Springer Proceedings in Physics **420**

Indexed by Scopus

The series Springer Proceedings in Physics, founded in 1984, is devoted to timely reports of state-of-the-art developments in physics and related sciences. Typically based on material presented at conferences, workshops and similar scientific meetings, volumes published in this series will constitute a comprehensive up to date source of reference on a field or subfield of relevance in contemporary physics. Proposals must include the following:

– Name, place and date of the scientific meeting
– A link to the committees (local organization, international advisors etc.)
– Scientific description of the meeting
– List of invited/plenary speakers
– An estimate of the planned proceedings book parameters (number of pages/articles, requested number of bulk copies, submission deadline).

Please contact:
For Americas and Europe: Dr. Zachary Evenson; zachary.evenson@springer.com
For Asia, Australia and New Zealand: Dr. Loyola DSilva; loyola.dsilva@springer.com

Hamid M. K. Al Naimiy · Hussein M Elmehdi ·
Ihsan A. Shehadi

Editors

Proceedings of the 14th Arabic Conference of the Arab Union for Astronomy and Space Sciences

AUASS-CONF23, 13–16 November 2023
Sharjah, United Arab Emirates

 Springer

Editors
Hamid M. K. Al Naimiy
University of Sharjah
Sharjah, Sharjah, United Arab Emirates

Hussein M Elmehdi
Applied Physics
University of Sharjah
Sharjah, United Arab Emirates

Ihsan A. Shehadi
University of Sharjah
Sharjah, Sharjah, United Arab Emirates

ISSN 0930-8989   ISSN 1867-4941 (electronic)
Springer Proceedings in Physics
ISBN 978-981-96-3275-6   ISBN 978-981-96-3276-3 (eBook)
https://doi.org/10.1007/978-981-96-3276-3

This work was supported by University of Sharjah.

This Springer imprint is published by the registered company Springer Nature Singapore Pte Ltd.
The registered company address is: 152 Beach Road, #21-01/04 Gateway East, Singapore 189721, Singapore

If disposing of this product, please recycle the paper.

Foreword

The field of astronomy and space sciences is one that has seen remarkable growth and development in recent years, thanks to technological breakthroughs and collaborative global efforts by the international scientific community to explore and unravel the mysteries of the universe. This is also reflected in the increased attention and a surge in interest in this field by scientists in the Middle East and Northern Africa region due to increased international collaborations where researchers in this region view exploring advancements in these fields as crucial. One prominent example of this international collaboration is the 14th Conference of the Arab Union of Space and Astronomy Sciences. This edited book, "Exploring Frontiers: Advances in Astronomy and Space Sciences," perfectly reflects the core themes of the manuscripts presented at this conference and published herein. This eBook probes into the fascinating world of space sciences and astronomy, highlighting their intricate connection exploring planetary science and astronomy at various wavelengths. Each chapter explores different frontiers of these intertwined fields and serves as a valuable foundation for researchers and astronomy enthusiasts to build upon and to increase the boundaries of our cosmic understanding. Topics such as astronomy, astrophysics, data analysis, deep space exploration, dark matter, instrumentations, planetarium design, remote sensing, cube satellite design, high-energy physics, electronic device characterizations for space applications, and much more are discussed in the book. This broad resource makes it an excellent reference for anyone interested in astronomy and related fields. This allows readers to stay current with the ever-evolving understanding of our cosmos. The field of astronomy and space sciences is perpetually evolving, with new discoveries constantly emerging and expanding the boundaries of our cosmic understanding. Continued investment in technology and international collaboration promises even more exciting discoveries in the years to come as we venture further into the cosmic frontiers. As we dig deeper into the cosmic frontiers, we are continually reminded of its complexity, vastness, and the endless opportunities for exploration and discovery.

Fawzi A. Ikraiam

Foreword

The manuscripts that are presented in this edited book cover many aspects of the current and future insights related to recent advances in astronomy, astrophysics and planetarium designs and technologies. This book also addresses the issues related to applied research in astronomy where data science and its components are utilized in the analysis of data collected using recent advances in Cube Satellite and potential use of AI and machine learning in assessing such data. With the advancements in computational and data sciences, astro and related sciences shall be evolved as is evident through the research work that is endorsed in the proceeding of this important conference. The papers cover a wide spectrum of space sciences from instrumentations, data collections and analysis, policies and legislation, sustainability, applications of computational tools and selected case studies in related fields. Hence, this edited book can be a valuable reference for applied research in astronomy in the MENA region and around the world.

Al-Mehdi M. Ibrahem

Foreword

The "Exploring Frontiers: Advances in Astronomy and Space Sciences" edited book shall be marked as a valuable reference for interested researchers and scientist in the Gulf and MENA region due to the quality of papers presented herein. The book tackles recent advanced topics in astronomy and astro science that range from deep space exploration of galaxies, stars and plants. In addition, some of the papers explore the possibility of extraterrestrial activities as based on data analysis and explorations. Planetarium designs and use to increase the public awareness about the importance of astronomy for the wellbeing of humanity in terms of sustainability, energy efficiency and productions, origin of life and remote sensing and communications. This book shall be of great interest to professionals and amateurs in the field of astronomy and astro sciences.

Maher Omar

Preface

"Exploring Frontiers: Advances in Astronomy and Space Sciences"

"Exploring Frontiers: Advances in Astronomy and Space Sciences" is a collection of peer-reviewed chapters that present the latest discoveries and advancements in the field of astronomy and space sciences. This volume offers readers a comprehensive overview of the cutting-edge research and technological innovations that are propelling us further in these fields.

The chapters published in this eBook are selected from the papers presented at the 14th Arab Conference on Astronomy and Space Sciences, held at the University of Sharjah, United Arab Emirates (UAE) from November 13–16, 2023. This conference came at a critical time when the world is witnessing unprecedented technological and scientific advances in astronomy and space sciences. These advances have led countries around the world, including the UAE, to identify these fields as prime areas in their strategic planning and future projects, with a focus on the exploration of planets and celestial bodies, the allocation of exploratory and scientific space flights, the development of communications technology and satellites, and the application of the latest space technology for peaceful uses and applications.

The chapters in this volume cover a wide range of topics, from the intricate dance of celestial bodies to the mysteries of dark matter and energy, from the exploration of distant planets and moons to the quest for extraterrestrial life. Each chapter not only delves into the scientific principles and findings but also highlights the collaborative efforts and interdisciplinary approaches that are essential in unraveling the complexities of the universe.

The conference aimed to bring together leading academics, scholars, and researchers in a stimulating hybrid setting to tackle the challenges encountered in recent years. The themes of the conference and topics include:

- Astronomy and Astrophysics
- Space Sciences and Technologies
  - Astronomy in Arabic and Islamic Cultures
- Education and Outreach
- Initiatives and Programs for the Sustainable Development Goals (SDG).

Over 100 papers were presented during the conference, and scholars were given the opportunity to submit their work for publication as a book chapter in this special volume. All chapters have gone through a rigorous blind review process by two experts in the respective field, in accordance with international standards, as well as similarity and plagiarism checks using AI tools and language reviews.

The twenty-one chapters published in this edited open-access eBook represent 21% of the total manuscripts submitted to the 14th Arab Conference on Astronomy and Space Sciences. These chapters cover case studies, models, and reviews of original collaborative research work conducted in more than 21 institutions in the Gulf and the

Middle East region, making this volume unique in presenting a comprehensive overview of the latest advances in the MENA region landscape during the past few years and providing an outlook on new evolving trends and discoveries, with an emphasis on technologically driven models.

The edited book is in line with the strategic goals of the University of Sharjah, which highlight the importance of scholarly activities, knowledge sharing, and exchange, as well as promoting collaboration among leading institutions and the scientific community at large at the national, regional, or international levels. The editorial board has worked diligently to ensure that the chapters contribute to the noble goal of disseminating knowledge among researchers and academics working in the fields of astronomy and space sciences.

Editorial Board

Contents

Transiting Exoplanets from Sharjah Astronomical Observatory (SAO-M47): The Exoplanet HAT-P-25 b Using L & V Filters

Mohammad F. Talafha[1]([✉]), Mashhoor A. Al-Wardat[1,2], Ammar E. M. Abdulla[1,2], and Hamid M. Al-Naimiy[1,2]

[1] Sharjah Academy of Astronomy Space Science and Technology, University of Sharjah, 27272 Sharjah, United Arab Emirates
mtalafha@shrjah.ac.ae

[2] Physics Department, Science College, University of Sharjah, 27272 Sharjah, United Arab Emirates

Abstract. In this study, we conduct a comparative analysis of observations carried out on the exoplanet HAT-P-25b at the Sharjah Astronomical Observatory (SAO). We have employed two distinct filters, namely, the Luminoso (L) and Visual (V) filters. Our research conducted aims to discern any variations in transit depth or exoplanet size resulting from the use of these different filters.

The primary focus of this study is to determine the exoplanet's size relative to its host star using the transit method. The application of different filters was expected to introduce subtle variations in size, influenced by factors such as the exoplanet's atmosphere. Notably, our findings reveal that the exoplanet's size appears larger when observed through the L filter compared to the V filter.

Throughout the analytical process, we employed the TRASCA model to determine the transit depth for each epoch. Fixed parameters, including the orbital period of the exoplanet (P, measured in days) and the transit duration (measured in minutes), were utilized in these calculations. Our results indicate that the transit depths observed with the L filter were greater than those with the V filter, measuring 0.0238 magnitudes and 0.0200 magnitudes, respectively. These values deviate from the reference result of 0.0204 magnitudes.

Keywords: Exoplanet · HAT-P-25 b · Transit depth · multi filter

1 Introduction

Over the last two decades, the identification of exoplanets, facilitated by the transit method, has swiftly transformed our comprehension of the cosmos, more than 60% of the discovered extrasolar planets were identified using this method (1) Moreover, there has been a rapid growth in the number of exoplanet candidates, a surge which was facilitated through surveys such as the TESS & Kipler satellite. The initial discovery of HAT-P-25b was made by Quinn (2012). Subsequent refinements to the physical and orbital parameters of the system were conducted by Wang (2018). (2) No photometric or

© The Author(s) 2025
H. M. K. Al Naimiy et al. (Eds.): AUASS-CONF 2023, SPPHY 420, pp. 1–7, 2025.
https://doi.org/10.1007/978-981-96-3276-3_1

spectroscopic studies have been conducted on the HAT-P-25 star-planet system other than the discovery paper and the paper by Wang et al. (2018). (3) In this study, we introduce new photometric follow-up observations using different filters. The parameters of the exoplanet, as listed in the discovery publication, report that it is a transiting extrasolar planet orbiting the $V = 13.19$, $G5$ dwarf star $GSC1788 - 01237$, with a period $P = 3.652836 \pm 0.000019d$, and transit duration $0.1174 \pm 0.0017d$. The host star has a mass of $1.01 \pm 0.03M_\odot$, and radius of $0.96 \pm^{0.05}_{0.04} R_\odot$, an effective temperature of 5500 ± 80 K, and metallicity $[Fe/H] = +0.31 \pm 0.08$. The planetary companion has a mass of $0.567 \pm 0.022M_J$, and radius of $1.190 \pm^{0.081}_{0.056} RJ$, yielding a mean density of 0.42 ± 0.07 gcm^{-3} (4). Also, the corrected transit depth found was (0.01626 ± 0.00035) using Sloan i band filter (2), observed by the Fred Lawrence Whipple Observatory (FLWO), as shown in Fig. 1. The TESS telescope has observed this target multiple times, as shown in Fig. 2. The multi-filter photometric observation was used to recognize the different responses for the transit depth. Observations in this spectral band are crucial for examining the presence of high hazes in the atmosphere. Rayleigh scattering gains significance at shorter wavelengths, potentially resulting in larger observed radii within this band. (5).

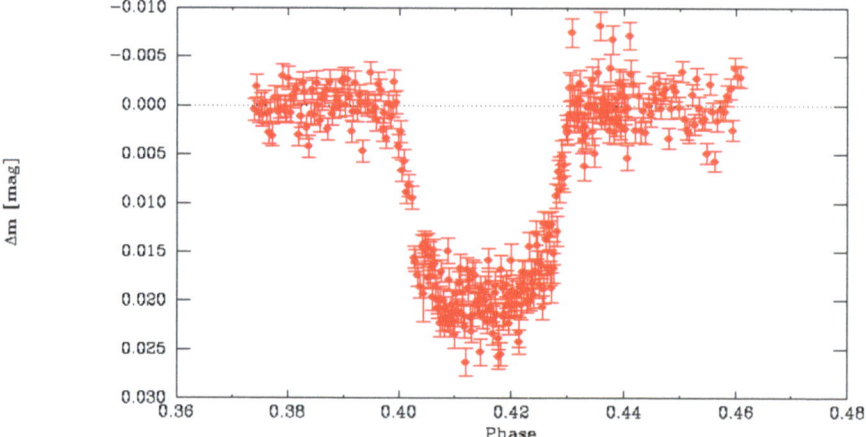

Fig. 1. The transit light curve for HAT-P-25b by Sloan i band filter (650 nm–85 m), using the Fred Lawrence Whipple Observatory (FLWO)

The resulting light curve is shown in Figs. 3 and 4. Although there is more dispersion in the data taken with the V filter compared to the L filter data, a flux depth of approximately 2% or more is observed. The best-fit of $Rp/R*$ values obtained are listed in Table 2.

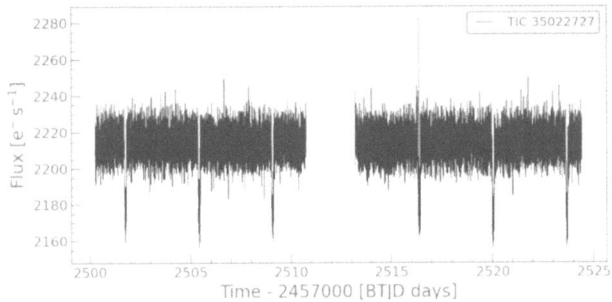

Fig. 2. Multi-epoch light curves by TESS space telescope observing HAT-P-25b found in sector 44, released in 2021.

2 Observations and Data Reduction

The observations were conducted at the Sharjah Astronomical Observatory (SAO-M47) employing a 431 mm aperture telescope equipped with two distinct optical filters: a Luminous filter L spanning the spectral range from 420 nm to 685 nm that provides 98% transmission over the entire visible spectrum from 420 to 685 nm, and a V filter covering the range of 500 nm to 700 nm. The V filter demonstrates a level of transmittance centered around its peak wavelength of about 530 nm. However, it allows only around 63% of the incident light at that specific wavelength to traverse through the filter. The SAO's optical system incorporates a CDK telescope configuration along with an $SBIGSTX - 16803CCD$ Camera $4K \times 4K$, delivering a field of view measuring $43\prime \times 43\prime$ and an angular pixel scale of 0.63 arcseconds per pixel.

Standard data processing procedures were applied to the compiled observations, involving dark frame subtraction, bias correction, and flat-field calibration. Aperture photometry was subsequently performed utilizing *MaximDL* software. The transit light curves were derived using differential photometry techniques (6).

The fundamental parameters of the exoplanet were determined based on the TRASCA (7) model, assuming a fixed radius for the host star denoted as $R\odot$ (solar radii), a fixed period for the exoplanet P (in days), and the transit duration (in minutes).

The observation commenced by consulting the Exoplanet Transit Database to ascertain the timing of observations and the altitude of the exoplanet. The initial observation, conducted with the Luminous filter, took place on November 8, 2021 (Julian Heliocentric Date: JHD 2459527.29170), featuring an exposure duration of 150 s during which a total of 73 individual images were acquired.

A year later, the second observation was performed utilizing the *V* filter on November 12, 2022 (Julian Heliocentric Date: JHD 2459896.22108), with a longer exposure time of 360 s. A total of 46 images were collected during this observation. All details of these observations are listed in Table 1 (Figs. 5 and 6).

Table 1. The observation parameters for each night of observation.

Parameters						Air mass	
Filter	Exposure sec	Binning	# Frames	1/SNR	Binning	Start	End
L	150	1	73	0.003	1 × 1	1.0658	1.3243
V	360	1	42	0.007	1 × 1	1.3082	1.0810

The analytical illustrations generated by the TRASCA Model depict consistency in outcomes regarding the planet's rotation period and transit duration when contrasted with other observations at the ETD site. Additionally, these illustrations show a clear disparity in the depth of the transit observed between the L and V filters.

3 Results

Utilizing the TRESCA model for data analysis and light curve generation, with fixed parameters including mid-transit and transit duration, we examined variations in transit depth. A comparative analysis of observations conducted with *L* and *V* filters revealed distinct differences in transit depth between them.

Moreover, transit depth (bottom) showed the transit depth of observations made by the L filter to be larger than ones by the *V* filter Fig. 3 and Fig. 4. On the other hand, the Top and middle plots show the remarkable agreement with reference results and correspondence in the calculated and observed results of observing the planet using the *V* filter and filter *L*, , and this is evident in all parameters (O-C, Duration time, and transit depth). Table 2 lists all the results of the observations that have been shared on the Exoplanet Transit Database website[1].

---

[1] http://var2.astro.cz/ETD/.

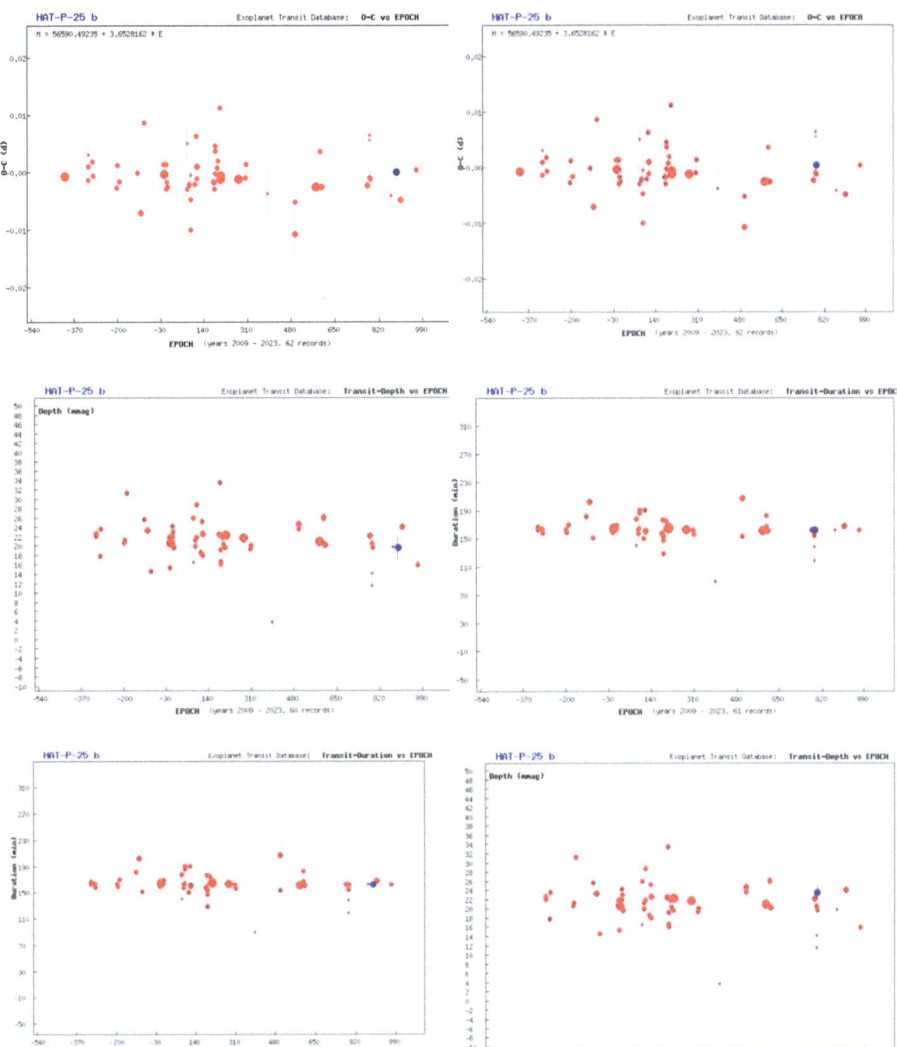

Fig. 3. The results of V filter observation for HAT-P-25b by TRESCA. Top: The fixed period of the exoplanet. Middle: Fixed result for the transit duration. Bottom: Depth of transit

Fig. 4. The results of L filter observations for HAT-P-25b. Top: The fixed period of the exoplanet. Middle: Fixed result for the transit duration. Bottom: Depth of transit.

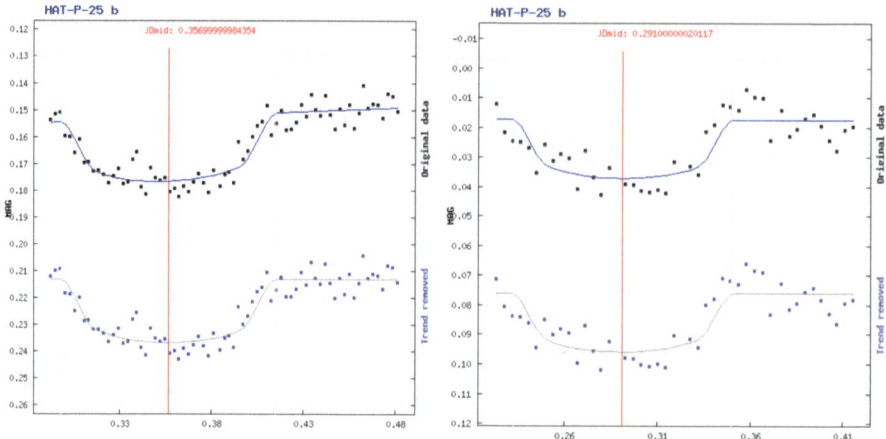

Fig. 5. Light curves in the L filter for HAT-P-25b.

Fig. 6. Light curve taken by V filter for HAT-P-25b.

Table 2. Results for HAT-P-25b observations by TRESCA model.

Filter	Date	Duration (min)	Epoch	Depth	R_p/R_*
Luminous (L)	8/11/2021	Fix	Fix	0.0238	0.1542
Visible (V)	12/11/2022	Fix	Fix	0.0200	0.1414
Reference (ETD)		169 min	---------	0.0204	0.1482

4   Conclusion

This study marks the inaugural presentation of a light curve of HAT-P-25b in the L filter. The significance of observations across various spectral bands lies in their ability to investigate the presence of high hazes in the atmosphere. This importance stems from the increased relevance of Rayleigh scattering at shorter wavelengths, potentially resulting in larger observed radii in the shortwave band (5). Transit duration and mid-transit were systematically fixed into the TRASCA Model as reference points to enable a rigorous comparison of transit depth outcomes between the two filters. The exciting results emphasize the importance of conducting more observations using various filters. These additional investigations are essential for detecting and quantifying the dynamic alterations in the depth of the light curve using different filters, and more photometric and transmission spectrum observations are needed.

References

Neilson, H.R., et al.: Limb darkening and planetary transits: testing center-to-limb intensity variations and limb-darkening directly from model stellar atmospheres. Astrophys. J. **845**, 12 (2017)

Wang, X.Y., et al.: Transiting Exoplanet Monitoring Project (TEMP). IV. Refined system parameters, transit timing variations, and orbital stability of the transiting planetary system HAT-P-25. Public. Astron. Soc. Pac. **130**(988) (2018)

Erdem, A., Öztürk, O.: New photometric observations of HAT-P-25b and WASP-11b/HAT-P-10b. In: AIP Conference Proceedings (2018)

Quinn, et al.: Transiting Exoplanet Monitoring Project (TEMP). IV. Refined system parameters, transit timing variations and orbital stability of the transiting planetary system HAT-P-25. Astrophys. J. **745**, 9 (2012)

Ricci, D., et al.: Multifilter transit observations of WASP-39b and WASP-43b with Three San-Pedro Mártir telescopes. Publ. Astron. Soc. Pac. **127**(948), 143 (2015)

Henry, G.W.: Techniques for automated high-precision photometry of sun-like stars. Astron. Soc. Pac. **111**(761) (1999)

Stanislav, P., Luboš, B., Ondřej, P.: Exoplanet transit database. Reduction and processing of the photometric data of exoplanet transits. New Astron. **15**(3), 297–301 (2010)

The Jewel in the Crown: Archiving and Analyzing Astronomical Spectro-Visual Binaries Big Data

Suzan Alnaimat[1], Raid Jameel[1], Ahmad Abushattal[2(✉)], Mashhoor Al-Wardat[3], and Mustafa H. Ahmed[1]

[1] Library Science Department, College of Arts, Al-Hussein Bin Talal University, Ma'an, Jordan
[2] Department of Physics, Al-Hussein Bin Talal University, P. O. Box 20, Ma'an 71111, Jordan
`ahmad.abushattal@ahu.edu.jo`
[3] Department of Applied Physics and Astronomy, College of Sciences, and Sharjah Academy for Astronomy, Space Sciences and Technology, University of Sharjah, 27272, Sharjah, United Arab Emirates

Abstract. This article discusses how Astroinformatics has transformed astronomical research by providing advanced data mining techniques to adapt to the challenges posed by large-scale data sets. Focusing on the realm of binary star systems, we employ Descriptive Data Mining to analyze and integrate data from three pivotal catalogs: The Sixth Catalog of Orbits of Visual Binary Stars, The Ninth Catalogue of Spectroscopic Binary Orbits, and The Fourth Catalog of Interferometry Measurements of Binary Stars. To conduct this study, we consolidated SB9 data from 2021, which records 4021 binary systems, and cross-referenced this data with orbits from 6COVBS and observations from 4CIMBS, which contains data on 66,225 resolved stars in 2020. Digital archiving helps create a comprehensive, unified database spanning multiple catalogs, which is the overall objective of this study. Binary star data can be compared using this unified database. Among these catalogs, we identified a total of 600 standard systems. In this discovery, spectroscopic and visual orbital solutions were integrated to provide a rich dataset for calculating the physical properties of binary stars. It sets a precedent not only for future research in Astroinformatics but also highlights the power of Astroinformatics in modern astronomy.

Keywords: Binary Star · Digital Archiving · Data Mining · Catalogues · visual · spectroscopic

1 Introduction

As a starting point, this work relies heavily on three basic concepts: binary stars, astronomical catalogs, and electronic archives. As we use huge astronomical data in this work, we will clarify the importance of complementarity and sharing between these three concepts. The binary star is a pair of stars whose components are bound by gravity to each other, and, for this reason, they orbit the common center of mass. One of these

H. M. K. Al Naimiy et al. (Eds.): AUASS-CONF 2023, SPPHY 420, pp. 8–20, 2025.
https://doi.org/10.1007/978-981-96-3276-3_2

movements around the center of mass can be considered to be fixed (the primary) and the other (the secondary) can be considered to be relative concerning the first. According to numerous estimates, more than half the stars we observe belong to binary systems or to complex systems that include three, four, or more components. Depending on how far the stars are separated, the orbital periods of binary stars can vary significantly [1–5]. However, depending on the distance between the two stars, the orbital periods of the binaries can vary from hours to centuries. The study of binaries and multiple stars is important in Astrophysics and Astrodynamics for different and interesting reasons but, especially because of their orbits and parallaxes, it is possible to obtain the values of the stellar masses.

To study the evolutionary tracks of stars, mass is a fundamental parameter. However, several other physical parameters may be interesting, including the size of the components, their parallax (dynamical and orbital parallax), luminosities, and exact orbital elements [6–8]. As opposed to binary systems, which function in a similar way to "shop windows," binary systems reflect a variety of physical processes: mass loss, mass exchange, component variability, relativistic processes, the Nova phenomenon, flares, x-Ray components, Wolf-Rayet components, pulsars, etc. A dynamical perspective includes looking at perturbations, searching for dark components like brown dwarfs and exoplanets, determining orbital methods, etc.

Numerous observatories equipped with large telescopes and refractors, especially in Europe and North America, were constructed at the beginning of the past century, with the primary objective of observing binaries [9–14]. Therefore, many relevant astronomers pointed to double stars as an interesting astronomical field involving very different lines of research. Traditionally and considering the technique used to discover and later study them, binaries are classified into three categories: visual, spectroscopic, and eclipsing [15–18].

Space missions and satellites send huge amounts of astronomical data worldwide every day to analyze and study these data in various fields of space science and astronomy. Consequently, it is one of the most common problems faced by researchers and specialists in the field [19, 20]. Organization, arrangement, and archiving of this vast astronomical data, and here comes the role of such a study, which solves these problems by archiving astronomical data electronically in a way that facilitates the analysis of it and reaches the desired results through the work of researchers. There are also many catalogs of astronomical tasks due to the multiplicity of these tasks. Individuals specialize in single stars, planets, binary stars, multiple stars, and galaxies [9, 13, 21–24].

In addition, physicists in space science face the challenge of integrating catalogs and searching for common characteristics and data. As a result, the problem of multiple catalogs has arisen, which is how to ensure that data is shared across them. Several catalogs are specialized in binary stars, which is the focus of this study since we will search for stars shared between them, archive them digitally, and prepare them for research in binary star research [15, 25].

The electronic and digital archiving of astronomical catalogs, which contain huge amounts of data and are specialized in binary stars, allows researchers to calculate astronomical parameters such as mass, orbital parallax, temperature, gravity, spectral type, absolute magnitude, and other astrophysical parameters [26–28]. The data deluge is transforming research practices across a wide range of scientific disciplines, including astronomical research daily. To process astronomical data and its variety, innovative algorithms and methods are essential.

There is no doubt that archives are valuable resources for astronomy, but statistics indicate they are even more important than most people realize. The archive is crucial to the science of survey projects in astronomy and space sciences. Even more unexpectedly, archival data contributes significantly to the science of astrophysical missions like Chandra, Spitzer, and ground-based observatories. Missions and observatories can generate a great deal of science with the help of archives. Establishing new projects should consider the value of archives [29, 30]. Mission and observatory design and operations decisions should take a robust archive into account not only when budgeting for future missions and observatories. We can increase our science output from our major investment in large projects by increasing both funding for archive users and archive centers especially to enable cross-archive, multiwavelength science.

Data-centric astronomy, which encompasses many interdisciplinary applications, has been recognized as a distinct academic research field called Astroinformatics. Astronomical surveys and catalogs are analyzed using various computational methods and software tools [31, 32]. Many tools are available to assist with handling data, such as data modeling, data mining, data access, digital astronomical databases, machine learning, statistics, and other specialized software. Astroinformatics focuses on discovering and studying binary stars from catalogs of binary stars and how this data can be mined [21, 33–35].

2   The Importance of the Digitalization of Astronomical Data

The significance of this study, which focuses on archiving astronomical data digitally, especially with binary stars, cannot be overstated [9, 10, 21, 35–37]. This study addresses the challenges associated with using and managing vast amounts of astronomical data, thereby advancing our understanding of the cosmos. Research in space science and astronomy can be conducted more effectively and comprehensively, collaboration can occur, and significant discoveries can be made. There are several key aspects of significance in this study:

- Astronomical data generated by space missions and satellites requires effective management, organization, and archiving to overcome the challenges posed by the vast amount of data. By establishing electronic archives, researchers and scientists can access and utilize valuable information more easily, facilitating their understanding of the universe.
- The research of astronomers covers many different topics, from individual stars to entire galaxies. Their data must be organized and archived efficiently to facilitate efficient research. This study simplifies data access and management, enabling researchers to concentrate on their research questions instead of dealing with data management issues.
- Celestial objects can be more comprehensively understood with integrated catalogs, according to this study. Identifying common characteristics and trends in data from multiple sources allows researchers to gain a deeper understanding of the universe.

- Various astrophysical parameters can be calculated based on the documentation and organization of data in this study. Astronomers use these parameters to understand how celestial objects move through the universe and where they are positioned.
- A key goal of this study is to facilitate collaboration among astronomers through the creation of electronic archives and the accessibility of data. The field of astronomy often benefits from collaboration because it brings together multiple perspectives and expertise.
- Accessible astronomical data offers research and educational benefits. This data can be utilized by students, educators, and science enthusiasts to learn about the universe, igniting an interest in space science and astronomy.

3 The Objective of Archiving and Analyzing Astronomical Spectro-Visual Binaries

Managing and archiving vast astronomical datasets have become increasingly crucial in space science and astronomy. As space missions and satellites generate copious amounts of data, digital archiving is essential for managing these datasets. We aim to facilitate data access, analysis, and sharing within the field, with an emphasis on binary stars. We present the main objectives of this study, each intricately linked to digital archiving. Main objectives:

- Comprehensive Digital Archiving System: Create a comprehensive digital archive system that enables space missions and satellites to store and manage astronomical data efficiently.
- Efficient data organization: Provide researchers with an easy-to-search, easy-to-retrieve digital organizational structure within the archive.
- Facilitate Data Analysis: Simplify the retrieval and analysis of digital data for researchers. Utilize digital tools to make data usage more efficient.
- Digital Catalogue Integration: Achieve seamless data sharing and cross-referencing within the digital archive by integrating multiple astronomical catalogs digitally.
- Binary Star Specialization: Provide digital archiving and analysis of binary star data, recognizing their importance in astrophysics. Data from specialized catalogs are compiled to create a binary star database.
- Digital Parameter Calculation: Establish digital methodologies for computing astronomical parameters from digital archives. Digital calibration ensures the accuracy of parameter calculations.
- Validation and testing: Analyze optically resolved spectroscopic binary systems within the digital archive with both spectroscopic and visual orbits to assess the efficacy of digital methodologies and systems. Ensure that the physical parameters calculated are accurate.
- Improved researcher accessibility: Provide easy access to digitally archived data within the digital archive for researchers and specialists in binary star research and other space science and astronomy fields. Utilize digital means to advance research using this data.

4   The Sources of Electronic Archiving

For decades, stellar dynamics research has been based on the meticulous study and cataloging of binary star systems. Our understanding of binary star systems has been significantly enhanced by the synergistic analysis of three pivotal catalogs. Combined, these catalogs provide a comprehensive view of the binary star universe, each with its own specialized focus and historical significance. To highlight advancements in binary star research, this study examines and integrates data from these sources. A comprehensive data analysis is also emphasized in the paper.

- *The Ninth Catalogue of Spectroscopic Binary Orbits (SB9):*
  A new catalog of 2386 spectroscopic binary orbits has been published by Batten and colleagues at http://sb9.astro.ulb.ac.be. Through three key applications, this catalog distinguishes itself from its predecessors: (1) Identifying the shortest periods in the H-R diagram, (2) Evaluating completeness based on the period distribution of SB1s and SB2s, and (3) Examining the relationship between periods and eccentricities (Pourbaix et al., 2004). Newly observed orbits can be submitted directly to SB9 by contributors worldwide. ADS bibliographic service references are also provided for each system in SB9, as a result of the increasing volume of data. However, SB9 provides multiple orbits, individual radial velocities, and references to the ADS bibliographic service for every new system.
  New orbits will be graded automatically while existing orbits will be rated from best (grade 5) to worst (grade 1). Similarly, for double-lined binaries, elements such as the center-of-mass velocity V0 must be determined separately for primary and secondary components since they represent the actual Keplerian motion rather than the radial velocity curve. Objects of interest can be located through SB9's search engine by common catalog identifiers (such as HD, BD, HIP), bibcodes, or coordinates [38].
- *The Sixth Catalog of Orbits of Visual Binary Stars (6COVB):*
  A major development in the study of visual double stars has occurred over the past thirty years with the development and refinement of interferometry. In contrast to speckle interferometry used on large telescopes, which is more precise (down to milliarcseconds), micrometry and other visual techniques played a less important role in orbit calculations by the time the Fourth catalog was published. Speckle interferometry has been available since 1970 but played only a small role in orbit calculations by then. Today, many orbits still rely only on speckle results, despite its popularity during the 1980s. Today, technologies such as the Navy Precision Optical Interferometry and long-baseline interferometry are acquiring the same maturity as speckle.
  It is becoming more common for multi-aperture telescope arrays to be used to study binaries previously observed by only spectroscopists. Spectral and visual observations will become increasingly similar as new interferometers become more sensitive to magnitude. Spectroscopic and visual data are increasingly being integrated into catalogs like this one. To create a new catalog, each orbit must be evaluated. Using a numerical scale (1 = definitive, 5 = indeterminate), it assesses orbital coverage, the number of observations, and the overall quality of observations in binary star

catalogs. In addition to qualitative assessments, observations have been graded numerically based on their cumulative experience in Astroinformatics: The Importance of Mining Astronomical Data in Binary Stars Catalogues [39].

- *The Fourth Catalog of Interferometry Measurements of Binary Stars (4CIMB):*
  In 1982, the Fourth Catalogue of Interferometric Measurements of Binary Stars was used to survey binary star observations made at Georgia State University's Center for High Angular Resolution Astronomy (CHARA). After a wide range of high-angular-resolution methods were used, the Speckle catalog included all astrometric and photometric data for binary stars (and single stars detected by duplicate surveys). It was also possible to obtain infrared speckle or imaging results, even though some of the results were not considered high resolution. This catalog includes bands of right ascension. The catalog is updated regularly, but statistics are only updated occasionally. Aside from astrometric data, this catalog includes 73,894 photometric observations [40].

5 Methodology

A vast amount of astronomical data is handled in this study, sourced from diverse spectroscopic and visual databases. A versatile, high-level programming language like Python is used to process and compile this data. How we approach these datasets is to organize and identify common elements, ensuring they are structured in a way that makes them easily accessible. Researchers in astrophysics and experts who study binary stars will be able to gain access to the data if this is done. Digital archiving of astronomical surveys and catalogs is made possible by Astroinformatics, an innovative, data-centered methodology. Access to astronomical data is facilitated by interoperability between various astronomical archives and data centers. Figure 1 illustrates the significance of digital archiving in connection with big data science and astronomy.

Using electronic and digital archives of astronomical catalogs, which are specialized in binary stars and contain huge amounts of data, researchers can calculate astronomical parameters like mass, orbital parallax, temperature, gravity, spectral type, and absolute magnitude.

Algorithms and methods that are innovative are required for the processing of astronomical data. In this study, descriptive data mining techniques are used to extract shared data from three binary star catalogs. In addition to these catalogs, the Ninth Catalogue of Spectroscopic Binary Orbits (9SBO), the Sixth Catalogue of Visual Binary Stars (6COVBS), and the Fourth Catalogue of Interferometry Measurements (4CIMBS) are also available. A total of 4021 binary systems are listed in the Ninth Catalog's most recent edition of 2021. Under the 6COVBS data, around 3000 binary systems will be organized. Additionally, 66,225 stars have been resolved in the most recent edition of the 4CIMBS.

The work involves processing and collecting astronomical data from a variety of spectroscopic and visual catalogs. Arrange them so they are easily accessible by sorting and identifying their commonalities. To make them accessible to astronomers and binary star specialists to benefit from them. Our work introduces Digital archiving (the use of Astroinformation resources to process astronomical surveys and catalogs) as a new

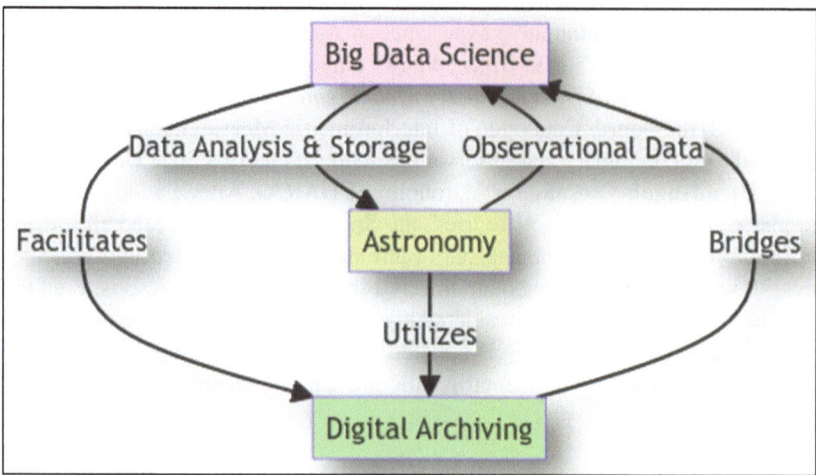

Fig. 1. The Significance of Digital Archiving in Connecting Big Data Science and Astronomy

data-oriented approach. It is easy to access astronomical data thanks to interoperability between different astronomical archives and data centers. Many catalogs organize and archive astronomical data. Physicists in space science must integrate these catalogs and search for common characteristics and data. By archiving astronomical data electronically, this study addresses these problems.

Both plain text and gzipped versions of the Fourth Interferometric Catalogue are available. There are two sections for each system listed: an identification line with catalog numbers, followed by individual measurements arranged according to the date of observation. The ID line format has been updated to accommodate longer names. Links to reference files are included as part of the observations, as are links to additional information contained in note files. There is also a connection between this catalog and the Sixth Orbit Catalogue, which contains astrometric and visual orbits. There is one notes file for all three catalogs: WDS, Interferometric, and Orbit. To make it easier to access, the notes file has been split into 24 smaller files due to its size.

In SB9, the number of stellar identifiers per system has been increased from a few to an unlimited number, while HD and HIP continue to be the primary sources for stellar identification. It now includes Declination and Right Ascension to the hundredth of a second to match other catalogs with different epochs and equinoxes than 1900.0. An overview of uncertainties is provided as well. The SB9 web interface allows you to search by catalog identifiers and coordinates to display specific orbits. The user can choose between several orbits if multiple systems fit the selection criteria, using a list of publication years and HTML links to open detailed information. Abstracts, coordinates, apparent magnitudes, and identifiers can be retrieved directly from ADS. Based on actual observations, orbit plots can also be generated through the interface. On the main page, researchers can download a compressed version of the SB9 database if they want a broader overview [38].

Approximately 10,000 binary star orbits are included in the Sixth Catalogue of Orbits of Visual Binary Stars. As interferometry advances, such as long-baseline interferometry, we have seen shorter periods and smaller semi-major axes in orbits observed with interferometry. There have been updates to the master file to accommodate formal errors and higher precisions, including the ability to quote periods in centuries, minutes, milliarcseconds, microarcseconds, and arcseconds, and whether to quote T0 in Julian dates or fractional Besselian years. The web catalog shows each orbit over two lines, with catalog names such as HD and Hipparcos displayed at the bottom of the page. Data from these three catalogs has been processed using astronomical data science and specialized software. Our unified database of visual spectroscopic binaries was formed by linking them and searching for stars that are common across all catalogs. Figure 2 summarizes this process.

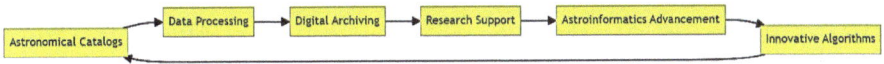

Fig. 2. Digital Archiving and Astroinformatics in Binary Star Research

6 Results

As part of this project, two pivotal fields are successfully merged: observational astronomy, which studies, monitors, and records data on various astronomical objects, and digital archiving, which processes and organizes this data for easier access by researchers. This research reveals a connection between precision-visual binaries, astronomical data, and data science. This study combined data from the Ninth Catalogue of Spectroscopic Binary Orbits (SB9), the Sixth Catalogue of Orbits of Visual Binary Stars (6COVBS), and the Fourth Catalogue of Interferometry Measurement of Binary Stars (4CIMBS) to generate spectroscopic binary orbits. Over 600 common systems were identified between these three catalogs as a result of this work.

Tables 1 and 2 of this subsection show the names of the stars used in the three considered catalogs, respectively, with their numerical identifiers, which are most popular among the three considered catalogs. The first table contains the star names that are common to the three catalogs under study, along with their respective numerical identifiers.

Binary star researchers use these systems to determine the orbits and physical characteristics of binary stars. This endeavor not only enhances our understanding of celestial bodies but also makes substantial contributions to astrophysics. Research and observations meticulously gathered by Al-Hussein Bin Talal University will be made available on the university's website. By disseminating this information, scientists and students worldwide will have access to this valuable dataset, allowing them to conduct further academic research and collaborate. University officials are committed to advancing knowledge and cultivating an open, accessible research culture through the initiative.

Table 1. The common binary stars from all catalogs (Part one) based on HD Identification.

28	16908	41040	91636	120803	157482	188307	217924
245	16909	41040	91962	120901	157482	188753	218568
895	17084	41116	92000	120901	157978	189340	218658
1273	17134	41380	92139	121370	157978	191589	218670
1976	17134	42954	92626	122742	158222	191854	219018
2057	17382	42995	93030	123299	158837	193216	219018
2070	18198	43821	93549	123999	160239	193216	219675
2261	18198	44762	94363	124547	160269	193554	221264
2333	21754	44780	95689	125337	160346	193797	221264
2343	22091	44896	97907	125351	160538	194152	221757
3196	22468	45088	97916	128141	160922	195725	221914
3196	22468	46101	98353	128620	162587	195987	222404
3266	22649	46407	98800	128642	162596	196524	223323
3443	23052	47703	98800	129132	162724	196574	223617
4775	23140	48914	99946	129132	163151	196795	223778
5408	23626	48915	100018	129333	163611	196795	224930
5408	23838	49293	101013	131511	163708	197433	234677
6118	24031	50310	101379	132756	163840	199870	237287
7331	26630	50337	102928	132813	165590	199939	237354
7374	26659	53299	103483	133621	165590	200580	284414
7483	26961	53424	103613	134320	165590	201626	332954
7640	26961	54563	104321	136138	166181	202275	
8054	27691	58368	104471	136504	166181	202447	

Table 2. The common binary stars from all catalogues (Part two) based on HD Identification.

7640	26961	54563	104321	136138	166181	202275
8054	27691	58368	104471	136504	166181	202447
8556	28363	58728	104471	137052	166208	202710
9021	28634	59717	105981	137687	167954	202908

(*continued*)

Table 2. (*continued*)

9053	28910	61421	105982	137909	168339	202908
9312	29095	61994	106400	138369	168532	202940
9313	29140	64096	106516	138525	169156	203345
9939	29140	64440	106760	138690	170000	203345
10009	29608	65339	107259	139006	170200	204075
10307	30869	66216	107259	139691	170547	205478
10516	31278	69148	108907	139691	170737	205767
10800	31964	71663	109011	140667	172831	206058
10800	32008	71663	109281	142267	173764	206644
11559	32068	73712	110024	142474	174457	206901
11613	32092	73752	110314	143275	175039	208816
11636	32850	74874	110555	144217	175515	210027
11753	33647	75958	110555	144253	176051	210647
12111	33856	76943	110743	145389	176411	211416
12534	34101	78362	110833	145849	178125	211594
12889	34318	78418	112048	147395	178593	212697
13161	34334	78515	112445	147508	179484	212754
13480	35155	79028	112914	147584	179558	212989
13480	35317	79910	112985	149414	179950	213429
13520	35877	81809	113449	150710	179950	214222
13611	35877	82674	113697	151613	181602	214511
13738	37013	83270	115955	151746	181615	214511
14214	37297	83270	116127	152751	183255	214608
15064	37393	85040	116458	152751	183536	214608
15096	37507	86590	116594	153597	184467	214850
15755	39587	88284	116656	154732	185082	216494
15777	40932	89758	118216	155410	185734	216598
15862	40932	90242	119458	155714	185936	216608
16458	40932	90242	119834	155937	186922	217580
16620	40932	90442	120539	156558	187076	217675
16739	41040	90537	120690	156635	187949	217792

7 Conclusions

We have successfully integrated observational astronomy with digital archiving, which organizes and processes this data to make it easier for researchers to access it. Digital archiving focuses on the study, monitoring, and recording of data on astronomical objects. Our research has established a key interface between astronomical data, data science, and precision-visual binaries, which has been made possible by digital archiving techniques. Our analysis was based on data from three major sources: spectroscopic binary orbit data from the Ninth Catalogue of Spectroscopic Binary Orbits (SB9), visual binary star data from the Sixth Catalogue of Visual Binary Star Orbits (6COVBS), and interferometry data from the Fourth Catalogue of Interferometry Measurement of Binary Stars (4CIMBS). Our research methodology, which used advanced big data programs to process these binary star catalogs, found more than 600 common systems among them.

Digital archiving was integral to this process. Additionally, it improved the accessibility and usability of data from these diverse sources. Data science and observational astronomy research should be carried out in joint research groups. In various astrophysical sciences, such as observational astronomy and astrobiology, such teams analyze observed data from different astronomical catalogs. This study demonstrates a progressive step in astrophysical research by combining digital archiving, data science, and observational astronomy, opening new avenues for investigation.

8   Recommendations and Future Work

It is crucial to leverage technological advancements and digital data in binary star research and astroinformatics. A brief overview of key recommendations and future directions in this area can be found in the following five points. As a result of these points, it has been highlighted that sophisticated data management, artificial intelligence, machine learning, user-friendly tools, a detailed study of binary star systems, as well as educational outreach and data preservation, are all extremely important. Astronomical research is more efficient and effective when these strategies are used to advance our knowledge of the cosmos.

- Advanced Data Management and Integration: Integration of various astronomical catalogs into a unified, easily accessible database for managing the increasing volume of astronomical data.
- AI and Machine Learning in Astronomy: Use artificial intelligence and machine learning in astronomy to enhance data analysis, including pattern recognition and predictive modeling.
- Collaboration Platforms and User-Friendly Analytical Tools: Develop tools that assist researchers across disciplines with data analysis and foster collaboration among astronomers and data scientists.
- Investigate binary star systems in-depth, focusing on their physical characteristics, formation, and evolution, and expanding research into related phenomena such as exoplanets and dark components using digital archives.
- Outreach to the public and education: Provide public and educational access to a portion of the data to encourage interest in the space sciences and preserve astronomical data for long-term research.

References

1. Al-Wardat, M., et al.: Physical and geometrical parameters of CVBS. XII. FIN 350 (HIP 64838). Astrophys. Bull. **72**, 24–34 (2017)

2. Docobo, J.A., et al.: Precise orbital elements, masses and parallax of the spectroscopic–interferometric binary HD 26441. Mon. Not. R. Astron. Soc. **469**(1), 1096–1100 (2017)
3. Duquennoy, A., Mayor, M.: Multiplicity among solar-type stars in the solar neighbourhood. II-Distribution of the orbital elements in an unbiased sample. Astron. Astrophys. **248**(2), 485–524 (1991). Research supported by SNSF, ISSN 0004-6361
4. Chen, X., Liu, Z., Han, Z.: Binary stars in the new millennium. Progr. Particle Nuclear Phys. 104083 (2023)
5. Hussein, A.M., et al.: Atmospheric and fundamental parameters of eight nearby Multiple stars. Astron. J. **163**(4), 182 (2022)
6. Abushattal, A.A., et al.: The 24 Aqr triple system: a closer look at its unique high-eccentricity hierarchical architecture. Adv. Space Res. (2023)
7. Docobo, J., et al.: Double Stars Inf. Circ (2018)
8. Docobo, J., Campo, P., Abushattal, A.: IAU Commiss. Double Stars **169**, 1 (2018)
9. Abushattal, A., Kraishan, A., Alshamaseen, O.: The exoplanets catalogues and archives: an astrostatistical analysis. Commun. BAO **69**(2), 235–241 (2022)
10. Abushattal, A.A.M.: The modeling of the physical and dynamical properties of spectroscopic binaries with an orbit: doctoral dissertation. Universidade de Santiago de Compostela (2017)
11. Taani, A., Abushattal, A., Mardini, M.K.: The regular dynamics through the finite-time Lyapunov exponent distributions in 3D Hamiltonian systems. Astron. Nachr. **340**(9–10), 847–851 (2019)
12. Taani, A., et al.: On the wind accretion model of GX 301–2. J. Phys. Conf. Ser. (2019). IOP Publishing
13. Taani, A., et al.: Probability distribution of magnetic field strengths through the cyclotron lines in high-mass x-ray binaries. arXiv preprint arXiv:2002.03011 (2020)
14. Taani, A., et al.: Jordan J. Phys. **13**(3), 243–251 (2020)
15. Abushattal, A.A., Docobo, J.A., Campo, P.P.: The most probable 3D orbit for spectroscopic binaries. Astron. J. **159**(1), 28 (2019)
16. Alameryeen, H., Abushattal, A., Kraishan, A.: The physical parameters, stability, and habitability of some double-lined spectroscopic binaries. Commun. BAO **69**(2), 242–250 (2022)
17. Algnamat, B., et al.: The precise individual masses and theoretical stability and habitability of some single-lined spectroscopic binaries. Commun. BAO **69**(2), 223–230 (2022)
18. Abushattal, A.A.M.: The modeling of the physical and dynamical properties of spectroscopic binaries with an orbit. Universidade de Santiago de Compostela (2017)
19. Abushattal, A., et al.: Extrasolar planets in binary systems (statistical analysis). J. Phys. Conf. Ser. (2019). IOP Publishing
20. Southworth, J.: Space-based photometry of binary stars: from Voyager to TESS. Universe **7**(10), 369 (2021)
21. Abushattal, A., Alrawashdeh, A., Kraishan, A.: Astroinformatics: the importance of mining astronomical data in binary stars catalogues. Commun. BAO **69**(2), 251–255 (2022)
22. Ochsenbein, F., Bauer, P., Marcout, J.: The VizieR database of astronomical catalogues. Astron. Astrophys., Suppl. Ser. **143**(1), 23–32 (2000)
23. Boubert, D., Everall, A.: A selection function toolbox for subsets of astronomical catalogues. Mon. Not. R. Astron. Soc. **510**(3), 4626–4638 (2022)
24. Rah, M., et al.: Unraveling the origins and development of the galactic disk through metal-poor stars. arXiv preprint arXiv:2402.07045 (2024)
25. Al-Tawalbeh, Y.M., et al.: Precise masses, ages, and orbital parameters of the binary systems HIP 11352, HIP 70973, and HIP 72479. Astrophys. Bull. **76**, 71–83 (2021)
26. Vavilova, I., et al.: Surveys, catalogues, databases, and archives of astronomical data. In: Knowledge Discovery in Big Data from Astronomy and Earth Observation, pp. 57–102. Elsevier, Amsterdam (2020)

27. Mickaelian, A., et al.: BAO plate archive project: digitization, electronic database and scientific usage. Commun. Byurakan Astrophys. Observ. **67**, 293–301 (2020)
28. Mickaelian, A., et al.: Armenian astronomical archives and databases. BAOJ Phys. **2**(008) (2017)
29. Ubertini, P., et al.: Future of space astronomy: a global road map for the next decades. Adv. Space Res. **50**(1), 1–55 (2012)
30. Ashby, M., et al.: The spitzer deep, wide-field survey. Astrophys J **701**(1), 428 (2009)
31. Abushattal, A.A., Loureiro, A.G., Boukortt, N.E.I.: Ultra-high concentration vertical homo-multijunction solar cells for CubeSats and Terrestrial applications. Micromachines **15**(2), 204 (2024)
32. Boukortt, N.E.I., et al.: Electrical and optical investigation of 2T–Perovskite/u-CIGS Tandem Solar Cells with~ 30% efficiency. IEEE Trans. Electron. Devices **69**(7), 3798–3806 (2022)
33. Brescia, M., et al.: Astroinformatics (2017)
34. Abushattal, A., Alrawashdeh, A., Kraishan, A.: Astroinformatics: the importance of mining astronomical data in binary stars catalogues. Commun. Byurakan Astrophys. Observ. **69**, 251–255 (2022)
35. Brescia, M.: Time domain astroinformatics. In: ML4Astro International Conference. Springer, Cham (2022)
36. Calderón, J.H., et al.: The digital archive of the photographic images of the Córdoba observatory plates collections. Astrophys. Space Sci. **290**, 345–351 (2004)
37. Kurtz, M.: The future of memory: archiving astronomical information. In: Symposium-International Astronomical Union. Cambridge University Press (1994)
38. Pourbaix, D., et al.: The ninth catalogue of spectroscopic binary orbits. Astron. Astrophys. **424**(2), 727–732 (2004)
39. Hartkopf, W.I., Mason, B.D., Worley, C.E.: The 2001 US naval observatory double star CD-ROM. II. The fifth catalog of orbits of visual binary stars. The Astron. J. **122**(6), 3472 (2001)
40. Hartkopf, W.I., McAlister, H.A., Mason, B.D.: The 2001 US naval observatory double star CD-ROM. III. The third catalog of interferometric measurements of binary stars. Astron. J. **122**(6), 3480 (2001)

New Insight Concerning Primordial Lithium Production

Tahani Makki[1($\boxtimes$)], Mounib El Eid[2], and Grant Mathews[3]

[1] Department of Physics, American University of Beirut (Alumna), Bliss Street, Beirut, Lebanon
trm03@mail.aub.edu
[2] Department of Physics, American University of Beirut (emeritus), Bliss Street, Beirut, Lebanon
meid@aub.edu.lb
[3] Department of Physics, Center for Astrophysics, University of Notre Dame, Notre Dame, South Bend, IN 46556, USA
gmathews@nd.edu

Abstract. To constrain the universe before recombination (380000 years after the Big Bang), we mostly rely on the measurements of the primordial abundances that indicate the first insight into the thermal history of the universe. The first production of light elements is obtained by the Big Bang Nucleosynthesis (BBN). The production of the elements D, ^{3}He, and ^{4}He during BBN matches well the observations; however, the production of lithium (^{7}Li) based on the Standard Big Bang Nucleosynthesis (SBBN) is found to be higher by about a factor of three than the observed abundance from metal-poor halo stars. This so-called "Cosmological Lithium Problem" is still elusive and needs to be resolved. One important attempt to resolve this problem is to invoke a non-standard description of the SBBN to decrease the lithium abundance. In our previous work, we encountered a problem that the decrease in the ^{7}Li abundance requires an increase in the deuterium abundance to maximum values that are not accepted by observations. In the present work, a decrease in the lithium abundance could be achieved without maximizing the deuterium abundance by modifying the time-temperature relation in the range $(4.3 - 9.1) \times 10^8$ K during the nucleosynthesis process. This range is crucial to reducing the strong correlation between lithium and deuterium production. The main conclusion of the present work is that the ^{7}Li abundance in the atmospheres of metal-poor stars cannot be analyzed without considering possible modifications to the primordial nucleosynthesis.

Keywords: Big Bang Nucleosynthesis · Lithium Problem · Entropy Modification · Non-Standard Physics

---

M. El Eid and G. Mathews—Contributing authors.

H. M. K. Al Naimiy et al. (Eds.): AUASS-CONF 2023, SPPHY 420, pp. 21–35, 2025.
https://doi.org/10.1007/978-981-96-3276-3_3

1 Introduction

The standard Big Bang nucleosynthesis (SBBN) is the production site of the light elements (D, ^{3}He, ^{4}He, ^{7}Li, ^{7}Be) before the formation of the first stars after the "dark age" which followed the recombination epoch, 380000 years after the Big Bang. The SBBN is a well-established theory since it depends on a single parameter the baryon-to-photon ratio (η) which has been well determined by the Planck satellite collaboration. The resulting abundances in the case of D and ^{4}He match the observed abundances in the atmospheres of very metal-poor halo stars. However, the abundance of ^{7}Li is found to be higher by at least a factor of three. This is termed as the "Cosmological Lithium Problem", which is not yet resolved. Various works [1–6, 8, 9, 28] have attempted to resolve that problem, and the main result of these investigations was that the decrease in the lithium abundance was linked to an increase in the deuterium abundance to values not compatible with observations in stars. The drawback of that link is that deuterium is easily destroyed in stars at temperatures exceeding 10^6 K. In our previous work [10], we attempted to resolve the lithium problem by varying the number of neutrinos, their chemical potentials, and their temperature. In addition, the effect of dark fluid and photon cooling with axion dark matter [11] was included. However, this approach failed to relax the strong correlation between deuterium and lithium production. In the present work, a decrease of the lithium abundance that matches closely the observations is possible without increasing the deuterium abundance to values contradicting the observations. As described in Sect. 4, this is obtained by modifying the time-temperature relation in the range $(4.3 - 9.1) \times 10^8$ K when adding a dark entropy component. Section 2 introduces the SBBN. In Sect. 3, various suggestions are outlined to resolve the lithium problem. In Sect. 4 and 5, we have introduced our new approach to have a better understanding of the decrease of the lithium abundance resulting from the modified SBBN. Section 6 summarizes the conclusion.

2 Standard Big Bang Nucleosynthesis (SBBN)

The SBBN model is described by the competition between the expansion time scale of the universe and the lifetime of the involved reaction rates. In particular, the weak interaction rates are sensitive to the expansion rate, so they will freeze out when their rates become less than the time scale of the universe. Figure 1 displays the network of the nuclear reactions, which we have used in the numerical simulations using a modified solver called "AlterBBN" [20]. This code requires the baryon-to-photon ratio η, the neutron lifetime τ_n, and the relativistic effective degrees of freedom N_{eff} as input. However, several modifications were required for the treatment of the non-standard Big Bang Nucleosynthesis (BBN). This concerns additional effective degrees of freedom (N_{eff}), neutrino degeneracy parameters (β_{ve}, $\beta_{v\mu}$, $\beta_{v\tau}$), and dark components. Table 1 includes the results obtained in this work using updated nuclear reaction rates, which lead to the lowest lithium production by the SBBN. These calculations are in agreement with previous works [22, 23]. Table 1 shows that the ratios of the abundances normalized to hydrogen are in good agreement with observations, except for lithium which is overproduced by the SBBN.

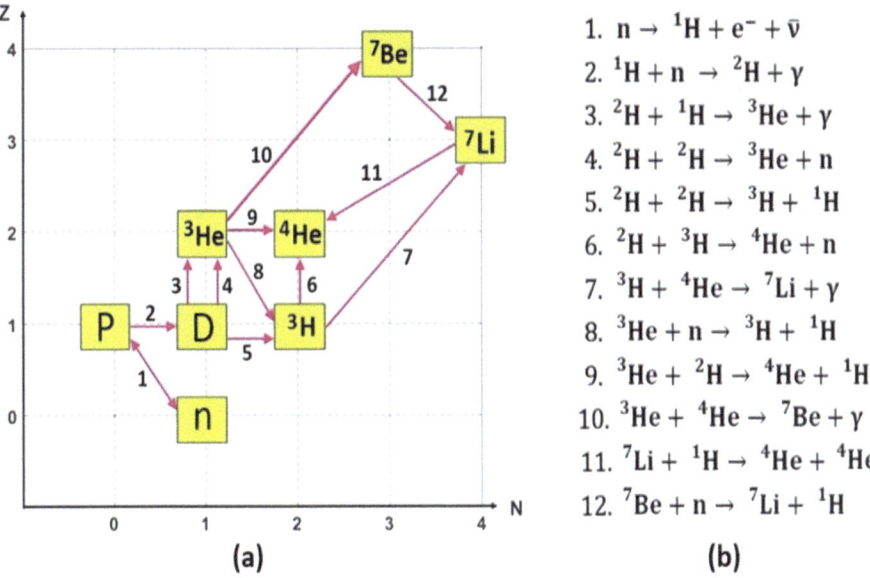

Fig. 1. Network of the most important nuclear reactions included during primordial nucleosynthesis.

Table 1. The SBBN abundance ratios normalized to hydrogen obtained in the present work and by several other works as indicated.

	This work	Pitrou et al. (2018) [23]	Cyburt et al. (2016) [22]	Observations
Yp	0.2461 ± 0.0002	0.24709 ± 0.00017	0.24709 ± 0.00025	0.2449 ± 0.0040 [24]
D/H $\times$ 10^5	2.653 ± 0.123	2.459 ± 0.036	2.58 ± 0.13	2.58 ± 0.07 [26]
^{3}He/H $\times$ 10^5	1.017 ± 0.053	1.074 ± 0.026	1.0039 ± 0.0090	1.1 ± 0.2 [25]
^{7}Li/H $\times$ 10^{10}	4.284 ± 0.378	5.623 ± 0.247	4.68 ± 0.67	$1.58^{+0.35}_{-0.28}$ [27]

3 Different Insights in the Cosmological Lithium Problem

An important observation indicating the discrepancy between the SBBN predictions and observations of the lithium abundance in the atmospheres of very metal-poor halo stars is called the Spite Plateau, as shown in Fig. 2. This plateau exhibits single values with small dispersion in the metallicity range $-3 \lesssim$ [Fe/H] $\lesssim -2$ and the abundance range $2.1 \lesssim$ A(Li) $\lesssim 2.4$ (Fig. 2 explains these quantities). As Fig. 2 shows, the lithium abundance predicted by the SBBN is higher by a factor of at least three compared with the plateau level. In addition, that figure depicts a significant dispersion of the abundances below [Fe/H] = −3.0, a "melting" of the plateau behavior. It is rather challenging to figure out one single reason to understand the behavior of the lithium abundances described above. Possible attempts are as follows.

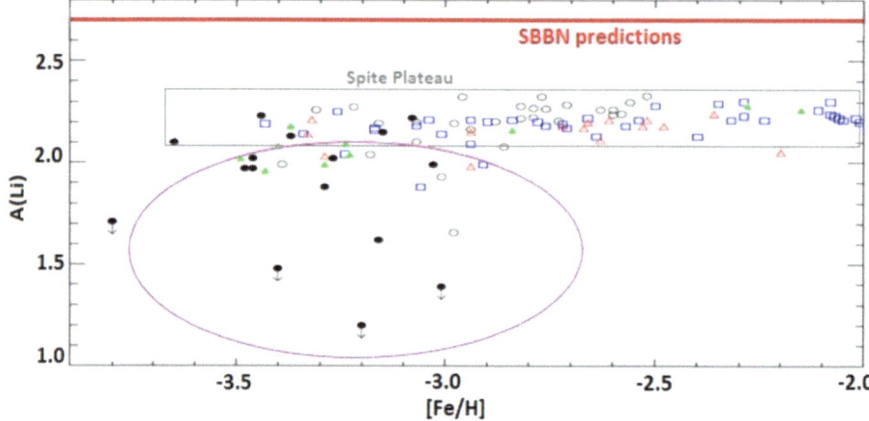

Fig. 2. Observed [30, 31] and predicted abundances A(Li) = log(^{7}Li/H) + 12 as a function of metallicity characterized by [Fe/H] = log(N_{Fe}/N_H)$_{star}$ − log(N_{Fe}/N_H)$_{Sun}$, where the N's are number densities. Note that [Fe/H] = 0 is solar. Notice the difference between the predicted lithium abundance and the observed ones especially at very low metallicity.

3.1 Stellar Evolution Aspects

The main question is: what is the origin of lithium in the stellar atmospheres? One would not necessarily expect that element in the stellar atmosphere to indicate the original composition. The existence of the plateau led to the view that it could represent the primordial values [29]. However, lithium is very fragile because it is destroyed at a temperature T ∼ 2.5 × 10^6 K. Given that the stellar atmosphere is colder, convective mixing is needed to possibly bring the lithium down to hotter layers. But how deep could this penetration be, called overshooting? This mechanism is a non-local convective mixing beyond the border of the convectively unstable region according to the "Schwarzschild criterion". The overshooting is commonly a parametrized process [32, 33]. The authors in Ref. [32] assumed that the SBBN is correct, and they used overshooting, microscopic diffusion, and residual mass accretion during the pre-main sequence evolution of very metal-poor stars of masses 0.57–0.80 M$_\odot$. Their approach led to obtaining the spite plateau, but it does not explain the dispersion below it. According to Ref. [34], they also suggested an astrophysical solution without modifying the SBBN. Such a solution needs to be supported by future observations and should consider the plateau level and the dispersion below it. Then, it is worth to include the non-standard effects we describe in the present work.

3.2 Nuclear Physics Aspects

The observed lithium abundances in metal-poor halo stars depend on the nuclear reaction rates involved in the SBBN, especially the reactions (10) and (12) listed in Fig. 1. Reaction (10) produces ^{7}Be, which enables the production of ^{7}Li via electron capture after the end of BBN. In addition, the reactions responsible for the production and destruction of deuterium can indirectly influence the final abundance of ^{7}Li since these two elements are directly related as shown in Ref. [12] and our recent calculations [10]. It is clear from the network shown in Fig. 1 that the three fundamental processes (strong nuclear, weak nuclear, and electromagnetic) indicate how fundamental the Big Bang Nucleosynthesis (BBN) is.

3.3 Modification of the SBBN

A third attempt is to investigate a modification of the SBBN. Various attempts have been proposed to resolve the lithium problem dealing with inhomogeneous nucleosynthesis [6, 7], the decay of massive particles during the SBBN [1, 3, 4], and modification of the Maxwell-Boltzmann distribution of nuclei during BBN [2]. Another approach to reducing the primordial lithium abundance was to assume photon cooling, the decay of long-lived X particles, and fluctuations of a primordial magnetic field [28]. In our previous work [10, 11], we considered a non-standard treatment of the SBBN to find out how far we can reduce the lithium abundance. We have included non-standard neutrino properties and dark components. We have been able to reduce the lithium abundance but at the expense of increasing the deuterium abundance, which was not compatible with observations. In the present work, we describe a reasonable way to avoid this drawback.

4 Modification of the Entropy During BBN

4.1 General Comments

Dark matter is a crucial component in the universe whose constituents have been the focus of many works for a long time. Different candidates of dark matter are proposed, such as gravitino dark matter [13], sterile neutrinos [14], axions [15], and unifying dark matter with dark energy [16]. Such particles may decay into other particles during BBN. For example, the decay of cold dark matter was examined to alleviate the tension between the actual Hubble constant (H_0) measured by cosmic microwave radiation and the one obtained from the distance ladder measurements from SNIa [17]. Gravitino decay modes produce hadronic and electromagnetic spectra, which interact with the background nuclei during BBN and affect their abundance [18]. However, that decay is constrained by the observed abundances of D and ^{4}He; then, for large gravitino mass (equal to or greater than 3 Tev) and a lifetime of around 103 s, the lithium problem is alleviated. It is beyond the scope of the present work to discuss all possibilities of modifying the SBBN (see Ref. [19] for details). In any case, possible suggestions have to produce the observed abundances of D and ^{4}He. In the next paragraph, we focus on investigating the impact of dark entropy on the resulting lithium abundance during BBN. We note that the adopted description of dark matter and entropy in the present work relies on the conventional model of dark matter; this is to illuminate its effect on the primordial lithium production. We think that more investigations are needed in that respect.

4.2 Variation of the Entropy During BBN

As we will see, it is important to vary the entropy content during BBN since this affects the energy conservation equation and consequently the time-temperature relation. Such a variation could result from the decay or interaction of some weakly interacting particles with the background radiation or any other source that could be modelled as a dark component. With the adiabatic expansion of the universe and no modification of its entropy content, the energy conservation equation reads:

$$\frac{d}{dt}\left(\rho_{tot}\,a^3\right) + P_{tot}\frac{d}{dt}\left(a^3\right) = 0, \tag{1}$$

where "a" is the scale parameter, ρ_{tot} and P_{tot} are the total energy density and the corresponding pressure of all constituents, namely, photons, neutrinos, baryons, electrons, and positrons. In BBN codes, Eq. (1) is transformed into the following relation:

$$\frac{d}{dT}\left(lna^3\right) = -\frac{\frac{d\rho_{tot}}{dT}}{\rho_{tot} + P_{tot}} \tag{2}$$

Having $\frac{d}{dT}(\ln a^3)$, the time-temperature relation is implemented in BBN codes as follows:

$$\frac{dT}{dt} = 3H/\frac{d}{dT}\left(lna^3\right), \tag{3}$$

where H is the Hubble parameter. Introducing the dark entropy into Eq. (1), it takes the following form:

$$\frac{d}{dt}\left(\rho_{tot}\,a^3\right) + P_{tot}\frac{d}{dt}\left(a^3\right) + T\frac{d}{dt}(f_D a^3) = 0 \tag{4}$$

In this equation, we represent the modification of the energy conservation equation by a function f_D that could take many forms. In this work, f_D is the temperature-dependent dark entropy density. This function is used to modify Eq. (2) which becomes:

$$\frac{d}{dT}\left(lna^3\right) = -\frac{\frac{d\rho_{tot}}{dT} + T\frac{df_D}{dT}}{\rho_{tot} + P_{tot} + Tf_D} \tag{5}$$

Varying the entropy content of the early universe was studied in our previous work [11]. However, in the present work, the variation of the entropy is restricted to the temperature range $T = (4.3 - 9.1) \times 10^8$ K. This modification will affect the final element abundances. We adopt the proposal in Ref. [21], where a unified fluid is adopted to describe dark energy and dark matter as two different aspects of the same component. This is represented by temperature-dependent components of dark energy density ρ_D and dark entropy s_D as follows:

$$\rho_D(T) = k_\rho \times \rho_{rad}(T_0) \times \left(\frac{T}{T_0}\right)^{n_\rho} \tag{6}$$

$$s_D(T) = k_s \times s_{rad}(T_0) \times \left(\frac{T}{T_0}\right)^{n_s} \tag{7}$$

where $\rho_{rad}(T_0)$ and $s_{rad}(T_0)$ are the radiation and entropy energy densities respectively at $T_0 = 1.0 \, \text{Mev} = 11.6 \, \text{GK}$, k_ρ is the ratio of the effective dark fluid energy density over the total radiation at T_0, and k_s is the ratio of the effective dark fluid entropy density over the total entropy density at T_0. The exponents n_ρ and n_s characterize the behaviors of dark energy density and entropy; for example, $n_s = 3$ corresponds to a radiation behavior, while $n_s = 1$ characterizes the entropy behavior that appears in reheating models. In this case, $f_D = s_D$ in Eq. (4). The effects of adding a dark entropy component in the temperature range $(4.3 - 9.1) \times 10^8$ K and from the beginning to the end of BBN are summarized in Table 2.

Table 2. Effect of dark entropy on light elements when inserted in the temperature range $T = (4.3 - 9.1) \times 10^8$ K and from the beginning till the end of BBN ($T = 0.001 - 100$ GK). We show the parameters of dark entropy as implemented in AlterBBN code and the resulted abundances of light elements.

Dark entropy parameters	Temperature range	Yp	D/H $\times 10^5$	^{7}Li/H $\times 10^{10}$
$n_s = 1$, $k_s = 10^9$	$T = (4.3 - 9.1) \times 10^8$ K	0.2516 ± 0.0005	2.690 ± 0.202	2.249 ± 0.374
$n_s = 1$, $k_s = 10^9$	$T = 0.001 - 100$ GK	0.2993 ± 0.0001	1.517 ± 0.081	10.06 ± 0.925
$n_s = 0.8$, $k_s = 9 \times 10^8$	$T = (4.3 - 9.1) \times 10^8$ K	0.2506 ± 0.0004	2.725 ± 0.216	2.285 ± 0.434
$n_s = 0.8$, $k_s = 9 \times 10^8$	$T = 0.001 - 100$ GK	0.2791 ± 0.0001	2.360 ± 0.112	5.725 ± 0.515
$n_s = 0.5$, $k_s = 10^{10}$	$T = (4.3 - 9.1) \times 10^8$ K	0.2531 ± 0.0006	3.091 ± 0.312	1.544 ± 0.374
$n_s = 0.5$, $k_s = 10^{10}$	$T = 0.001 - 100$ GK	0.3001 ± 0.0001	2.974 ± 0.126	4.734 ± 0.420
Standard Big Bang Nucleosynthesis (SBBN)		0.2461 ± 0.0002	2.653 ± 0.123	4.284 ± 0.378

The results in Table 2 indicate that using the description of the dark entropy as in Eq. (7), the decrease of the lithium abundance while keeping the helium and deuterium abundances compatible with the observations ($n_s = 1$, $k_s = 9 \times 10^9$) is possible only if we apply Eq. (7) in the temperature range $T = (4.3 - 9.1) \times 10^8$ K. The consequence of this restriction alters the time-temperature relation and affects the nuclear reaction rates shown in Fig. 1. Other combinations of n_s and k_s are possible to achieve similar results ($n_s = 0.8$, $k_s = 9 \times 10^8$). However, if we want to reach the lower level of the plateau, or even go below it, we must relax the constraint on deuterium. In his case, we

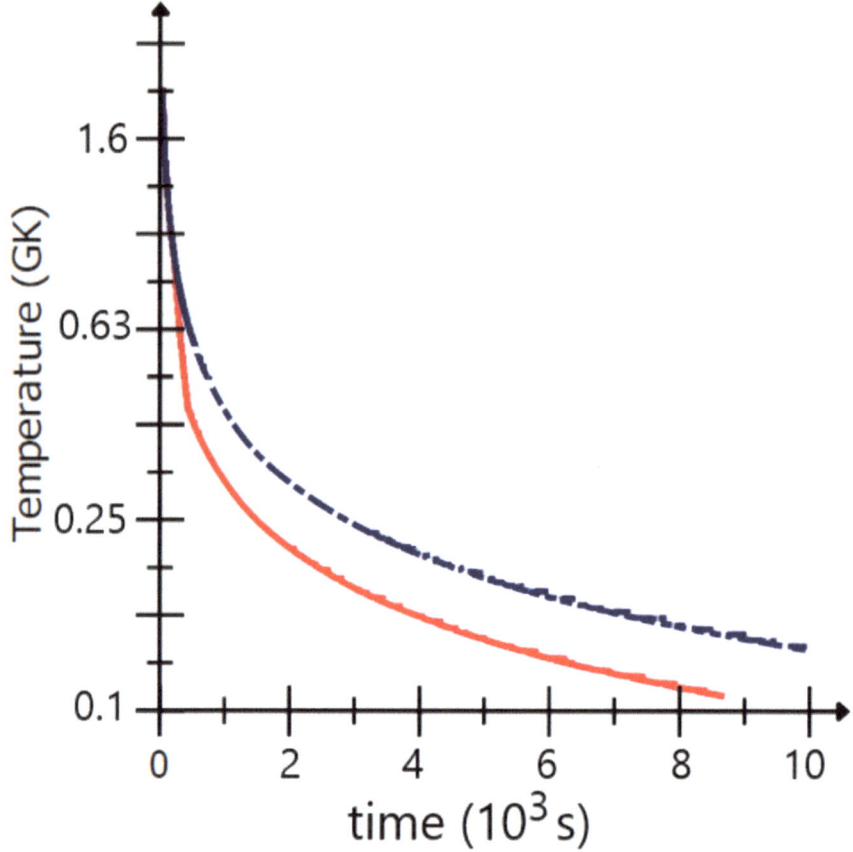

Fig. 3. The variation of the temperature as a function of time during the SBBN (blue dashed line) and when including the dark entropy component in the temperature range $T = (4.3 - 9.1) \times 10^8$ K (red solid line).

can argue that this increase in deuterium could be accepted because deuterium is fragile and can be destroyed in stars. For the same set of parameters, adding the dark entropy from the beginning till the end of BBN ($T = 0.001 - 100$ GK) violates the observational constraints on D and ^{4}He. We emphasize that the resulting lithium abundance is the sum of ^{7}Li and ^{7}Be; more precisely, ^{7}Be has a crucial role in determining the lithium abundance since it represents more than 90% of the final lithium abundance. In the following, we present the abundances of light elements as a function of temperature resulting from the present treatment.

4.3 Effect of Dark Entropy on the ^{4}He Abundance

Figure 4 shows the helium abundance as a function of temperature. Inserting the dark entropy component during BBN does not alter the shapes of the abundance profile. Including this entropy component during the whole BBN epoch ($T = 0.001 - 100$

GK) increases the helium abundance, which contradicts the observations. However, the helium abundance agrees with the value from the SBBN and observations if the dark entropy is applied in the range T = $(4.3 - 9.1) \times 10^8$ K. To understand this result, we note the following. Helium is affected by two important stages: (i) freeze-out of weak interaction reactions ($n + \nu_e \leftrightarrow p + e^-, n + e^+ \leftrightarrow p + \overline{\nu}_e, n \leftrightarrow p + e^- + \overline{\nu}_e$), (ii) deuterium bottleneck (its maximum) during BBN. Since the dark entropy causes a faster decrease of the temperature (see Fig. 3) during BBN, the deuterium abundance reaches its bottleneck at a higher temperature or earlier (see green dotted curve in Fig. 5), which leads to an increase in the helium abundance (green dotted curve in Fig. 4). This relation between helium and deuterium bottleneck is given explicitly in Ref. [12] where the final abundance of helium is proportional to $\exp(-t_N/\tau_n)$, with t_N is the time of the bottleneck and τ_n is the neutron lifetime. However, restricting the dark entropy in the adopted range does not affect the helium abundance (red dashed line curve in Fig. 4). This is because the time of the deuterium bottleneck is not shifted although the abundance of the bottleneck is higher than that of the SBBN (red dashed line in Fig. 5). In addition, the freeze-out temperature is not affected because it happened at higher temperatures, so the ^{4}He abundance remains in the acceptable range as Fig. 4 shows.

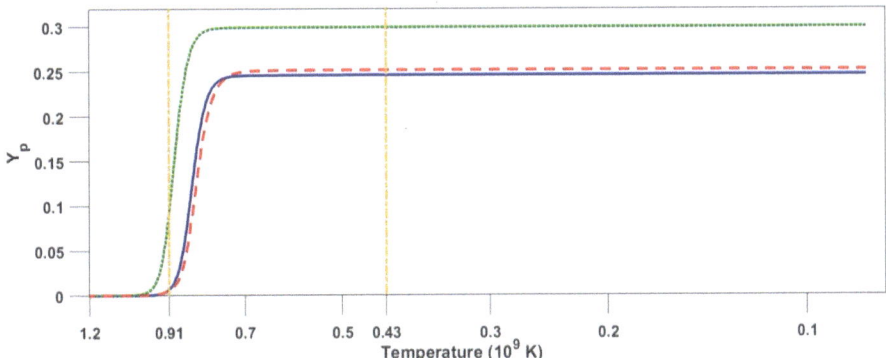

Fig. 4. The abundance of helium-4 based on the SBBN (blue solid line), when including the dark entropy during the BBN epoch (green dotted line), and when including the dark entropy in the temperature range T = $(4.3 - 9.1) \times 10^8$ K (red dashed line).

4.4 Effect of Dark Entropy on the Deuterium Abundance

It is emphasized that the modification of the entropy content took place before two important phases: (i) the deuterium bottleneck (its maximum), and (ii) when the neutron abundance becomes comparable to that of deuterium. As seen in Fig. 5, the deuterium abundance increases at the bottleneck but converges to its standard value below 0.43 GK (the red dashed curve converges to the blue solid curve). It is clearly shown in Table 2 that the final deuterium abundance 2.690×10^{-5} is approximately equal to the resulting abundance of the SBBN within the uncertainties. This result is due to the efficiency of the reaction rates shown in Fig. 6.

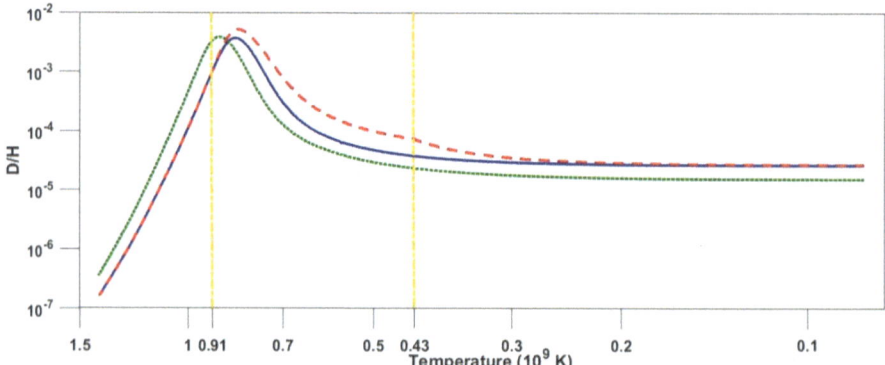

Fig. 5. The deuterium abundance based on the SBBN (blue solid line), when including the dark entropy during the BBN epoch (green dotted line), and when including the dark entropy in the temperature range $T = (4.3 - 9.1) \times 10^8$ K (red dashed line).

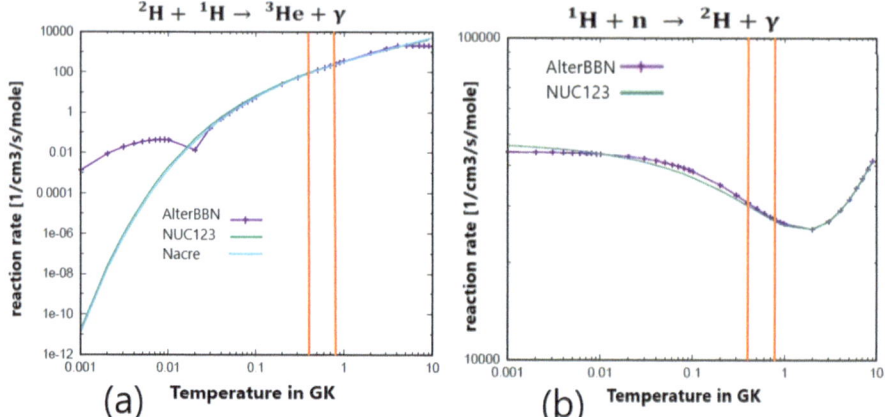

Fig. 6. Example of two reaction rates responsible for the destruction (a) and production (b) of deuterium. The vertical lines represent the restricted temperature range.

4.5 Effect of Dark Entropy on ^{7}Be and ^{7}Li Abundances

The lithium abundance decreases significantly to 2.249×10^{-10} (see Table 2) to match closely the observed plateau level. This is due to adding the dark entropy in the restricted temperature range $T = (4.3 - 9.1) \times 10^8$ K, which leads to a faster decrease in the temperature shown in Fig. 3. Figure 7 shows the abundance of ^{7}Be as a function of temperature. A similar profile is obtained in all three cases: (a) SBBN, (b) dark entropy applied in the range of BBN, and (c) in the restricted temperature range we have adopted. In case (c), the resulting lower abundance reflects the behavior of the reaction rates as shown in Fig. 8.

The inspection of Fig. 8a reveals that the production of ^{7}Be through the reaction ^{3}He + ^{4}He → ^{7}Be + γ is not enough compared to the destruction by the reaction ^{7}Be + n → ^{7}Li + p (Fig. 8b). The present investigation indicates the importance of the

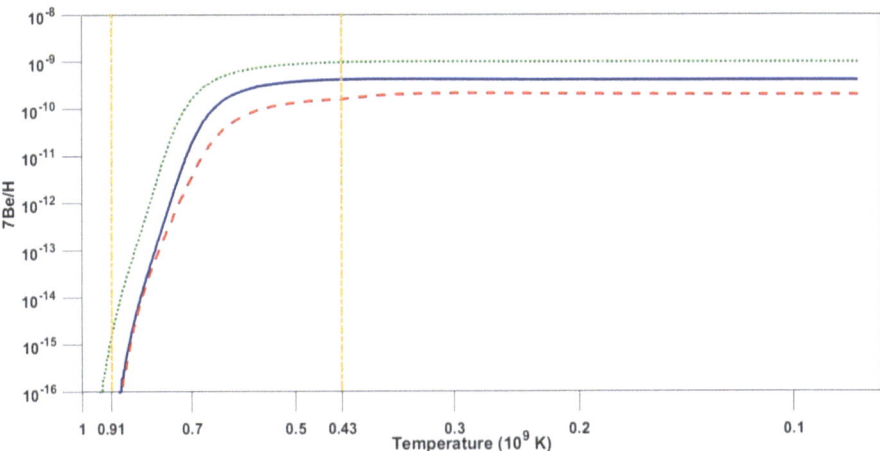

Fig. 7. The abundance of beryllium based on the SBBN (blue solid line), when including the dark entropy during the BBN epoch (green dotted line), and when including the dark entropy in the temperature range $T = (4.3 - 9.1) \times 10^8$ K (red dashed line).

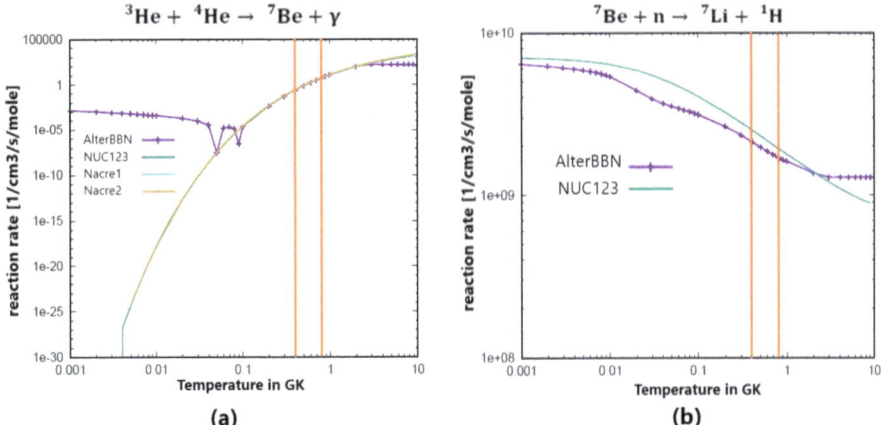

Fig. 8. The main reaction rates responsible for the production (a) and destruction (b) of ^{7}Be. The vertical lines represent the restricted temperature range.

time evolution of the temperature and its impact on the reaction rates in determining the abundances of ^{7}Be and consequently ^{7}Li. Then, the desired decrease in the abundance of ^{7}Li is achieved without affecting the observational constraints on helium and deuterium.

5 Another View of the Modification of the Time-Temperature Relation

In Sect. 4, we modified the energy conservation equation given in Eq. (1), which affected the time-temperature relation. Here, we will directly perturb the time-temperature relation to emphasize that altering the energy conservation equation by a function f_D implies a modification of the time-temperature relation. In this way, we could reach a substantial decrease in the lithium abundance without increasing the deuterium abundance using different parametrizations. The perturbation is introduced by a function g(T) added to Eq. (3) in the range $T = (4.3 - 9.1) \times 10^8$ K, so that Eq. (3) becomes:

$$\frac{dT}{dt} = 3H / \frac{d}{dT}\left(lna^3\right) + 3H \times g(T), \qquad (8)$$

Interesting to find an expression to this function, for example by a linear fitting. Our trial yields:

$$g(T) = 1.64 - 3.72T \qquad (9)$$

Modifying the time-temperature relation in this way leads to the results in Table 3 showing that the lithium abundance has significantly decreased without altering the abundances of helium and deuterium from the SBBN.

Table 3. The effect of modifying the time-temperature relation of the early universe by a function g(T) in the temperature range $T = (4.3 - 9.1) \times 10^8$ K.

	SBBN	using g(T)
Yp	0.2461 ± 0.0002	0.2549 ± 0.0008
D/H $\times 10^5$	2.653 ± 0.123	2.624 ± 0.119
^{7}Li/H $\times 10^{10}$	4.284 ± 0.378	2.204 ± 0.222

6 Summary and Conclusion

The standard Big Bang nucleosynthesis (SBBN) is a fundamental process in cosmology responsible for the formation of light elements up to lithium (^{7}Li), which needs extended work to resolve the discrepancy between its predicted lithium abundance and the observed abundance in very metal-poor halo stars. To resolve this so-called "Cosmological Lithium Problem", we need an extension of the SBBN before drawing conclusions based on stellar modelling. In our previous works [10, 11], we have implemented non-standard treatments of BBN by varying neutrino properties and adding dark components. We have achieved a decrease in the lithium abundance but only combined with an increase of the deuterium abundance, which does not match the accurate observations. In the present work, we have invoked an additional treatment based on modifying the energy conservation equation by adding a dark entropy term (see Sect. 4.2). This resulted (Sect. 4 for details) in relaxing the tight connection between deuterium and lithium. With this approach, we figured out that the dark entropy term with its used parametrized form (Sect. 4.2) cannot be applied through the whole temperature range of BBN, but in a restricted range $T = (4.3 - 9.1) \times 10^8$ K. This restriction allows to reduce the lithium abundance to match closely the observations and satisfy the constraints on helium and deuterium (see Table 2). Additionally, as seen in Sect. 5, we find a perturbation term g(T) describing the behavior of the time-temperature relation in the restricted range above. Finally, we emphasize that including the effect of the dark entropy component during BBN, which affects the time-temperature relation and relevant nuclear reactions, is a crucial issue in resolving the cosmological lithium problem. Stellar evolution modelling cannot simply assume that the SBBN is just correct.

References

1. Goudelis, A., et al.: Light particle solution to the cosmic lithium problem. Phys. Rev. Lett. **116**, 211303 (2016)
2. Hou, S.Q., et al.: Non-extensive statistics to the cosmological lithium problem. ApJ **834**, 165 (2017)
3. Jedamzik, K.: Neutralinos, big bang nucleosynthesis, and 6Li in low-metallicity stars. Phys. Rev. D **70**, 083510 (2004)
4. Jedamzik, K.: Did something decay, evaporate, or annihilate during big bang nucleosynthesis? Phys. Rev. D **70**, 063524 (2004)
5. Kusakabe, M., et al.: Review on effects of long-lived negatively charged massive particles on Big Bang Nucleosynthesis. Int. J. Mod. Phys. E **26**(08), 1741004 (2017)
6. Mathews, G.J., et al.: Coupled baryon diffusion and nucleosynthesis in the early universe. ApJ **358**, 36 (1990)
7. Nakamura, R., et al.: Big-Bang nucleosynthesis: constraints on nuclear reaction rates, neutrino degeneracy, inhomogeneous and Brans–Dicke models. Int. J. Mod. Phys. E **26**(8), 1741003 (2017)
8. Scherrer, R.J., Turner, M.S.: Primordial nucleosynthesis with decaying particles. II. Inert Decays. ApJ. **331**, 33–37 (1988)
9. Scherrer, R.J., Turner, M.S.: Primordial nucleosynthesis with decaying particles. I. Entropy-producing decays. ApJ. **331**, 19–32 (1988)
10. Makki, T.R., El Eid, M.F., Mathews, G.J.: A critical analysis of the Big Bang Nucleosynthesis. MPLA **34**, 1950194 (2019)
11. Makki, T.R., El Eid, M.F., Mathews, G.J.:Impact of neutrino properties and dark matter on the primordial lithium production. Int. J. Mod. Phys. E **28**(08), 1950065 (2019)
12. Physical Foundations of Cosmology, Cambridge University Press, New York, ISBN 9780521563987 (2005)
13. Steffen, F.D.: Gravitino dark matter and cosmological constraints. JCAP **2006**(09), 001 (2006)
14. Abazajian, K.: Sterile neutrino hot, warm, and cold dark matter. Phys. Rev. D **64**, 023501 (2001)
15. Duffy, L.D., van Bibber, K.: Axions as dark matter particles. New J. Phys. **11**(10), 105008 (2009)
16. Arbey, A.: The unifying dark fluid model. AIP Conf. Proc. **1241**, 700–707 (2010)
17. Vattis, K., Koushiappas, S.M., Loeb, A.: Dark matter decaying in the late Universe can relieve the H_0 tension. Phys. Rev. D **99**, 121302 (2019)

18. Luo, F.: Thesis dissertation, The Effects of Supersymmetric Particle Decays and Annihilations on Big-Bang Nucleosynthesis, UNIVERSITY OF MINNESOTA (2012)
19. Menestrina, J.L., Scherrer, R.J.: Dark radiation from particle decays during big bang nucleosynthesis. Phys. Rev. D **85**, 047301 (2012)
20. Arbey, A.: AlterBBN: a program for calculating the BBN abundances of the elements in alternative cosmologies. Comput. Phys. Commun. **183**, 1822–1831 (2012)
21. Arbey, A., Mahmoudi, M.: SUSY constraints, relic density, and very early universe. JHEP **2010**, 51 (2010)
22. Cyburt, R.H., Fields, B.D., Olive, K.A.: Big bang nucleosynthesis: present status. Rev. Mod. Phys. **88**, 015004 (2016)
23. Pitrou, C., et al.: Precision big bang nucleosynthesis with improved Helium-4 predictions. Phys. Rept. **754**, 1–66 (2018)
24. Aver, E., et al.: The effects of He I λ10830 on helium abundance determinations. JCAP **2015**(07), 011 (2015)
25. Bania, T.M., et al.: The cosmological density of baryons from observations of 3He+ in the Milky Way. Nature **415**, 54 (2002)
26. Cooke, R.J., et al.: Precision measures of the primordial abundance of deuterium. ApJ **781**, 31 (2014). Cooke, R.J., et al.: The primordial deuterium abundance of the most metal-poor damped Lyα system. ApJ **830**, 148 (2016)
27. Sbordone, L., et al.: The metal-poor end of the Spite plateau. A&A **522**, A26 (2010)
28. Yamazaki, D.G., et al.: The new hybrid BBN model with the photon cooling, X particle, and the primordial magnetic field. Int. J. Mod. Phys. E **26**(8), 1741006 (2017)
29. Brian, D.: Fields: the primordial lithium problem. Annu. Rev. Nucl. Part. Sci. **61**, 47–68 (2011)
30. Iocco, F.: The lithium problem, a phenomenologist's perspective. arXiv: http://arxiv.org/abs/1206.2396v2 (2012)
31. Bonifacio, P., Sbordone, L., Caffau, E., et al.: Chemical abundances of distant extremely metal-poor unevolved stars. A&A **542**, A87 (2012). Sbordone, L., Bonifacio, P., Caffau, E., et al.: A&A **522**, A26 (2010). Aoki, W., Barklem, P.S., Beers, T.C., et al.: ApJ **698**, 1803 (2009). Hosford, A., Ryan, S.G., García Pérez, et al.: A&A **493**, 601 (2009). Asplund, M., Lambert, D.L., Nissen, et al.: ApJ **644**, 229 (2006). González-Hernández, J.I., Bonifacio, P., Ludwig, H.-G., et al.: A&A **480**, 233 (2008)
32. Bressan, A., et al.: Uncertainties in stellar evolution models: convective overshoot. In: ASSP, vol. 39, p. 25, Springer, Cham (2015)
33. Xiaoting, F., et al.: Lithium evolution in metal-poor stars: from pre-main sequence to the Spite plateau. MNRAS **452**, 3256 (2015)
34. Brian, D.F., Olive, K.A.: Implications of the non-observation of 6Li in Halo Stars for the Primordial 7Li Problem. JCAP **2022**(10), 078 (2022)

Neutrinos from Type-II Supernovae

Leen Binchi$^{(\boxtimes)}$ and Amine Ahriche

Department of Applied Physics and Astronomy, University of Sharjah, Sharjah, UAE
leenbinchi@gmail.com

Abstract. In our work we investigate the stellar surface fluxes of different neutrino flavors, both with and without matter effects, while aiming to consider the varying density of the progenitor star to reveal intriguing variations in their behavior. By examining both normal and inverted neutrino mass hierarchies in studies of neutrino fluxes, a behavioral difference is detected which highlights the significance of the chosen mass hierarchy for studies in understanding the neutrino oscillation phenomena. This difference in flux behavior also serves as a clue to probe the mass hierarchy; within the energy range of 13–19 meV, the neutrino mass hierarchy can be determined as normal or inverted by detecting excess or missing events in the non-electronic neutrinos. We also examine the level crossing diagrams of neutrinos to illustrate how different neutrino masses and energies affect the transformation of neutrino flavors as they propagate through matter with constant density. We determine the resonance regions (both high and low) with the highest flipping probability using the level crossing diagrams. This allows us to estimate where within a star we are most likely to find neutrino flavor conversions.

Keywords: Neutrinos · Supernovae · Adiabatic Conversion

1 Introduction

With the continuously growing understanding of neutrinos and their properties, their application in various research fields has expanded. Neutrinos have been frequently observed to contribute to or explain energy-related interactions in astronomical objects. They have also started playing an increasingly significant role in astrophysical studies by serving as a method for studying astrophysical phenomena as an alternative to electromagnetic methods. Supernovae are among the objects that can be observed through their neutrino emissions, as was the case with supernova 1987a. This supernova is unique for several reasons, including the fact that its progenitor star was a blue giant and that it is the nearest supernova to Earth that has ever been observed. This supernova was the first to be observed through its neutrino burst before its electromagnetic burst [14].

In February 1987, The Japanese Kamiokande-II and the Irvine-Michigan Brookhaven (IMB) Detector both detected neutrino emissions coming from the Large Magellanic Cloud where the supernova was later seen to take place. This observation confirmed the fact that not only do neutrinos leave the supernova before the electromagnetic burst occurs, but that neutrinos may truly in fact play a key role in the supernova explosion process. More accurately, neutrinos are believed to participate in the core collapse

H. M. K. Al Naimiy et al. (Eds.): AUASS-CONF 2023, SPPHY 420, pp. 36–53, 2025.
https://doi.org/10.1007/978-981-96-3276-3_4

mechanism of the progenitor star. This information amplifies the importance of studying supernova neutrinos in order to further understand supernovae [13].

It is generally understood that neutrinos have the ability to change flavors. However, what truly highlights the significance of these supernova neutrinos is the fact that the mechanism for flavor switching can be influenced by the density of their medium or a changing density profile. This, coupled with their extremely long mean free path compared to photons, underscores the importance of these particles. The impact of a density profile on the flavor flipping mechanism, known as the MSW effect, allows us to dissect the different inner layers of the progenitor star [9, 17]. Although, the observed supernova neutrino spectrum at underground detectors can be significantly affected by non-standard neutrino interactions [1].

In this paper, we delve into the neutrino emissions detected during SN1987A, with a focus on analyzing the energy spectra, neutrino flavor oscillation, and resonance regions. Additionally, we inspect neutrino level crossing schemes, flux, and compare different distribution functions used to estimate neutrino flux. Finally, we examine the predicted and observed behavior of neutrinos in both normal and inverted mass hierarchy schemes, considering a general density profile for the star.

2 Supernova Neutrino Energies

Neutrinos that have been unleashed after being trapped in the core travel through the star's envelope while preserving the initial energy distribution. The average energy is dictated by the temperature at the neutrino-sphere. Initially, supernovae neutrino spectra were thought to possess a Maxwell-Boltzmann distribution ($\phi \propto E_v^2/e^{Ev/T}$) or a Fermi-Dirac distribution ($\phi \propto E_v^2/(e^{Ev/T} + 1)$), or even a quasi-thermal distribution ($\phi \propto E_v^\gamma e^{-(\gamma+1)Ev/E0}$) [15].

These behaviors, however, do not necessarily align with the experimental neutrino spectra detected [10]. This discrepancy arises from contributions made by factors such as neutrino flavor oscillations, the MSW effect, and the varying mass distribution of the supernova's layers. A more precise approximation of the neutrino spectra, known as the power-law distribution, will be introduced later. The neutrino spectrum of SN1987A in Fig. 1 displays individual and combined datasets from the IMB and Kamiokande detectors, considering their target and cross sections. The shaded region represents a Fermi-Dirac spectrum with an average energy of $E_0 = 15, MeV$ and a luminosity of $L = 2.2 \times 10^{53}, erg$ [21]. In terms of individual neutrino flavor mean energies, both Kamiokande II and IMB detectors measured electron and muon neutrinos energies to be [8]

$$\begin{aligned} <E_{ve}> &= 11.0 \pm 1.2 MeV, \\ <E_{v\mu}> &= 17.3 \pm 2.2 MeV. \end{aligned} \tag{1}$$

The detection of this event were dominated by anti-electron neutrinos in both Kamiokande II and IMB rather than electron neutrinos due to anti-electron neutrinos having a higher interaction cross section than electron neutrinos. This is because they experience both elastic scattering and inverse beta decay interactions rather than just elastic scattering like electron neutrinos do. The fact that electron neutrinos would most likely be detected in a different flavor also plays a role in the lack of their detection in comparison to the detection of muon and tau neutrinos. This however does not change the fact that the neutrino burst after the core bounce is dominated by electron neutrinos [11].

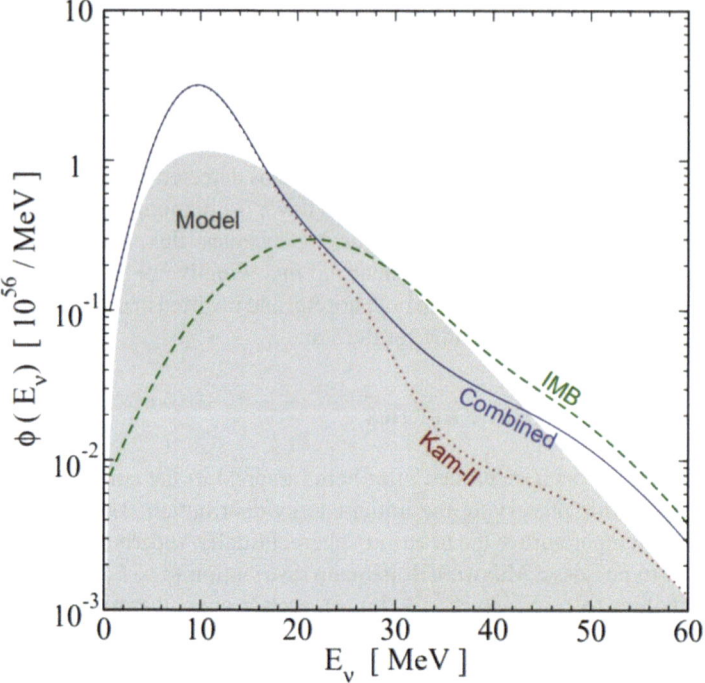

Fig. 1. SN1987A Neutrino energy spectrum [21].

Immediately after the core bounce, the matter density below the shock wave decreases dramatically allowing the neutrinos to escape freely, producing the phenomenon of neutronization. Furthermore, the processes of electron capture and the neutrino burst lead to a reduction in the lepton number in the core which leads to the ability of all neutrino flavors to be produced [10, 20]. The effect of these processes can be seen in the behavior of neutrino luminosity with time in Fig. 2. When comparing the behavior of the 2 spectra, it's been understood that the cooling stage spectra results in a narrower curve in comparison to that of the thermal neutrinos. The narrowing of the curve is referred to as "pinching" and is a consequence of 2 processes. The first process is the decrease of the temperature in the supernova as the radius increases. The second process is the decrease in density alongside the temperature at a rate faster than the common $\frac{1}{r}$. The normalized Fermi-Dirac distribution function in terms of energy and pinching parameter η can be described as

$$f_{\nu\beta}(E, \eta_{\nu\beta}) = \frac{1}{1 + exp[E/T_{\nu\beta} - \eta_{\nu\beta}]}, \qquad (2)$$

where the neutrino temperature is considered and is linearly proportional to the mean energy when the pinching parameter $\eta_{\nu\beta} = 0$. When the pinching parameter however is $\eta_{\nu\beta}l = 0$, the mean energy is then proportional to the square of the temperature.

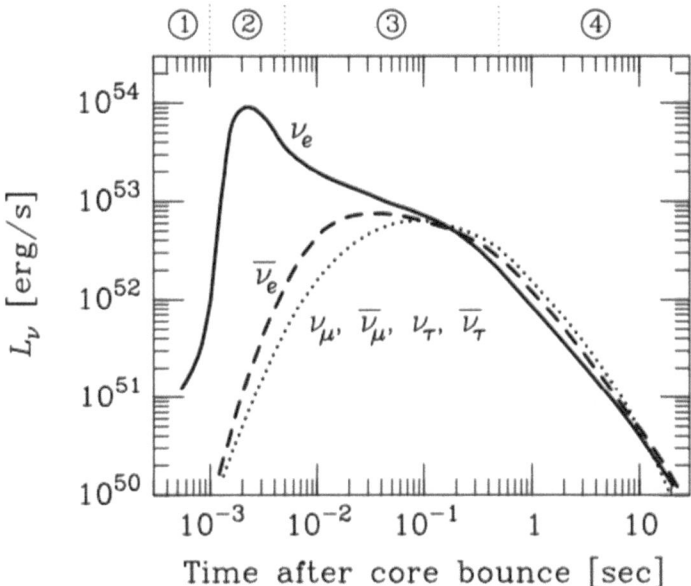

Fig. 2. Luminosities of neutrinos and antineutrinos versus time on 4 stages. 1: core collapse and bounce; 2: shock propagation and neutrino burst; 3: matter accretion and mantle cooling; 4: neutron star cooling [20]

3   Level Crossing Schemes and Neutrino Flux

Neutrino level crossings arc synonymous to flavor oscillations which we now know occur because neutrinos have non zero masses. To analyze how the neutrino flavor mixing behavior changes with the changing density of a star, we must understand the density distribution. The density within a star decreases the further away we move from the core, in other words the greater the distance traveled outwards, the more the density decreases [8]. Accordingly, neutrinos exhibit 2 resonances following the MSW effect called H and L (which stand for High matter density and Low matter density) which are regions at which the flavor conversion mechanism is at its greatest efficiency [19].

In supernovae, the high resonance density regions, which corresponds to the square mass difference of atmospheric neutrinos ($\Delta m^2_{atm} = \Delta m^2_{32} \approx 10^{-3}$ eV with $\sin^2 2\theta = 0.8$ − 1), is approximately $\rho_H = 10^3 \rightarrow 10^4$ g/cm^3 [12]. The low density resonance regions however, have densities that range depending on the mixing angle and are characterized by the square mass difference of solar neutrinos ($\Delta m^2_{sol} = \Delta m^2_{21} \approx 10^{-6} - 10^{-5} eV^2$

for the respective small, or large, mixing angles) where [3]

$$\rho_L = \begin{cases} 5 - 15\left(g/cm^3 \right) & Small(\theta) \\ 10 - 30\left(g/cm^3 \right) & Large(\theta) \end{cases} \tag{3}$$

An accurate calculation of the resonance region density can be conducted using

$$\rho_{res} \approx \frac{1}{2\sqrt{2}G_F} \times \frac{\Delta m^2}{E} \times \frac{m_N}{Y_e} \times \cos 2\theta. \tag{4}$$

where Y_e is the electron fraction, m_N is the proton mass, and G_F is the Fermi constant.

In a level crossing diagram drawn with respect to energy v.s. electron density, such as that in Fig. 3 it can be understood that higher electron densities equate to regions close to or inside of the core. Decreasing electron densities equate to moving away from the core and towards the outer layers of the star [6]. With our main concern being the normal mass hierarchy, in Fig. 3 it is illustrated how there exist regions of resonance at low and high densities at which flavor flipping occurs. Focusing on positive electron density regions, it can be seen that at low matter density resonance, flavor switching between electron and muon neutrinos is most probable. While at high matter density, the flavor switching of electron and tau neutrinos is most probable. There exists a relation between the neutrino level crossings and their observed fluxes. The flux of a particular flavor of neutrinos can be affected by level crossings. Detectors built to detect a particular flavor of neutrinos could end up detecting a lot less than expected due to these level crossings and the change of a neutrino's flavor from the desired one to another flavor. In cases where sources produce neutrinos of all 3 flavors however, detectors would detect different values from what is expected for each flavor also due to the level crossings.

The distinction between the normal and inverted mass hierarchies when considering the neutrinos under discussion can be attributed to the sign of Δm_{32}^2. The sign determines the type of mass hierarchy. The order is [4]

$$\Delta m_{32}^2 > 0; \, m_3 > m2, m1, \tag{5}$$

$$\Delta m_{32}^2 < 0; \, m2, m1 > m3. \tag{6}$$

When situation (5) is satisfied, the spectrum follows a normal mass hierarchy, while when (6) is satisfied, the spectrum follows an inverted mass hierarchy.

The densities of star cores are much greater in comparison to the kind of densities discussed during level crossings (ρ_H, and ρ_L). At such extreme densities, neutrino level crossings or flavor oscillations are suppressed and do not occur, instead the neutrino flavor coincides with the eigenstates in the medium ($\nu_{1m}, \nu_{2m}, \nu_{3m}$); In the normal hierarchy we find that

$$\nu_{1m} = \nu_\mu, \, \nu_{2m} = \nu_\tau, \, \nu_{3m} = \nu_e, \, \bar{\nu}_{1m} = \bar{\nu}_e,$$
$$\bar{\nu}_{2m} = \bar{\nu}_\mu, \, \bar{\nu}_{3m} = \bar{\nu}_\tau.$$

While for the inverted hierarchy we see that

$$\nu_{1m} = \nu_\mu, \, \nu_{2m} = \nu_e, \, \nu_{3m} = \nu_\tau, \, \bar{\nu}_{1m} = \bar{\nu}_\tau,$$
$$\bar{\nu}_{2m} = \bar{\nu}_\mu, \, \bar{\nu}_{3m} = \bar{\nu}_e.$$

$$\rho_{res} \approx \frac{1}{2\sqrt{2}G_F} \times \frac{\Delta m^2}{E} \times \frac{m_N}{Y_e} \times \cos 2\theta.$$

(4)

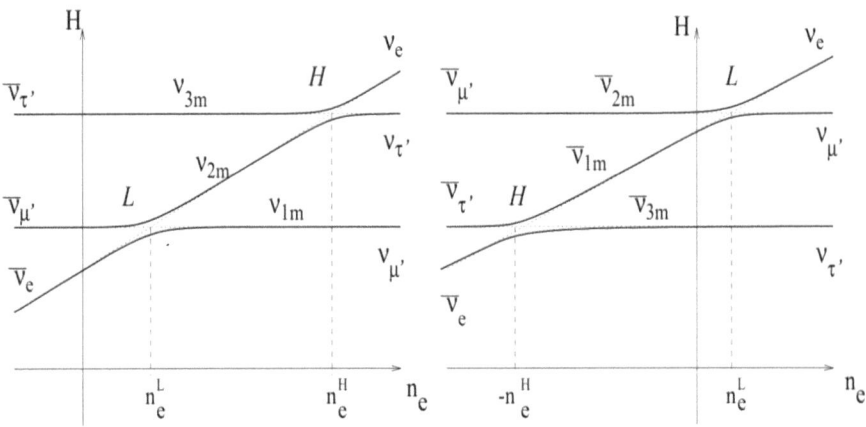

Fig. 3. The level crossing diagrams for normal (left) and inverted (right) mass hierarchy. The horizontal dotted lines represent the neutrinos' energies while the vertical dotted lines represent the electron densities at which resonance occurs.

These level crossing diagrams have been plotted to highlight the role of the absolute mass and the energy. As can be seen from the plots in Fig. 4, the value of the absolute mass drastically changes the shape of the plot in relation to at what masses' ranges do you encounter the least mass square difference yielding to different resonances. An important thing to note is that the logarithmic scale of the masses allows us to see the vast difference in values between the minimal square difference at low and high resonances. The plots were obtained using the Hamiltonian matrix.

$$H = \frac{1}{2E} \begin{pmatrix} m_{ee}^2 + A_{CC} & m_{e\mu}^2 & m_{e\tau}^2 \\ m_{\mu e}^2 & m_{\mu\mu}^2 & m_{\mu\tau}^2 \\ m_{\tau e}^2 & m_{\tau\mu}^2 & m_{\tau\tau}^2 \end{pmatrix}$$

(7)

We can now take our understanding of resonance regions a step further by attempting to roughly locate how far away from the core we can expect high and low resonances to take place. To do so, we must use the generalized electron density [7]

$$\rho_e = \rho_0 \left(\frac{r}{r_0}\right)^{-3}$$

(8)

And at resonance, the electron density at resonance is, with the use of the non adiabatic gamma parameter

$$\gamma = \frac{\Delta m^2}{2E} \frac{\sin^2 2\theta}{\cos 2\theta} \frac{1}{\frac{1}{\rho_e}\frac{d\rho_e}{dx}}$$

(9)

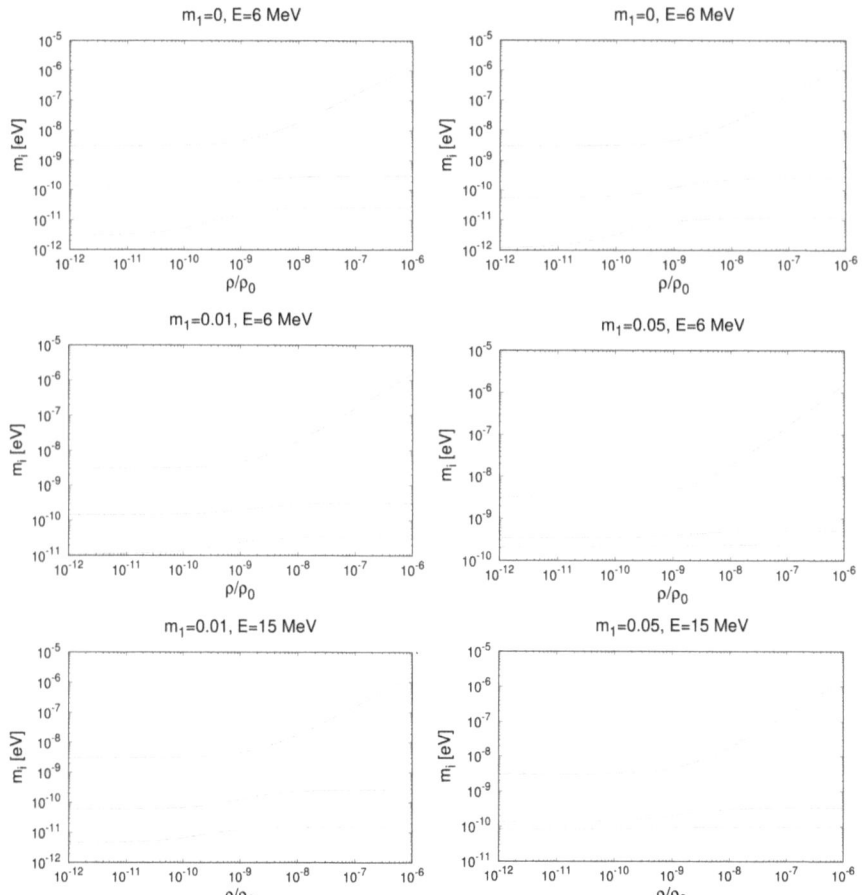

Fig. 4. The level crossing diagrams for different masses of m_1, the absolute neutrino mass, and different energies where $\rho_0 = 3.5 \times 10^{10} g/cm^3$.

to simplify,

$$r_{res} = r_0 \left(\frac{\Delta m^2 m_N \cos 2\theta}{2\sqrt{2}\rho_0 E G_F} \right)^{-1/3}. \tag{10}$$

This now allows us to find the resonance regions in terms of distance.

Figure 5 shows the role of energy in determining how far away from the core the resonance regions exist. The higher the energy, the further away from the core we find the resonances. As expected, and is true, the distance of the high resonance regions are always much closer to the core than low resonance due to the necessity of a highly dense environment. A key note to point out in this plot is that due to the logarithmic scale used to describe the distance in km, the high resonance regions never exist at $r = 0$ km because the resonance regions where flavor flipping occur MUST be located outside of the core which is roughly 10–16 km in radius. To further extend the relation between neutrino

$$\gamma = \frac{\Delta m^2}{2E} \frac{\sin^2 2\theta}{\cos 2\theta} \frac{1}{\frac{1}{\rho_e} \frac{d\rho_e}{dx}} \tag{9}$$

to simplify,

$$r_{res} = r_0 \left(\frac{\Delta m^2 m_N \cos 2\theta}{2\sqrt{2}\rho_0 EG_F} \right)^{-1/3}. \tag{10}$$

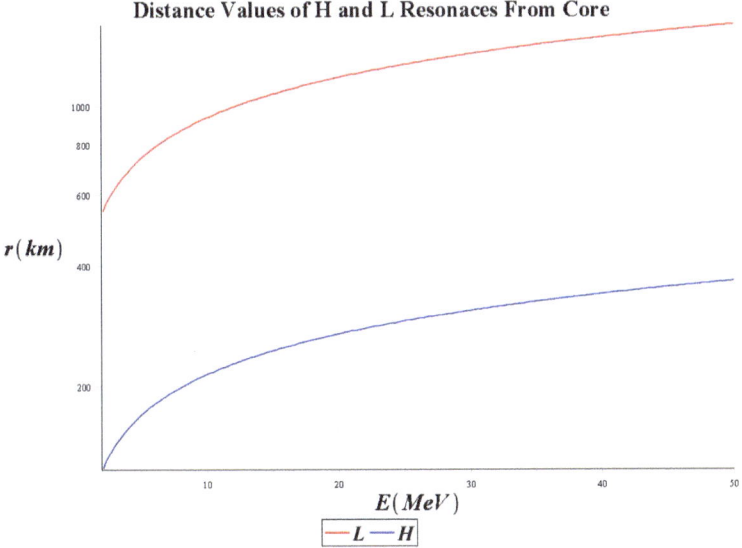

Distance Values of H and L Resonaces From Core

$r(km)$

$E(MeV)$

— L — H

Fig. 5. The resonance distance for the regions H and L in function of the neutrino energy.

fluxes and level crossings, the level crossing diagrams such as those in Fig. 3 can be used to write an association between the flux of each neutrino eigenmass and their survival probability, where the probability is minimal at resonance due to the amplification of the flavor oscillation mechanism. In the normal mass hierarchy, these relations take the following form

$$F_1 = P_H P_L F_{e0} + (1 - P_L)F_{\mu 0} + P_L(1 - P_H)F_{\tau 0},$$
$$F_2 = P_H(1 - P_L)F_e^0 + P_L F_\mu^0 + (1 - P_L)(1 - P_H)F_\tau^0, \tag{11}$$
$$F_3 = (1 - P_H)F_{e0} + P_H F_{\tau 0}.$$

with $\overline{P}_L = 0$, where adiabaticity is assumed to be complete.

The form of the flavor fluxes for both normal and inverted mass hierarchies is

$$F_\alpha = \sum_i |U_{\alpha i}|^2 F_i, \overline{F}_\alpha = \sum_i |U_{\alpha i}|^2 \overline{F}_i, \tag{12}$$

The total flux of all non-electron neutrinos and non-electron antineutrinos is then the sum of the fluxes of $\nu_\mu, \nu_\tau, \overline{\nu}_\mu, \overline{\nu}_\tau$. This is made simpler when we consider that all

4 fluxes of the other 2 flavors and their anti particles can be represented by the same generic F_x flux such that

$$F_x = F_\mu + F_\tau + \overline{F}_\mu + \overline{F}_\tau \qquad (13)$$

Which can then be summarized so that the fluxes of all neutrinos in the normal hierarchy, dependent on P_H and P_L and using (12) are

$$F_e = [|U_{e1}|2P_H P_L + |U_{e2}|2P_H(1 - P_L) + |U_{e3}|2(1 - P_H)]F_{e0}$$
$$+ \left[|U_{e1}|^2(1 - P_L P_H) + |U_{e2}|^2(1 - P_H(1 - P_L)) + |U_{e3}|^2 P_H \right] F_x^0, \qquad (14)$$

$$\bar{F}_e = |U_{e1}|^2 \bar{F}_e^0 + \left(1 - |U_{e1}|^2 \right) F_x^0,$$

$$F_x = F_\mu + F_\tau + \bar{F}_\mu + \bar{F}_\tau$$
$$= \left[\left\{ |U_{\mu 1}|^2 + |U_{\tau 1}|^2 \right\} P_H P_L + \left\{ |U_{\mu 2}|^2 + |U_{\tau 2}|^2 \right\} P_H(1 - P_L) \right.$$
$$+ \left\{ |U_{\mu 3}|^2 + |U_{\tau 3}|^2 \right\}(1 - P_H) \right] F_e^0 + \left\{ |U_{\mu 1}|^2 + |U_{\tau 1}|^2 \right\} \bar{F}_e^0$$
$$+ \left[2 - \left\{ |U_{\mu 1}|^2 + |U_{\tau 1}|^2 \right\} P_L P_H + \left\{ |U_{\mu 2}|^2 + |U_{\tau 2}|^2 \right\}(1 - P_H + P_H P_L) \right.$$
$$+ \left\{ |U_{\mu 3}|^2 + |U_{\tau 3}|^2 \right\} P_H \right] F_x^0 \qquad (15)$$

Using the relation $|U_\mu|^2 + |U_\tau|^2 = 1 - |U_e|^2$, F_x can be simplified to

$$F_x = \left[\left\{ 1 - |U_{e1}|^2 \right\} P_H P_L \right.$$
$$+ \left\{ 1 - |U_{e2}|^2 \right\} P_H(1 - P_L)$$
$$+ \left\{ 1 - |U_{e3}|^2 \right\}$$
$$\times (1 - P_H)] F_e^0 + \left\{ 1 - |U_{e1}|^2 \right\} \bar{F}_e^0$$
$$+ \left[2 - \left\{ 1 - |U_{e1}|^2 \right\} P_L P_H \right.$$
$$+ \left\{ 1 - |U_{e2}|^2 \right\}(1 - P_H(1 - P_L))$$
$$+ \left\{ 1 - |U_{e3}|^2 \right\} P_H \right] F_x^0$$

For the inverted mass hierarchy in the level crossing diagram, we find that fluxes are;

$$F_1 = F_\tau^0$$
$$F_2 = P_L F_e^0 + (1 - P_L) F_\mu^0$$
$$F_3 = (1 - P_L) F_e^0 + P_L F_\mu^0$$
$$\overline{F}_1 = (1 - P_H) \overline{F}_e^0 + P_H \overline{F}_\tau^0$$
$$\overline{F}_2 = P_H \overline{F}_e^0 + (1 - P_H) \overline{F}_\tau^0$$

$$\overline{F}_3^0 = \overline{F}_\mu^0.$$

and so we can find the flavor fluxes using (12) to be

$$F_e = \left[|U_{e2}|^2 P_L + |U_{e3}|^2 (1 - P_L)\right] F_e^0$$
$$+ \left[|U_{e1}|^2 + |U_{e2}|^2 (1 - P_L)\right.$$
$$\left. + |U_{e3}|^2 P_L\right] F_x^0$$

$$\overline{F}_e = \left[|U_{e1}|^2 (1 - P_H) + |U_{e2}|^2 P_H\right] \overline{F}_e^0$$
$$+ \left[|U_{e1}|^2 P_H + |U_{e2}|^2\right.$$
$$\left. \times (1 - P_H) + |U_{e3}|^2\right] F_x^0,$$

$$F_\mu + F_\tau + \overline{F}_\mu + \overline{F}_\tau = \left[\left(1 - |U_{e2}|^2\right) P_L + \left(1 - |U_{e3}|^2\right)(1 - P_L)\right] F_e^0$$
$$+ \left[|U_{e1}|^2 \times (1 - P_H) + |U_{e2}|^2 P_H\right] \overline{F}_e^0$$
$$+ \left[3 - |U_{e2}|^2 + \left[|U_{e2}|^2 - |U_{e3}|^2\right]\right.$$
$$\left. \times P_L + \left[|U_{e2}|^2 - |U_{e1}|^2\right] P_H\right] F_x^0$$

3.1 Flavor Flipping Probabilities

The level crossings diagram Fig. 3 displays the two resonance regions where flavor flipping are most likely to occur, they are the regions where the mass difference between the potentially conversing flavors is minimal. The flip probabilities (or the probability that a flavor change will take place, commonly denoted as P_f where f stands for "flip") are related to the adiabatic parameter γ through their definition

$$P_f = e^{-\frac{\pi}{2}\gamma}. \tag{16}$$

Using the gamma definition (9), we simplify the statement with the generalized electron density (8), where

$$\frac{1}{\frac{1}{\rho_e}\frac{d\rho_e}{dr}} = \frac{r}{3} \tag{17}$$

We then use the relation (10). The purpose of this simplification is an attempt to find the energy at non adiabatic regions (E_{na}) such that an alternative definition to flipping probabilities (16) is found, which according to literature is

$$P_f = \left(\frac{E_{na}}{E}\right)^{2/3}. \tag{18}$$

Equating as necessary, we find that

$$
\frac{\pi}{2}\gamma = \frac{\pi}{2}\frac{\Delta m^2}{2E}\frac{\sin^2 2\theta}{\cos 2\theta}\frac{1}{\frac{3}{r_{res}}}
$$

$$
= \frac{\pi}{2}\frac{\Delta m^2}{6E}\frac{\sin^2 2\theta}{\cos 2\theta}\left(\frac{\Delta m^2 m_N \cos 2\theta}{2\sqrt{2}\rho_0 E G_F}\right)^{-1/3} \quad r_0 = \left(\frac{E_{na}}{E}\right)^{2/3} \tag{19}
$$

With further simplifications,

$$
E_{na} = \left(\frac{\pi}{2}\right)^{3/2}\Delta m^2 \frac{\sin^3 2\theta}{\cos^2 2\theta}\left(\frac{\sqrt{2}}{108}\frac{\rho_0}{m_N}G_F\right)^{1/2} r_0^{3/2} \tag{20}
$$

For many of the upcoming constructed plots, the P_H and P_L values were obtained using the Eq. (18). In the case of inverted mass hierarchy, where the square mass difference is negative as well as the resonance density, we see that in the formula (19) the negatives cancel out and the Energy at non adiabaticity remains positive. When the adiabatic parameter is larger than 1, the region is adiabatic and the crossing probability is close to 0 and close to no flavor switches take place.

4 Results and Discussion

4.1 Supernovae Neutrino Fluxes and Distributions

The same could hence be said about the fluxes of the neutrinos of a given flavor that coincide with the fluxes of the medium eigenstates (still at very high densities). For normal hierarchy; neutrinos: $F_{1m}^0 = F_{2m}^0 = F_x^0, F_{3m}^0 = F_e^0$; and for anti neutrinos $\overline{F}_{3m}^0 = \overline{F}_{2m}^0 = F_x^0, \overline{F}_{1m}^0 = F_e^0$. While for the inverted hierarchy; neutrinos: $F_1{}^0_m = F_3{}^0_m = F_x^0, F_2{}^0_m = F_e^0$; and for anti neutrinos $\overline{F}_{1m}^0 = \overline{F}_{2m}^0 = F_x^0, \overline{F}_{3m}^0 = F_e^0$ [5].

Neutrino fluxes are the number of neutrinos that pass through a given area per unit time. They can be measured using neutrino detectors through the neutrinos interaction with matter in the detector. The spectral flux of neutrinos can be found using the inverse square distance relation and the luminosity flux as well as the average energy and a neutrino distribution function following the form of a Fermi-Dirac distribution just like the spectral behavior discussed before. The initial flux function written in terms of the mean neutrino energy and time is (without considering flavor conversions due to matter effects) (2)

$$
F_{\nu\beta}^0(E, t) = \frac{L_{\nu\beta}(t)}{4\pi D^2}\frac{f_{\nu\beta}}{< E_{\nu\beta} >}, \tag{21}
$$

written in units of $\frac{1}{MeV \cdot cm^2}$ for a neutrino of flavor β.

The behavior of neutrinos in a neutrino flux diagram could be described by a few distributions. The most general of which is the The Fermi-Dirac distribution function (2), which assumes that the neutrinos follow a thermal spectrum behavior and can be described in terms of energy and pinching parameter η [2], where the neutrino tempera-ture is considered. When narrowing down a function to describe the supernova neutrino flux, we must go back to when a comparison between the Fermi-Dirac distribution func-tion and other functions was made. It was stated that a Fermi-Dirac distribution isn't precise enough to model supernova neutrino spectra due to the role the MSW effect plays which the function doesn't consider. The case in which one can use such distribution is limited to the behavior of the neutrino spectrum behaving like a thermal spectrum. It is more general to use a power law distribution to describe the neutrino spectrum of supernova neutrinos. The normalized power law distribution takes the form [10];

$$f_{\nu\beta}(E) = \frac{(\alpha+1)^{(\alpha+1)}}{\Gamma(\alpha+1) <E>} \left(\frac{E}{<E>}\right)^{\alpha} e^{-(\alpha+1)E/<E>}, \tag{22}$$

where Γ is the Euler Gamma function, and α is a pinching parameter that can be described in terms of the energy and average energy. The most common literature value for α is one we also use generalized as $\alpha = 3$ for all flavors' fluxes.

$$\frac{1}{1+\alpha} = \frac{<E^2> - <E>^2}{<E>^2}. \tag{23}$$

The flux could then be described as [16]

$$F_{\nu\beta}^{0}(E, t)$$
$$= \frac{L_{\nu\beta}(t)}{4\pi D^2} \frac{(\alpha+1)^{(\alpha+1)}}{\Gamma(\alpha+1) <E_{\nu\beta}>^2}$$
$$\times \left(\frac{E}{<E_{\nu\beta}>}\right)^{\alpha} e^{-(\alpha+1)E/<E_{\nu\beta}>} \tag{24}$$

A comparison can be conducted on the accuracy of both the Power Law Distribution (22) and the Fermi Dirac Distribution (2) in representing supernovae neutrino fluxes.

Figure 6 shows a comparison between the two distributions and extracted detected data from SN1987a originally fitted [18]. The data portrays neutrinos with a luminosity identical to the one used in expressing the normalized distributions, $L = 2.2 \times 10^{53} ergs$. Although the distributions have been re-scaled for easier visual comparison, it can clearly be seen that the power law distribution does a much better job at modeling the behavior of supernovae neutrinos than the Fermi-Dirac distribution. What is interesting about this plot is that it proves that the neutrinos obtained from the SN1987a explosion were in fact non-thermal in nature.

In our plot Fig. 7, the comparison was conducted by plotting the initial neutrino fluxes of all flavors using both distributions, normalized, with the same initial conditions such as luminosity values. Using the common pinching parameter of the power law function $\alpha = 3$, and the abandonment of the pinching parameter η in the Fermi-Dirac distribution, it becomes evident how vastly different each distribution presents the neutrino fluxes. The most interesting of these behaviors Fig. 7 is the evolution of the electron neutrino flux (F_e^0, red). It appears that after some very small energy, the portrayal of electron neutrinos becomes much more significant in the power law distribution than in the Fermi-Dirac distribution.

Anti-electron Neutrino Comparison

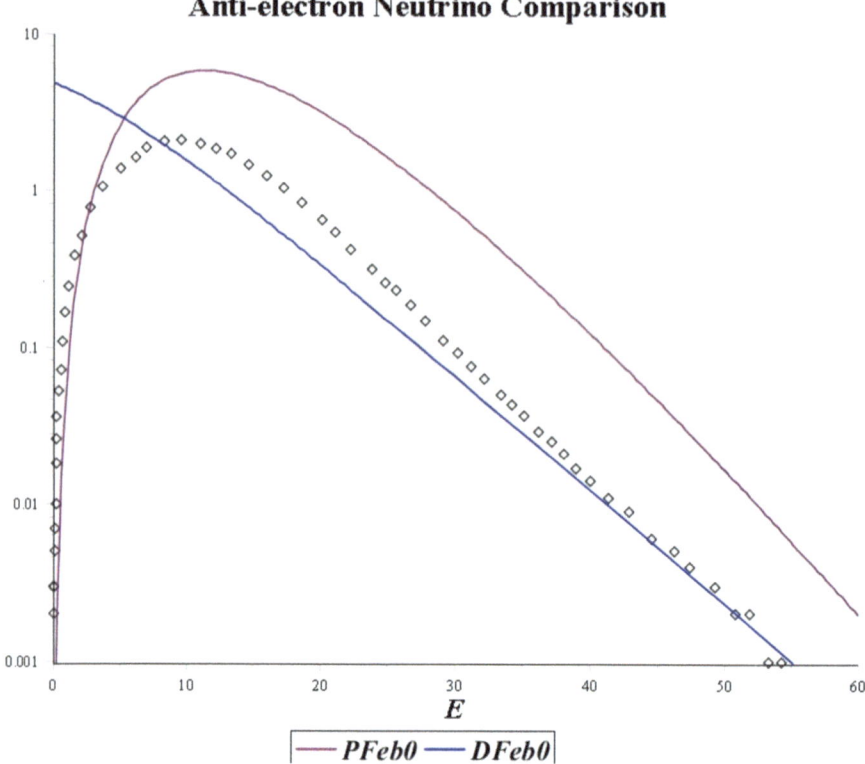

Fig. 6. Flux versus Energy plot, the SN1987a detected electronic anti-neutrinos Versus power law (PF) and Fermi-Dirac (without pinching parameters) distributions (DF).

4.2 Supernova Neutrino Fluxes and Mass Hierarchies

The flux comparison shown in Fig. 8 was conducted using the Power Law distribution. We find in the graph patterns what we've initially expected. In the normal hierarchy, with and without flavor conversions, electron neutrinos occupy the majority of neutrinos on the surface of supernovae up until a certain energy where that majority is then overtaken by either anti electron neutrinos or the "x" combination of other flavors and anti-flavors. The most significant effect we must note in the graph is the behavior of the green plots "F_x"; due to flavor conversions, we find a decrease in the flux of electron and anti electron neutrinos and an increase in "F_x". Eventually, at high energies we find that "F_x" becomes most prominent.

In the case of inverted mass hierarchies, a difference in behavior can immediately be seen in comparison to the normal hierarchy plot. Most prominent is the difference in behavior of (F_x) where in the inverted mass hierarchy, the increase in its flux is much lesser in comparison to the normal hierarchy behavior. The behavior of the electron neutrino flux (red) in both plots is also interesting, it can be seen that in the inverted hierarchy Fig. 8 the "post conversions" electron neutrino flux remains greater than the initial anti electron neutrino flux (blue) up until a certain energy. While in the normal

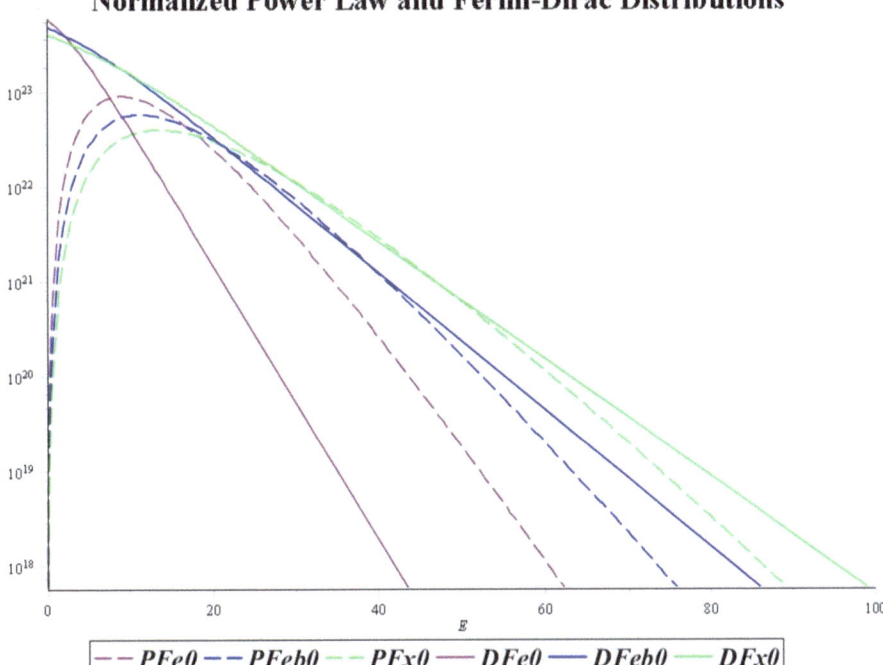

Fig. 7. Flux versus Energy Plot, initial neutrino fluxes using normalized power law (PF) and Fermi-Dirac (without pinching parameters) distributions (DF). Where PFe0 (initial electron neutrino power law flux), PFeb0 (initial anti-electron neutrino power law flux), PFx0 (sum of other neutrino flavors initial power law fluxes), DFe0 (initial electron neutrino Fermi-Dirac flux), DFeb0 (initial anti-electron neutrino Fermi-Dirac flux), DFx0 (sum of other neutrino flavors initial Fermi-Dirac fluxes) are plotted.

hierarchy Fig. 8, that electron neutrino flux remains almost always smaller than the flux of initial anti electron neutrinos.

For an easier and more direct comparison between the Normal Mass Hierarchy and the Inverted Mass Hierarchy with the initial fluxes, we can form a plot such as that in Fig. 9. It is expected to see how insignificantly both Hierarchies come into play regarding the anti electron neutrino flux. The shape of the flux curve remains almost the same with only the maximum flux varying. This is unlike the minimal but present reaction of the electron neutrino flux to the Hierarchies.

Initially, we observe the same behavior and curve with a flux lower than the initial flux, but after some energy point around $E \approx 20 MeV$, we can see an increase in the electron neutrino flux in comparison to the initial. The most prominently influenced neutrino flux is the group (F_x). Although initially both hierarchies lead to higher fluxes in comparison to the initial flux, the peak of the flux curve and its decay varies depending on the hierarchy in an obvious manner. More interestingly, at around the same energy region $E \approx 20 MeV$, both the normal and inverted mass hierarchies lead to fluxes below the initial flux at said energies and higher. To observe these comparisons in a more

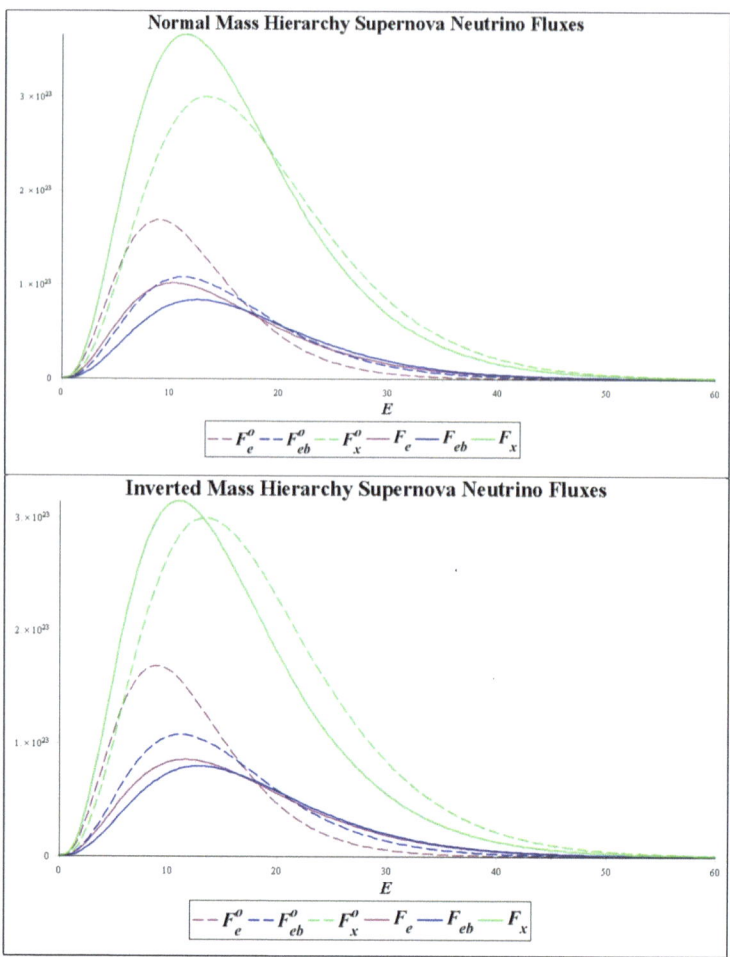

Fig. 8. Flux versus Energy Plots of flavor fluxes in normal and inverted mass hierarchy with matter effects (F) and initial fluxes without matter effects (F^0) comparisons.

sophisticated or clear manner, we can present the difference between the normal mass hierarchy fluxes and initial fluxes as well as the inverted mass hierarchy and initial fluxes difference 9. Such a diagram allows us to not only understand the progressing behavior of the converted fluxes with changing energy, but also helps us learn about a key feature.

The fraction of energy range highlighted by a thicker line shows a significant region which we can use to determine the type of hierarchy we have only from the obtained difference between converted and initial fluxes, given of course that the neutrinos energies are limited to within that range which can be approximated to ($13MeV < E < 19MeV$). For instance, if the flux difference were to be positive then a Normal Mass Hierarchy was observed. If the difference was negative however, then an Inverted Mass Hierarchy was observed. It is also interesting to note how insignificant the Hierarchy actually is when it comes to only the electron and anti electron neutrino fluxes. Unlike the useful

F_x relations, the maximum difference between the Inverted and Normal Mass hierarchies is found at nearly the same energy for both electron and anti electron neutrinos, which is still insignificant compared to the impact experienced by F_x. Figure 9 shows the difference in fluxes from normal and inverted mass hierarchies.

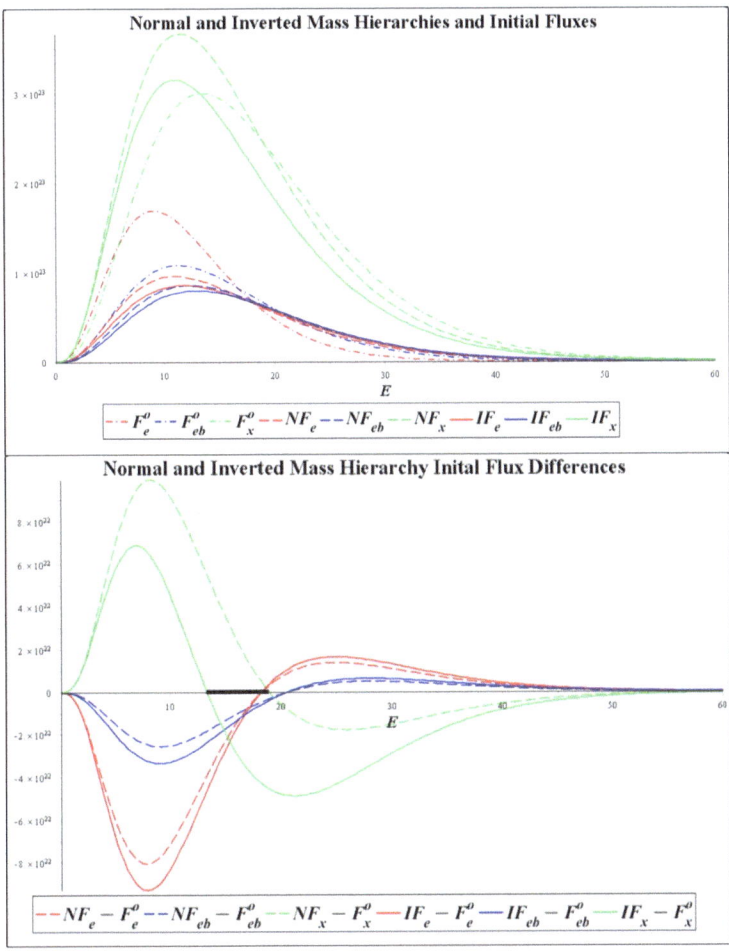

Fig. 9. Flux versus Energy plots comparison of normal mass hierarchy, inverted mass hierarchy, and initial fluxes and their differences from each hierarchy.

5 Conclusion

Through our studies, we graphed neutrino fluxes using the power law distribution and the density profile of $\rho_e = \rho_0(r/r_0)^{-3}$. We also conducted a comparison between the power law distribution and the Fermi-Dirac distribution in an attempt to observe which is more accurate in representing neutrino fluxes. We compared the produced fluxes to experimental data for a more confident conclusion.

Both normal and inverted mass hierarchies were studied through neutrino fluxes. The difference in resonance regions with the hierarchy portrayed differences in behavior due to changing transition probabilities.

We have also compared the difference in initial and final fluxes and examined the window of energies through which the sign of the flux difference alone could distinguish the type of hierarchy experienced. This happens in an energy window of the approximate range of ($13\ meV < E < 19\ meV$).

We have considered the future of supernova neutrino detectors. These future detection promise much more detailed and thorough data. With an increase in amount of neutrinos expected to be detected due to a better understanding of neutrino interactions. An important property to anticipate is the detection of electron neutrinos that was not possible in previous neutrino detectors.

References

1. Ahriche, A., Mimouni, J.: Supernova neutrino spectrum with matter and spin flavor precession effects. JCAP **11**, 004 (2003). https://doi.org/10.1088/14757516/2003/11/004
2. Chen, Y.M., Sen, M., Tangarife, W., Tuckler, D., Zhang, Y.: Core-collapse supernova constraint on the origin of sterile neutrino dark matter via neutrino self-interactions. 2022(11), 014 (2022). https://doi.org/10.1088/14757516/2022/11/014
3. Dasgupta, B., Dighe, A.: Collective three-flavor oscillations of supernova neutrinos. Phys. Rev. D **77**, 113002 (2008). https://doi.org/10.1103/PhysRevD.77.113002
4. Dighe, A.S., Keil, M.T., Raffelt, G.G.: Identifying earth matter effects on supernova neutrinos at a single detector. J. Cosmol. Astropart. Phys. **2003**(06), 006 (2003). https://doi.org/10.1088/1475-7516/2003/06/006
5. Dighe, A.S., Smirnov, A.Y.: Identifying the neutrino mass spectrum from a supernova neutrino burst. Phys. Rev. D **62**, 033007 (2000). https://doi.org/10.1103/PhysRevD.62.033007
6. Duan, H., Fuller, G.M., Qian, Y.Z.: Collective neutrino oscillations. Annu. Rev. Nucl. Part. Sci. **60**(1), 569–594 (2010). https://doi.org/10.1146/annurev.nucl.012809.104524
7. Esposito, S., Mangano, G., Miele, G.: Supernova neutrino energy spectra and resonant transition effects. Nucl. Phys. B Proc. Suppl. **66**(1), 265–268 (1998). https://doi.org/10.1016/S0920-5632(98)00051-6. https://www.sciencedirect.com/science/article/pii/S092056329800 00516. Proceedings of the XVI Workshop on Weak Interactions and Neutrinos
8. Janka, H.T.: Explosion mechanisms of core-collapse supernovae. Annu. Rev. Nucl. Part. Sci. **62**(1), 407–451 (2012). https://doi.org/10.1146/annurev-nucl-102711-094901
9. Janka, H.T.: Neutrino-driven explosions. In: Alsabti, A.W., Murdin, P. (eds.) Handbook of Supernovae, p. 1095 (2017). https://doi.org/10.1007/978-3-319-21846-5_109
10. Janka, H.T.: Neutrino emission from supernovae. In: Handbook of Supernovae, pp. 1575–1604. Springer (2017). https://doi.org/10.1007/978-3-319-21846-5_4
11. Kajita, T.: Atmospheric neutrinos and discovery of neutrino oscillations. Proc. Japan Acad. Ser. B Phys. Biol. Sci. **86**, 303–321 (2010). https://doi.org/10.2183/pjab.86.303
12. Lunardini, C., Smirnov, A.Y.: Neutrinos from sn1987a: flavor conversion and interpretation of results. Astroparticle Phys. **21**(6), 703–720 (2004). https://doi.org/10.1016/j.astropartphys. 2004.05.005. https://www.sciencedirect.com/science/article/pii/S0927650504000921
13. Mezzacappa, A., Messer, O.: Neutrino transport in core collapse supernovae. J. Comput. Appl. Math. **109**(1), 281–319 (1999). https://doi.org/10.1016/S0377-0427(99)00162-4. https://www.sciencedirect.com/science/article/pii/S0377042799001624

14. Nomoto, K., Shigeyama, T.: Supernova 1987A: constraints on the theoretical model. In: Kafatos, M., Michalitsianos, A.G. (eds.) Supernova 1987A in the Large Magellanic Cloud, pp. 273–288 (1988)
15. Raffelt, G.G.: Stars as laboratories for fundamental physics: the astrophysics of neutrinos, axions, and other weakly interacting particles (1996)
16. Saez, M.M., Mosquera, M.E., Civitarese, O.: Neutrino interactions in liquid scintillators including active-sterile neutrino mixing. Int. J. Mod. Phys. E **31**(03) (2022). https://doi.org/10.1142/s0218301322500239
17. Smirnov, A.Y.: The MSW effect and matter effects in neutrino oscillations. Physica Scripta Volume T **121**, 57–64 (2005). https://doi.org/10.1088/0031-8949/2005/T121/008
18. Vissani, F., Pagliaroli, G.: The diffuse supernova neutrino background: expectations and uncertainties derived from SN1987a. Astronomy Astrophysics **528**, L1 (2011). https://doi.org/10.1051/0004-6361/201016109
19. Wolfenstein, L.: Neutrino oscillations in matter. Phys. Rev. D **17**, 2369–2374 (1978). https://doi.org/10.1103/PhysRevD.17.2369
20. Xing, Z.Z., Zhou, S.: Neutrinos in particle physics, astronomy and cosmology (2011)
21. Yüksel, H., Beacom, J.F.: Neutrino spectrum from SN 1987a and from cosmic supernovae. Phys. Rev. D **76**(1) (2007). https://doi.org/10.1103/physrevd.76.083007

Estimation of the Transit Time of Multiple Coronal Mass Ejection Events

Firas Al-Hamdani$^{(\boxtimes)}$ ⓘ and Doha Al-Feadh ⓘ

Basrah University, Basra 61004, Iraq
firas.balbool@uobasrah.edu.iq

Abstract. We study the transit time (T.T) of triple events where the first two coronal mass ejections (CME) are associated with solar energetic particle events (SEP), while the third one is not. Our study focuses on the propagation of these CMEs and their associated shocks to estimate the best acceleration equation for T.T calculations and to determine the minimum error of the T.T of the CMEs. Using different models, we estimated the T.T of very high-speed CMEs and compared them with the empirical shock arrival model (ESA). Our results show that the acceleration cessation distance (ACD) in the ESA model yields the lowest possible T.T errors for the first event (0.7–0.5 AU) compared to other equations, and a distance of 0.7 AU for the second event. We also reveal minimum errors for the third event with a distance of 0.6 AU. In conclusion, these events exhibit different responses to the models used, even when they have very high speeds.

Keywords: Coronal mass ejections · Initiation and propagation · interplanetary shock wave

1 Introduction

Coronal mass ejections (CMEs) play a crucial role in space weather prediction, often resulting in the most intense geomagnetic disturbances on Earth (e.g., Gopalswamy et al. [1], Manoharan and Mujiber Rahman [2]; Shanmugaraju et al. [3]). These phenomena have been reported to affect Earth's atmosphere in various ways, with geomagnetic storms primarily caused by CMEs directed towards Earth. The direction, speed, magnitude, density, orientation, and magnetic field strength of CMEs near Earth are key determinants of their geo-effectiveness. Initially accelerated by the coronal magnetic field, CMEs may experience further acceleration mechanisms during their journey through space (Michalek et al. [5]).

Moreover, CMEs and solar flares are considered the most significant manifestations of solar activity, with their kinematics depending on initial velocity and solar wind conditions. The interaction between CMEs and their internal energy, as well as their interaction with each other, leads to constantly changing propagation dynamics between the Sun and Earth (e.g., Manoharan [6]).

Furthermore, coronal and/or CME-driven shocks produce solar radio type II bursts by accelerating electrons and transforming plasma oscillations into radio waves. Solar

H. M. K. Al Naimiy et al. (Eds.): AUASS-CONF 2023, SPPHY 420, pp. 54–62, 2025.
https://doi.org/10.1007/978-981-96-3276-3_5

burst data confirming type II emission in the solar wind at radial locations as far out as 20 Rs-30 Rs are well-documented (Al-Hamadani et al. [7]). Numerous studies and models have been developed to anticipate acceleration processes, with some focusing on forecasting the arrival of interplanetary (IP) shocks caused by rapid CMEs (Smart and Shea [8]; Smith [9]). However, these models have demonstrated varying degrees of precision (e.g., Gopalswamy et al. [10]).

Gopalswamy et al. [11] introduced an empirical model based on observations from 1996 to 2000 to predict CME arrival time at 1 AU. They established an efficient acceleration equation (a = 2.193 − (0.0054 × u)), coinciding with the CME starting speed (u), and calculated travel time with minimal errors using an ESA model. Michalek et al. [5] examined the transit time (T.T) of 83 CMEs near Earth, deriving an effective acceleration equation (a = 4.11 − (0.0063 × u)). Additionally, they obtained another relationship for effective acceleration (a = 3.35 − (0.007 × u)) for very slow and very fast events, assuming no acceleration cessation distance (ACD) between the Sun and Earth.

In this study, we focus on triple events in sequence where solar energetic particle (SEP) increases are associated with the first two CMEs, while the third shows no signs of a SEP enhancement. Our objective is to investigate the acceleration processes of multiple solar eruptions during their transition in IP space. The selection criteria for the analyzed events are detailed further in Pohjolainen et al. [14].

2 Data Analysis

We analyzed a set of CMEs observed by *Solar and Heliospheric Observatory/Large Angle and Spectrometric Coronagraph* (SOHO/LASCO). These events show three distinct features: a quick halo-type CME, a propagating shock evident in radio emission, and powerful flares occur in the same active region (AR) one after the other. All CMEs associated with M and X class X-ray flares, the details of CMEs, shock wave arrival time and T.T are listed in the Table 1. Where the first three columns refer to the CME date, time and linear speed respectively. Next four columns show date, time, speed and the actual T.T of IP shock. The T.T of the CMEs and their shocks found from the time various between the first apparent of CMEs in the SOHO/LASCO and IP shock/ *interplanetary coronal mass ejections* (ICME) arrival time in *Advanced Composition Explorer* (ACE) and WINDs spacecraft. The shock wave details can be obtained by ACE/WIND spacecraft, one can calculate the speed of the shock wave as in the Table 1.

The ideal and vastly used empirical production model is the *empirical CME arrival* (ECA) and ESA models which is developed by Gopalswamy and his collaborated. We find the T.T by using Gopalswamy et al. (2001) and Kim, Moon and Cho (2007) model as stated below:

$$T_1 = \frac{-u + \sqrt{u^2 + 2ad_1}}{a} \tag{1}$$

$$T_2 = \frac{d_2}{-u + \sqrt{u^2 + 2ad_1}} \tag{2}$$

Table 1. CMEs and shock wave characteristics

CME			IP Shock/ACE				IP Shock/WIND			
Date	Time UT	Speed Km/s	Date	Arrival Time hh:mm	Shock speed Km/s	T.T hours	Date	Arrival Time hh:mm	Shock speed Km/s	T.T hours
24.11.00	05:30	1289	26.11.00	05:00	462	47.5	26.11.00	05:32	520	48.03
24.11.00	15:30	1245	26.11.00	11:24	632	43.9	26.11.00	11:43	666	44.21
09.04.01	15:54	1192	11.04.01	13:14	613	45.16	11.04.01	14:09	673	46.25
10.04.01	05:30	2411	11.04.01	15:28	731	33.96	11.04.01	16:18 18:18	828 830	34.8
11.04.01	13:31	1103	13.04.01	07:06	914	41.58				
22.08.05	01:31	1194	24.08.05	05:45	518	52.23	24.08.05	05:35	591	51.56
22.08.05	17:30	2378	24.08.05	08:20	795	38.83	24.08.05	08:24	525	38.9
23.08.05	14:54	1929	25.08.05	13:00		46.1				

The T.T formulated as $(T.T = T_1 + T_2)$, where T_1 is a travel time till to ACD or acceleration distance d_1. Because the assumption is that the CME subjected to a constant acceleration or deceleration process during travel to the Earth with constant speed, T_2 is the travel time after the distance d_1 up to 1 AU. From the initial linear speed of CME we can found the effective IP acceleration by using different acceleration equations as below:

$$a = 2.193 - (0.0054 \times u) \tag{3}$$

$$a = 4.11 - (0.0063 \times u) \tag{4}$$

$$a = 3.35 - (0.007 \times u) \tag{5}$$

3 Models

Table 2 discusses the variations between the real arrival time of IP shock and T.T which is found from above equations with various acceleration distances. Where T.T error value is less than (6h of all equations) for the second CME at the first event, and less than (7h of all equations) for the first CME at the second event.

At the cessation acceleration distance (0.5 AU) the error in the T.T is minimum at Eq. 5 for all events compared with the other equations, as shown in Fig. 1.

The T.T profile was found from ESA model for all events with ACD (0.7) AU, drown to compared with the other acceleration equations and distances. Where we used the same way with the other acceleration equations(Eq. 4 and Eq. 5) to find the T.T. Table 2 explain that at 0.7 AU the ACD produced minimum errors of T.T for Eq. 3

Table 2. T.T difference for various ACD. The bold values exhibit the smallest variation ($<$7 h) between the actual and estimated T.T, ΔT equal to the difference between actual T.T of IP shocks and estimate T.T.

Date	ΔT (hours) for Eq. (3)			ΔT(hours) for Eq. (4)			ΔT(hours) for Eq. (5)		
CME	0.7 A.U	0.6 A.U	0.5 A.U	0.7 A.U	0.6 A.U	0.5 A.U	0.7 A.U	0.6 A.U	0.5 A.U
24.11.00	**4.89**	**6.55**	8.19	7.63	8.69	9.82	**0.26**	**3.27**	**5.85**
24.11.00	**−0.84**	**0.97**	**2.75**	**2.41**	**3.50**	**4.67**	**−5.77**	**−2.47**	**0.31**
09.04.01	**−1.31**	**0.71**	**2.67**	**2.68**	**3.81**	5	**−6.61**	**−2.93**	**0.11**
10.04.01	14.56	14.89	15.27	14.47	14.81	15.20	13.44	13.96	14.52
11.04.01	−10.78	−8.33	**−6.02**	**−5.14**	**-3.99**	**−2.76**	−16.65	−12.27	−8.73
22.08.05	**4.09**	**6.12**	8.06	8.06	9.18	10.38	**−1.81**	**2.47**	**5.51**
22.08.05	18.33	18.68	19.07	18.25	18.60	19.01	17.17	17.72	18.30
23.08.05	19.78	20.37	21	19.94	20.49	21.1	17.89	18.85	19.83

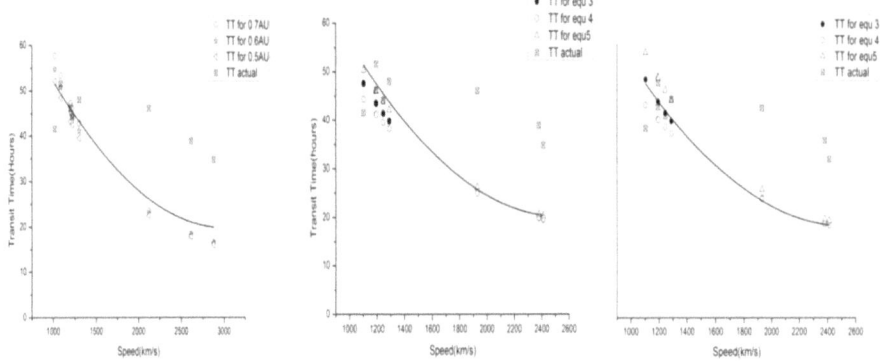

Fig. 1. The T.T of various acceleration distances. The left panel at 0.7AU, the middle panel at 0.6 AU and the right panel at 0.5 AU with different acceleration equations. The solid line indicates to the T.T from ESA model with cessation acceleration distance 0.7 AU.

to the events (1 and 2) comparing with the other equations, same thing applied to 0.6 AU acceleration distance. For the third event all ACD gives minimum error at Eq. 5, because this equation used either for very slow speed or very high speed events. The above procedure for estimate T.T was repeated but with the linear speed of CMEs at distances 20 Rs as explain in Table 3 to make a comparison of T.T with different speed.

Then we make graphically compression between the T.T obtained from the second order speed of CMEs reported in the SOHO/LASCO catalogue and final speed and used them in Eq. 3 with $d_1 = 0.7$ AU. From the Table 3, the first event T.T error is less than (9 h) at first CME, and less (5 h) for second CME. Second event errors less than (6 h)for first CME, and about (10–19 h) for the other CMEs, same result for the third CME but with the error limits in the T.T about (21 h) in the last two CMEs. Approximately the same result we can get for the final speed (right panel in Fig. 2) because there is no big differences between the second order speed at final distance and second order speed at 20 Rs distance in this events.

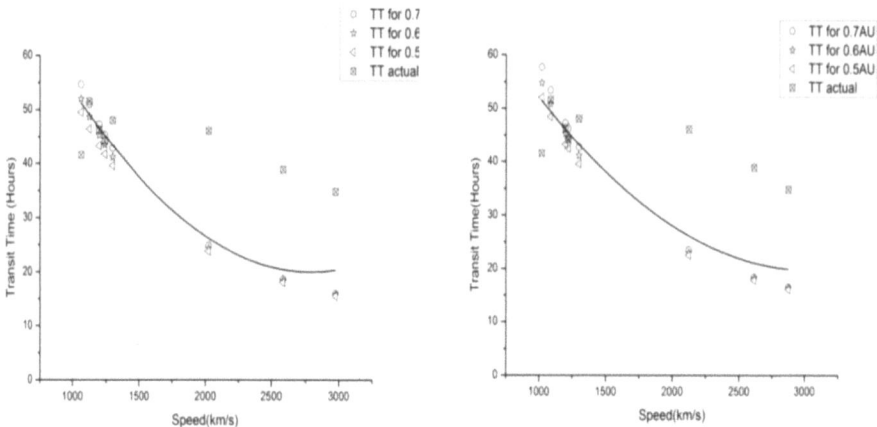

Fig. 2. Theestimated T.T using second order speed 20 Rs (left panel), and second order speed at final distance (right panel). Solid line refers to the ESA model for ACD 0.7.

4   Comparison with Other Models

We made a comparison between the actual T.T with different models such as; (1) Schwenn et al. [15], expansion speed model (2) Kim, Moon and Cho [16] model, and (3) Vrsnak et al. [12], *drag based model* (DBM). The expansion model based from the relation:

$$T_{arr} = 203 - 20.77 \times ln(V_{CME}),$$

where T_{arr} (in hours) is the T.T of linear speed V_{CME} in (km/s) of CME. Kim, Moon and Cho [16] indicate that the constant IP acceleration in the ESA model does not apply for all CME events, so that thy derived linear fit to the relationship between shock T.T (in hours) and V_{CME} in (km/s) during solar maximum, $T_{arr} = 76.86 - 0.02 \times (V_{CME})$. The DBM based on the motion of equation on CME where the drag acceleration /deceleration have quadratic dependence on therelative speed between CMEs and the background solar wind.

Table 4 and Fig. 3, reveals the variation between different models and actual T.T, for DBM we calculate the T.T based on basic solar wind speed parameters 450 km/s and the drag parameter (γ) as 0.2×10^{-7} km/s. The drag model gives in general the better result for minimum errors of T.T comparing with other models and its close to the real observed T.T, this result can be seen obviously in the Fig. 3. Where left side of Fig. 3

Table 3. The T.T differences for various ACD at 20 Rs. The bold values show the minimum difference (<7 h) between the actual and estimated T.T. ΔT equal to the difference between the actual T.T of IP shocks and the estimate T.T.

Date	ΔT (hours) for Eq. (3)			ΔT(hours) for Eq. (4)			ΔT(hours) for Eq. (5)		
CME	0.7 A.U	0.6 A.U	0.5 A.U	0.7 A.U	0.6 A.U	0.5 A.U	0.7 A.U	0.6 A.U	0.5 A.U
24.11.00	**5.18**	**6.55**	8.19	7.63	8.69	9.82	**0.26**	**3.27**	5.85
24.11.00	−0.84	**0.97**	2.75	**2.41**	3.5	4.67	−5.77	−2.47	0.31
09.04.01	−1.31	0.719	2.67	2.89	4.01	5.20	−6.27	−2.64	0.38
10.04.01	14.56	14.89	19.31	18.74	18.96	19.21	18.19	18.49	18.83
11.04.01	−13.08	−10.42	−7.94	−6.55	−5.41	−4.18	−19.14	−14.44	10.69
22.08.05	**0.56**	**2.90**	5.10	5.71	6.85	8.08	−5.16	−0.97	2.42
22.08.05	20.22	20.50	20.82	20.10	20.40	20.74	19.72	19.70	2.18
23.08.05	21.22	21.74	22.30	21.29	21.79	22.35	19.52	20.37	21.24

Table 4. Variations between the projected T.T for different models and the real IP shock T.T. The bold values show the minimum difference (<7 h) between the actual and estimated T.T.

Date	Expansion model	Kim model	Drag based model
24.11.00	**−6.22**	**−3.05**	4.25
24.11.00	−10.76	**−7.5**	−0.37
09.04.01	−9.62	**−6.77**	0.65
10.04.01	**−6.44**	6.16	3.37
11.04.01	−15.90	−13.22	−5.9
22.08.05	**−4.28**	**−1.42**	6
22.08.05	**−2.63**	9.6	7.23
23.08.05	**0.22**	7.82	10.66

show different T.T models with the linear speed of CMEs while right side with linear speed at distance 20 Rs.

When the CMEs emanating from the disk center or closed to it, there are projection effects at these disc-centered events. (Gopalswamy et al. [11]; Gopalswamy et al. [17]; Michalek et al. [5]; Shanmugaraju et al. [17]; Nicewicz and Michalek (2014)). To get rid from the projection effects we will use Gopalswamy et al. [11] Eqs. (6, 7) and Nicewicz and Michalek [18] Eq. 8.

$$v = u * \left| \frac{\sin \varphi + 1}{\sin \alpha + \sin \varphi} \right| \tag{6}$$

$$\varphi = arccos(\cos\lambda\cos\theta) \tag{7}$$

Table 5. The T.T following projected adjustment in the CME speed with the use of formulas (6, 7). The bold values show the minimum difference (>7 h) between the actual and estimated T.T.

Date	ΔT (hours) for Eq. (3)			ΔT(hours) for Eq. (4)			ΔT(hours) for Eq. (5)		
CME	0.7 A.U	0.6 A.U	0.5 A.U	0.7 A.U	0.6 A.U	0.5 A.U	0.7 A.U	0.6 A.U	0.5 A.U
24.11.00	20.40	21.05	21.74	20.64	21.25	21.90	18.31	19.40	20.48
24.11.00	14.96	15.69	16.47	15.34	16	16.71	12.63	13.87	15.06
09.04.01	15.78	16.58	17.42	16.28	16.98	17.74	13.26	14.62	16.94
10.04.01	21.10	21.25	21.42	20.97	21.13	21.32	20.59	20.81	21.05
11.04.01	**5.78**	6.91	8.07	**6.9**	7.87	8.81	**2.39**	**4.38**	**6.20**

Table 6. T.T following the adjustment of the projection in the CME speed by using Eq. (8). The bold values show the minimum difference (<7 h) between the actual and estimated T.T.

Date	ΔT (hours) for Eq. (3)			ΔT(hours) for Eq. (4)			ΔT(hours) for Eq. (5)		
CME	0.7 A.U	0.6 A.U	0.5 A.U	0.7 A.U	0.6 A.U	0.5 A.U	0.7 A.U	0.6 A.U	0.5 A.U
24.11.00	11.27	12.46	13.68	12.64	13.55	14.52	7.72	9.83	11.75
24.11.00	28.24	28.44	28.67	28.10	28.32	28.57	27.54	27.85	28.18
09.04.01	9.49	10.68	11.90	10.86	11.77	12.74	**5.94**	8.05	9.97
10.04.01	18.83	19.26	19.26	18.69	18.91	19.16	18.13	18.44	18.77
11.04.01	**1.10**	**2.56**	**4.02**	**3.21**	**4.22**	**5.28**	**−3.08**	**−0.45**	**1.84**

$$V = \upsilon LASCO + \upsilon LASCO * 0.23 \qquad (8)$$

where V is the corrected speed and u is the linear speed of the CME, α is the cone angle 360, λ and φ are the latitude and the longitude can be obtain from are location, v LASCO is the linear speed of CME. The T.T errors after using correction speed are given in Table 5 for Gopalswamy et al. [11] correction speed and Table 6 for Nicewicz and Michalek [18] speed. Form the Table 4 the T.T reduced for all equation for the last CME of second event and same thing is occur by using Nicewicz and Michalek [18] speed. Nevertheless, as indicated by Zhao and Dryer [19], Several factors, including the following, might contribute to the uncertainty in the T.T calculation: (1) the IP coronal mass ejections (ICMEs) 3D morphology, (2) the impact of the solar wind background, (3) variation in the ICMEs propagation, (4) the fluctuations in the solar wind and in homogeneities (5) T.T interactions with other structures, etc.

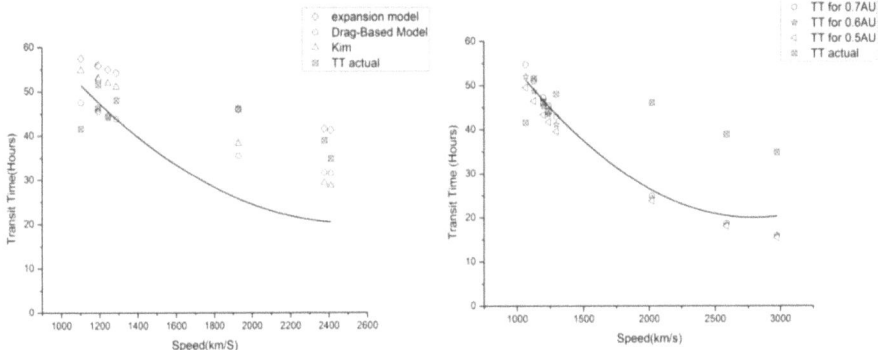

Fig. 3. Comparison between observed T.T and various T.T in different models. Solid line indicate ESA model for 0.7 AU ACD. Where the left side show different T.T models with the linear speed of CMEs while right side with linear speed at distance 20 Rs.

5   Conclusions

In this study, we focused on the transit time (T.T) of triple events involving coronal mass ejections (CMEs), where two CMEs were associated with solar energetic particle (SEP) enhancements, while the third CME did not exhibit enhancement in SEP observation. Radio emission observations confirmed that all these events had shocks propagating in the interplanetary (IP) medium (Pohjolainen, Al-Hamadani, and Valtonen [14]).

To investigate the propagation of these events in the IP medium, we calculated the T.T for each event. Linear CME speeds were used in various acceleration equations to evaluate the IP accelerations. These IP acceleration values were then applied in the ESA model with different acceleration distances. We also applied the same process using linear speed at 20 Rs distance to study the effects of speed differences on the estimated T.T. However, no distinct changes were observed in the T.T.

Our analysis revealed that the T.T errors were minimum for acceleration Eq. (5) in the ESA model with an acceleration cessation distance (ACD) of (0.7–0.5) AU for event (1) compared to other equations. Similarly, Eq. 3 with a distance of (0.7) AU yielded minimum errors for event (2), while Eq. 5 showed minimum errors for event (3) with a distance of (0.6) AU.

Furthermore, we estimated the T.T after correcting for the projection effect (correction speed), and found that the minimum errors occurred in Eq. 5 after using the Nicewicz and Michalek [18] equation.

Based on these conclusions, we can estimate the T.T for the third CME of event (1) using Eq. 5 with an acceleration distance of (0.5) AU to be (64.84 h), with the shock arrival time on 26th November 2000 at (16:50 min). The Double Bayesian Model (DBM) expected the CME arrival at the target (date and time) on 26th November 2000 at (23 h:57 min) with a T.T of 49.86 h, and the impact speed at the target (at 1 AU) to be 635 km/s

References

1. Gopalswamy, N., et al.: Astrophys. J. **572**, L103 (2002)
2. Manoharan, P.K., Mujiber Rahman, A.: J. Atmos. Solar-Terr. Phys. **73**, 671 (2011)
3. Shanmugaraju, A., Syed Ibrahim, M., Moon, Y.-J., Mujiber Rahman, A., Umapathy, S.: Solar Phys. **290**, 1417 (2015)
4. Gosling, J.T., McComas, D.J., Phillips, J.L., Bame, S.J.: J. Geophys. Res. **96**, 7831 (1991)
5. Michalek, G., Gopalswamy, N., Lara, A., Manoharan, P.K.: Astron. Astrophys. **423**, 729–736 (2004)
6. Manoharan, P.K.: Sol. Phys. **235**, 345–368 (2006)
7. Al-Hamadani, F., Pohjolainen, S., Valtonen, E.: SoPh. **292**, 127 (2017)
8. Smart, D.F., Shea, M.A.: J. Geophys. Res. **90**, 183 (1985)
9. Smith, Z., Dryer, M.: Solar Phys. **129**, 387 (1990)
10. Gopalswamy, N., et al.: J. Geophys. Res. **103**, 307 (1998)
11. Gopalswamy, N., Lara, A., Yashiro, S., Kaiser, M.L., Howard, R.A.: Geophys. Res. **106**, 29207–29218 (2001)
12. Vrsnak, B., et a.: Solar Phys. **285**, 295 (2013)
13. Vrsnak, B.: Solar Phys. **289**, 339 (2014)
14. Pohjolainen, S., Al-Hamadani, F., Valtonen, E.: Solar Phys. **291**, 487 (2016)
15. Schwenn, R., dal Lago, A., Huttunen, E., Gonzalez, W.D.: AnnalesGeophysicae. **23**, 1033 (2005)
16. Kim, K.-H., Moon, Y.-J., Cho, K.-S.: J. Geophys. Res. **112**, A05104 (2007)
17. Gopalswamy, N., Lara, A., Manoharan, P.K., Howard, R.A.: Adv. Space Res. **36**, 22892294 (2005)
18. Nicewicz, J., Michalek, G.: Adv. Space Res. **54**, 780786 (2014)
19. Zhao, X., Dryer, M.: Space Weather **12**, 448 (2014)

A Comprehensive Study of the Physical and Geometrical Characteristics of the Close Visual Binary System HIP 45571

Motasem J. Alslaihat[1,4](✉), Hatem S. Widyan[1], Mashhoor A. Al-Wardat[1,2,3,4](✉), Awni M. Kasawneh[1,2,3,4], Diala M. Taneenah[1,4], and Abdullah M. Hussein[1,4]

[1] Department of Physics, Al Al-Bayt University, Mafraq 25113, Jordan
Alabbade47@gmail.com, malwardat@sharjah.ac.ae
[2] Department of Applied Physics and Astronomy, University of Sharjah, Sharjah 27272, UAE
[3] Space Sciences and Technology, Sharjah Academy for Astronomy, University of Sharjah, Sharjah 27272, UAE
[4] Arab Union for Astronomy and Space Sciences, Amman 11941, Jordan

Abstract. In this paper, we estimated the physical and geometrical characteristics of the visually close binary stellar system Hip 45571, using "Al-Wardat's method for analyzing binary and multiple stellar systems". We estimated the physical properties of the components of the system for the four measured parallax given by Gaia and Hipparcos, which gives a dynamical mass sum ranges between 2.43 and 2.52 solar mass using the new orbital parameters following Tokovinin's dynamical method.

The method used is a spectrophotometrical computational technique that employs Kurucz plane-parallel line-blanketed model atmospheres for single stars. These model atmospheres are used to construct the synthetic spectral energy distributions (SED) of each component and for the entire system. To ensure the method's accuracy, we apply the fit between synthetic and observational photometry under different filters, including the recently published Gaia DR3 measurements. The positions of the components on the H-R diagram and the evolutionary tracks were used to estimate their masses and ages.

We found that the system consists of 2.24 Gyr two F2.5 IV and F3.5IV subgiant components with $T_{\text{eff}}^{A} = 6800$ K, $T_{\text{eff}}^{B} = 6700$ K, $\log g_A = 4.19$ m/s^2 $\log g_B = 4.33$ m/s^2, $R_A = 1.77 R_\odot$, $R_B = 1.34 R_\odot$, $L_A = 6.01 L_\odot$, $L_B = 3.25 L_\odot$ and, $Z_* = 0.011$.

Depending on the masses estimated by Al-Wardat's method, a new parallax value of 28.72 ± 0.30 mas was obtained. Which lies between the values given by DR2 and DR3. This research underscores the importance of precision and reliability in employing these methods and measurements in a dynamic context, deepening our understanding of such stellar systems.

Keywords: binary stars · binaries · HIP 45571 · HD 80671

© The Author(s) 2025
H. M. K. Al Naimiy et al. (Eds.): AUASS-CONF 2023, SPPHY 420, pp. 63–74, 2025.
https://doi.org/10.1007/978-981-96-3276-3_6

1 Introduction

In the vast expanse of the cosmos, many stellar systems are not solitary entities but rather intricate arrangements of multiple star systems. These systems typically involve two gravitationally linked stars, orbiting around a shared center of mass. Understanding the intricacies of binary and multiple stellar systems (BMSSs) is pivotal for delving deeper into stellar astrophysics. By unraveling the characteristics, behaviors, and developmental pathways of BMSSs, we unlock valuable insights that propel our understanding of the universe forward [1].

Scientists estimate various physical and geometrical properties, such as effective temperature, radius, and apparent and absolute magnitudes, to understand binary stars. This is done through computational methods as well as photometric and astrometric observations. Binary stars are important because they provide essential data on mass-luminosity and mass-radius relations (MLR and MRR), respectively, which can only be obtained from binary systems Kallrath & Milone, 2009).

This research delves into a comprehensive investigation of HIP 45571, a stellar entity within the vast celestial expanse. It embarks on a meticulous journey aimed at unraveling its fundamental properties. This pursuit involves precise analysis of its orbital elements and thoroughly examining its physical characteristics. It is a dedicated effort to deeply grasp its inherent attributes within astrophysical phenomena, facilitating a nuanced understanding of the star, its cosmic environment, and its cosmic significance.

The system was recently analyzed by Suhail Masda as a solar metallicity ($Z = 0.019$) binary system. We are reanalyzing the system using the same method, but this time utilizing Gaia DR3 observations and focusing on more details. We considered the mentality of the system as $Z = 0.011$ as reported by [4] and [5] (See Table 1). Furthermore, the Multiple Star Catalog shows that the system is a quadruple one. It consists of a close visual binary AB with a period of 3.445 years (the main focus of this work), and another binary CaCb located 18.834 arcseconds away. The CaCb binary has a separation of 0.557 arcseconds and an orbital period of 86.2266 years. We anticipate the system hierarchy to align with the description provided (see Fig. 1).

In this study, we leveraged four parallaxes obtained from two prominent space missions: the Hipparcos mission, which boasts two published catalogues brimming with relevant system data, and the Gaia mission, known for its three comprehensive data releases. To unravel the orbital solution of the system, we turned to Tokovinin's dynamical method, esteemed for its ability to provide thorough solutions in determining orbital parameters [6].

In estimating the system's physical characteristics, we employed "Al-Wardat's method for analyzing binary and multiple stellar systems (BMSSs)", a sophisticated computational spectrophotometric technique. This method seamlessly builds synthetic spectral energy distributions (SEDs) using the plane-parallel line-blanketed model atmospheres of Kurucz's (ATLAS9) with observational data, thereby unveiling the parameters of the individual components [7–12].

It is worth mentioning that Al-Wardat's method has been used to analyze many binary and multiple stellar systems, for example, see [7, 8, 10, 13–25].

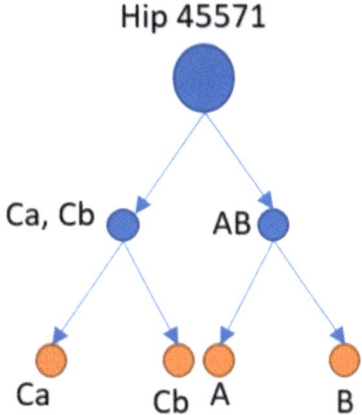

Fig. 1. The expected hierarchy of the quadruple system HIP 45571.

2 Basic Information of HIP 45571

HIP 45571 (HD 80671) has four different measured parallaxes: the first sourced from the Hipparcos 1997 catalogue ($\pi_{HIP97} = 29.83 \pm 0.60$ mas), followed by the Van Leeuwen catalogue published in 2007 ($\pi_{HIP2007} = 30.64 \pm 0.70$ mas), the third from Gaia data release 2 (DR2) ($\pi_{DR2} = 28.36 \pm 0.28$ mas) [26], and the last from Gaia data release 3 (DR3) ($\pi_{DR3} = 29.46 \pm 0.53$ mas) [5].

The binary system has F5V spectral type according to SIMBAD astronomical database, its effective temperature suggested by [27] is $T_{eff} = 6618$K.

Table 1. The basic data of HIP 45571

Parameter	HIP 45571	Reference
α	$09^h\ 17^m\ 17^s.2424584176$	SIMBAD*
δ	$-68°\ 41'\ 22''.535261160$	SIMBAD*
VJ	5.380	[30]
BT	5.871 ± 0.03	[31]
VT	5.438 ± 0.03	[31]
(B-V) HIP old	0.415 ± 0.02	[31]
π HIP (1997)	29.830 ± 0.60	[31]
π HIP (2007)	30.640 ± 0.70	[32]
π_{DR2}	28.36 ± 0.28	[26]
π DR3	29.46 ± 0.53	[5]
$[Fe/H]$	-0.25	[5]
$[Fe/H]$	-0.21 ± 0.05	[4]

The previous orbit parameters of [28] along with modified ones are listed in Table 3. The selection of the stellar system HIP 45571 is based on its expected nature to be a subgiant binary, despite being classified as a main-sequence binary in SIMBAD references. Exploring subgiant stars is crucial due to the scarcity of research in this field. Therefore, we should aim to enhance our understanding and carry out more precise investigations into these celestial systems.

The primary observational data for HIP 45571 is listed in Table 1, along with the corresponding references. These data were sourced from the SIMBAD astronomical database, HIP 97 and HIP 2007 catalogues, Gaia DR2 and DR3. To find the orbital elements, we utilize the relative positional measurements of the two components given in the Fourth Interferometric Catalogue [29].

3 Analysis

3.1 Orbital Solution

As we mentioned before the system was solved by Tokovinin and his team in 2015 [28], Tokovinin's dynamical method is a computer code that can be used to solve the orbits of binary and triple systems. It can be run under the IDL platform [33]. We used the relative position measurements: angular separations (ρ) in (arcsecond), position angles (θ) in (deg), and the relative position measurements of the system's residuals (Δρ) in (arcsecond). The new data provided in the fourth interferometric catalogue assisted us in slightly modifying the orbital solution of the system. Figure 2 shows a comparison between the orbit obtained in this work and the previous one. The two solutions are highly consistent.

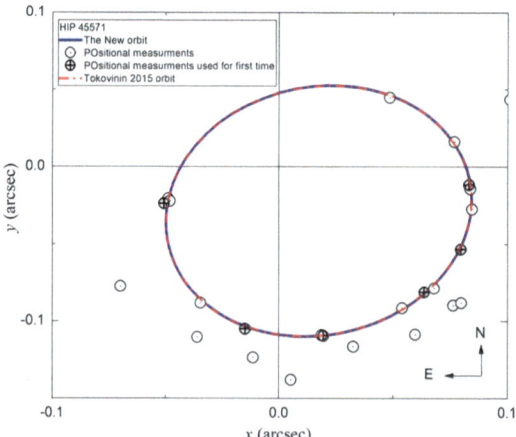

Fig. 2. Comparison between the previous orbit [28] and orbit from this work, where we used 5 new relative position measurements (labeled in crossed circles).

3.2 Physical Parameters

To estimate the physical parameters of HIP 45571, we used Al-Wardat's method, this method used primary calculation to build synthetic energy distribution (SED) for each component using ATLAS9 [34], the SED produced by ATLAS9 was used to build SED for the entire system using Al-Wardat's combination program. As a double-check, we used synthetic photometry from Al-Wardat's method to compare the synthetic magnitude and the color indices with the observational ones.

The primary calculation includes the difference in visual magnitude ($\Delta m_v = 0.64$) form [29], the apparent visual magnitude for component A and component B (m_v^A and m_v^B) of the binary star HIP 45571 calculated using the Eq. (1) for component A:

$$m_v^A = m_v + 2.5 log (1 + 10^{-0.4\Delta m_v}) \tag{1}$$

For component B we used the following equation:

$$m_v^B = m_v^A + m_v \tag{2}$$

Next, we calculated the absolute magnitude using the following equation:

$$M_v = m_v + 5 - 5log(d) - A_v \tag{3}$$

where d is the distance calculated depending on the parallaxes from the past mentioned catalogue and A_v is the interstellar extinction coefficient taken from [35]. To calculate the distances using the parallax we used Eq. (4):

$$d = \frac{1}{\pi} \tag{4}$$

We used Lang Table [36] to obtain the preliminary values of the effective temperatures (T_{eff}), the bolometric correction (BC), Masses (m) of the components and the spectral types. Using the absolute magnitude that we calculated from Eq. 3 and the bolometric correction (BC), we can find the preliminary absolute bolometric magnitude (M_{bol}) according to the following equation:

$$M_{bol} = M_v + BC \tag{5}$$

we calculated the luminosity (L) using the following equation:

$$M_{bol}^* - M_{bol}^\odot = -2.5log(\frac{L^*}{L^\odot}) \tag{6}$$

To calculate Radii (R) we used:

$$\frac{R_*}{R_\odot} = (\frac{L_*}{L_\odot})^{\frac{1}{2}} \times (\frac{T_{eff}^\odot}{T_{eff}^*})^2 \tag{7}$$

The final physical parameter we calculated is the gravitational acceleration ($log(g)$):

$$log(g) = log\left(\frac{M_*}{M_\odot}\right) + log\left(\frac{R_*}{R_\odot}\right) + 4.43 \tag{8}$$

Given $T_\odot = 5777$ K, $R_\odot = 6.69 \times 10^8$ m, and $M_{bol}^\odot = 4^m.75$ magnitudes, we utilize solar metallicity model atmospheres generated by ATLAS 9 with the preliminary parameters. The total energy flux from a binary star, created by the net luminosities of components a and b, situated at a distance d (in parsecs) from Earth, can be calculated as follows [7, 14]:

$$F_\lambda d^2 = H_\lambda^A R_A^2 + H_\lambda^B R_B^2 \tag{9}$$

Rearranging

$$F_\lambda = (R_A^2/d^2)^2 [H_\lambda^A + H_\lambda^B \cdot \left(\frac{R_B}{R_A}\right)^2] \tag{10}$$

H_λ^A and H_λ^B represent the fluxes from the unit surface of each component, with F_λ denoting the total spectral energy density of the system. R_A and R_b stand for the radii of the primary and secondary components in solar units. Synthetic photometry is a common method in various astronomical applications. Al-Wardat's method was employed to synthesize Spectral Energy Distributions, aiding in the calculation of magnitudes and color indices for the binary system.

The method also has a calibration procedure whereby the available observational data was compared with synthetic magnitudes of the complete star system. The accuracy of the results obtained is determined by the level of precision in the observations. Usually, a certain degree of error is acceptable to reach an agreement. The process of fitting colour measurements is used to determine the temperature of a star. By fitting the magnitude differences between the components, it is possible to estimate their relative brightness. Additionally, by fitting visual magnitude measurements between synthetic and observational data, it is possible to measure the overall brightness of the system. These comparative studies significantly contribute to our understanding of binary systems and their stellar components by testing and refining the models produced by this computational method.

The output in Fig. 3 shows the results of the synthetic components (A and B) and overall SEDs of the system Hip 45571 using Al-Wardat's Method and the new parallax ($\pi_{Thiswork} = 28.72 \pm 0.30$ mas).

Stellar metallicity is an important parameter in the analysis of any star, as it describes the abundance of elements heavier than hydrogen and helium in it and is measured by comparing the amount of iron (Fe) to hydrogen (H) in the star on a logarithmic scale relative to the Sun according to the following Equation:

$$\left[\frac{Fe}{H}\right] = \log_{10} \frac{[N_{Fe}/N_H]_{star}}{[N_{Fe}/N_H]_{sun}} \tag{11}$$

where N_{Fe}, N_H are the number of iron and hydrogen atoms in a unit of volume. The solar Metallicity was taken as $Z_\odot = 0.0196 \mp 0.0014$ [37].

Gaia DR3 gives the metallicity of Hip 45571 as $\left[\frac{Fe}{H}\right] = -0.25$ (See Table 1). Which means $Z_{Hip45571} = 0.011$. As a part of the method, we tested the positions of the components on the isochrons of different metallicities, and we found it fit the measured metallicity (see Fig. 4). According to Gaia DR3, the metallicity of Hip 45571 is given in

Table 1 as $\left[\frac{Fe}{H}\right] = -0.25$, i.e. $Z_{Hip45571} = 0.011$. As a part of the method's procedures, the positions of the components were tested on the isochrons of different metallicities. It was found that the measured metallicity fit the isochrons of $Z_{Hip45571} = 0.011$, as shown in Fig. 4.

Table 2. Modified orbital parameters of the HIP 45571 along with the previous solution.

Elements	Units	[28]	Modified elements (This work)
P	[year]	3.445	3.445 ± 0.005
a	[arcsec]	0.0888	0.0888 ± 0.0001
T	–	2013.376	2013.376 ± 0.001
e	–	0.455	0.4550 ± 0.0008
Ω	[deg]	154.6	154.6 ± 0.2
ω	[deg]	117.2	117.20 ± 0.21
i	[deg]	140.7	140.70 ± 0.12
$RMS(\theta)$	[deg]	–	0.85
$RMS(\rho)$	[mas]	–	0.0009

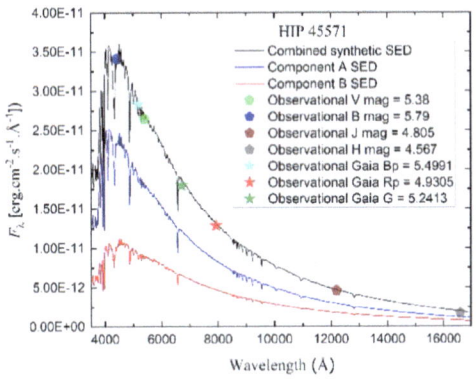

Fig. 3. The best fit for the spectral energy distribution (SED) of the Hip 45571 system using observational magnitudes from various sources.

The positions of the system components on the evolutionary tracks and isochrons are strong proof of the reliability of "Al-Wardat's method" as it shows that using the parallax measurements of Hipparcos and Gaia gives the exact solutions and masses. But as part of "Al-Wardat's method," when we combine the orbital solution with estimated masses, we are able to obtain a new dynamical parallax.

Kepler's third law was employed to determine the dynamical mass sum.

$$M_{dyn} = M_A + M_B = \frac{a^3}{\pi^3 P^2} M_\odot \qquad (12)$$

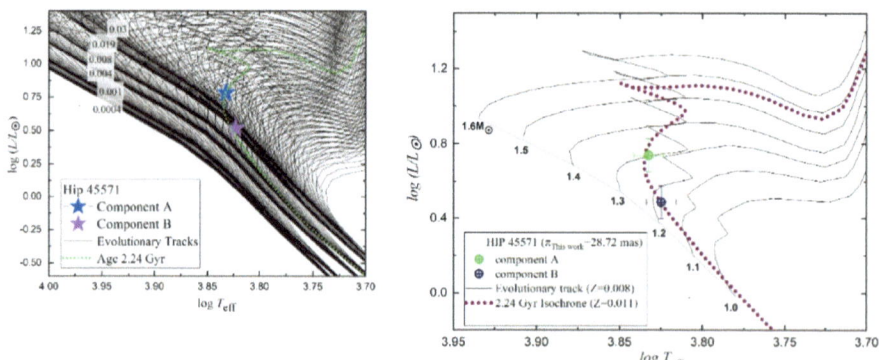

Fig. 4. The positions of the system's components on the isochrones and HR diagram and evolutionary tracks of different metallicities as of Girardi et al., 2000. It shows that the system's metallicity Z lies between 0.008 and 0.019.

where M_A represents the mass of the primary component, M_B represents the mass of the secondary component, a denotes the semi-major axis in arcseconds, π signifies the parallax in arcseconds and P represents the orbital period in years.

The formula gives the formal error in the total dynamical mass:

$$\frac{\sigma M_{Dyn}}{M_{Dyn}} = \sqrt{9\left(\frac{\sigma_\pi}{\pi}\right)^2 + 9\left(\frac{\sigma_a}{a}\right)^2 + 4\left(\frac{\sigma_P}{P}\right)^2} \tag{13}$$

Now, the important point is finding the system's exact masses and exact parallax.

The method proposed by Al-Wardat provides solutions that are not significantly affected by differences in parallax. Therefore, it is a reliable way to obtain precise masses for individual components. These masses and other estimated parameters are listed in Table 4. By using these masses and modified orbital parameters (Fig. 1), a new parallax value can be obtained as π this work $= 28.72 \pm 0.30$ mas.

Table 3. Comparison between dynamical masses and Al-Wardat's method mass sum of HIP45571.

Parallax (π)(mas)	Dynamical mass sum ($M_\odot$)		Al-Wardat's method mass sum	
	[28]	This work		
			With Z $= 0.008$	With Z $= 0.0019$
$\pi_{Hip 97} = 29.83$	2.2228	2.22 ± 0.13	2.48 ± 0.15	2.78 ± 0.15
$\pi_{Hip2007} = 30.64$	2.5113	2.05 ± 0.14	2.46 ± 0.15	2.74 ± 0.15
$\pi_{DR2} = 28.36$	2.58667	2.59 ± 0.08	2.51 ± 0.10	2.83 ± 0.15
$\pi_{DR3} = 29.46$	2.30761	2.31 ± 0.13	2.50 ± 0.15	2.80 ± 0.15
π this work $= 28.72 \pm 0.30$	2.49 ± 0.07	2.49 ± 0.07	2.48 ± 0.10	2.81 ± 0.15

Table 4. The estimated physical parameters for primary and secondary components of HIP 45571.

Parameter	Units	Comp. A	Comp. B
$T_{eff} \pm \sigma_{T_{eff}}$	[K]	6800 ± 80	6680 ± 70
$R \pm \sigma_R$	$[R_\odot]$	1.74 ± 0.09	1.35 ± 0.08
$\log g \pm \sigma_{\log g}$	$\left[\frac{m}{s^2}\right]$	4.19 ± 0.11	4.50 ± 0.13
$L \pm \sigma_L$	$[L_\odot]$	5.78 ± 0.20	3.23 ± 0.10
$M_v \pm \sigma_{M_v}$	[mag]	2.81 ± 0.13	3.43 ± 0.14
$M_{bol} \pm \sigma_{M_{bol}}$	[mag]	2.80 ± 0.08	3.47 ± 0.08
M	$M_\odot$	1.30 ± 0.15	1.18 ± 0.13
Sp. Type	--	F2.5IV	F3.5IV
Age	[Gyr]	2.24	

4 Conclusion

The purpose of this study was to enhance our understanding of the binary star system called HIP 45571. We achieved this by analyzing its characteristics and physical properties. To refine the orbital parameters, we used Tokovinin's dynamical method in the ORBITX code. This process led to the acquisition of accurate computed orbital elements that are considered to be reliable.

We utilized "Al-Wardat's Method for analyzing BMSSs" to determine physical parameters such as distances, radii, effective temperatures, and gravity. The alignment between synthetic and observational magnitudes and colour indices supported the accuracy of SED predictions, reaffirming the reliability of parallax measurements.

These are the key takeaways that we have concluded on:

- A modified orbit with more accurate orbital elements was achieved by adding five additional relative position measurements (listed in Table 2).
- The physical parameters of the system's components (Table 5) were estimated based on the best fit between synthetic Spectral Energy Distributions (SEDs) and observed photometry.
- We determined that the system comprises F2.5IV and F3.5IV solar-type subgiant stars with solar metallicity $Z_* = 0.011$. This conclusion is drawn from analyzing the physical properties of the system's components and their positions relative to evolutionary and age tracks (2.24 Gyr).
- We have determined that the system consists of two subgiant stars, assuring its metallicity $Z_* = 0.011$, which is a little bit smaller than that of the sun. We arrived at this conclusion by analyzing the system's components' physical properties and positions relative to evolutionary tracks and isochrons. We estimate that the system is approximately 2.24 billion years old.
- The analysis resulted in estimating the masses of system components. These estimates, along with the modified orbital elements, produced a new parallax ($\pi_{Thiswork} =$

28.72 ± 0.30 mas). It is worth noting that this value matches closely with the measurements provided by Gaia DR2. (π_{DR2} = 28.36 mas), and DR3 (π_{DR3} = 29.46 mas), Hipparcos 1997 ($\pi_{HIP\,97}$ = 29.83 mas), and Hipparcos 2007 ($\pi_{HIP\,2007}$ = 30.64 mas).

This comprehensive approach improved our understanding of HIP 45571. It showcased our success in achieving research objectives while emphasizing the importance of precision and reliability when employing these methods and measurements in a dynamic context.

Acknowledgments. "Al-Wardat's method for analyzing binary and multiple stellar systems" with its codes, as well as the Fourth Interferometric Catalogue of Binary Stars, SIMBAD database, Sixth Catalog of Orbits of Visual Binary Stars, and ORBITX code, have all been used in his work.

References

1. Raghavan, D., et al.: A survey of stellar families: multiplicity of solar-type stars. Astrophys. J. Suppl. Ser. **190**(1), 1–42 (2010). https://doi.org/10.1088/0067-0049/190/1/1
2. Kallrath, J., Milone, E.F.: Eclipsing Binary Stars: Modeling and Analysis. In: Astronomy and Astrophysics Library. Springer, New York (2009). https://doi.org/10.1007/978-1-4419-0699-1
3. Al-Wardat, M.A., Widyan, H.S., Al-thyabat, A.: Complex analysis of the stellar binary HD25811: a subgiant system. Publ. Astron. Soc. Aust. **31**, e005 (2014). https://doi.org/10.1017/pasa.2013.42
4. Gáspár, A., Rieke, G.H., Ballering, N.: The correlation between metallicity and debris disk mass. Astrophys. J. **826**(2), 171 (2016). https://doi.org/10.3847/0004-637X/826/2/171
5. Collaboration, G., et al.: Gaia Data Release 3: Summary of the content and survey properties (2022). https://doi.org/10.48550/arXiv.2208.00211
6. Tokovinin, A., Mason, B.D., Hartkopf, W.I., Mendez, R.A., Horch, E.P.: Speckle interferometry at soar in 2015. Astron. J. **151**(6), 153 (2016). https://doi.org/10.3847/0004-6256/151/6/153
7. Al-Wardat, M.A.: Physical parameters of the visually close binary systems Hip70973 and Hip72479. PASA **29**, 523–528 (2012)
8. Al-Wardat, M.A.: Spectral energy distributions and model atmosphere parameters of the quadruple system ADS11061. Bull Spec. Astrophys. Obs **53**, 51–57 (2002)
9. Al-Wardat, M.A., Hussein, A.M., Al-Naimiy, H.M., Barstow, M.A.: Comparison of Gaia and Hipparcos parallaxes of close visual binary stars and the impact on determinations of their masses. Publ. Astron. Soc. Aust. **38**, e002 (2021). https://doi.org/10.1017/pasa.2020.50
10. Masda, S.G., Al-Wardat, M.A., Pathan, J.M.: Orbital and physical parameters of the close binary system GJ 9830 (HIP 116259). Res. Astron. Astrophys. **19**(7), 105 (2019). https://doi.org/10.1088/1674-4527/19/7/105
11. Al-Wardat, M.A., El-Mahameed, M.H., Yusuf, N.A., Khasawneh, A.M., Masda, S.G.: Physical and geometrical parameters of CVBS XI: Cou 1511 (HIP 12552). Res. Astron. Astrophys. **16**(11), 166 (2016). https://doi.org/10.1088/1674-4527/16/11/166
12. Kurucz, R.: ATLAS9 stellar atmosphere programs and 2 km/s grid. ATLAS9 Stellar Atmosphere Programs 2 Kms Grid Kurucz CD-ROM No 13 Camb. Mass Smithson. Astrophys. Obs. vol. 13 (1993). http://adsabs.harvard.edu/abs/1993KurCD..13.....K. Accessed 12 Nov 2016

13. Abu-Alrob, E.M., Hussein, A.M., Al-Wardat, M.A.: Atmospheric and fundamental parameters of the individual components of multiple stellar systems. Astron. J. **165**(6), 221 (2023)

14. Al-Wardat, M.A.: Model atmosphere parameters of the binary systems COU1289 and COU1291. Astron. Nachrichten **328**, 63–67 (2007). https://doi.org/10.1002/asna.200610676

15. Al-Wardat, M.A., et al.: Speckle interferometric binary system HD375; Is it a sub-giant binary? Astrophys. Bull. **69**, 58–66 (2014)

16. Al-Wardat, M.A., El-Mahameed, M.H., Yusuf, N.A., Khasawneh, A.M., Masda, S.G.: Physical and geometrical parameters of CVBS XI: Cou 1511 (HIP 12552). Res. Astron. Astrophys. **16**, 166 (2016). https://doi.org/10.1088/1674-4527/16/11/166

17. Al-Wardat, M.A., Hussein, A.M., Al-Naimiy, H.M., Barstow, M.A.: Comparison of Gaia and Hipparcos parallaxes of close visual binary stars and the impact on determinations of their masses. Publ. Astron. Soc. Aust. **38** (2021). https://doi.org/10.1017/pasa.2020.50

18. Al-Wardat, M.A., Widyan, H.S., Al-thyabat, A.: Complex analysis of the stellar binary HD25811: a subgiant system. PASA **31**, 5 (2014)

19. Al-Wardat, M.A., Widyan, H.S., Al-thyabat, A.: Complex analysis of the stellar binary HD25811: a subgiant system. PASA **31**, e005 (2014). https://doi.org/10.1017/pasa.2013.42

20. Al-Wardat, M.A., Widyan, H.: Parameters of the visually close binary system Hip11253 (HD14874). Astrophys. Bull. **64**, 365–371 (2009)

21. Hussein, A.M., Abu-Alrob, E.M., Alkhateri, F.M., Al-Wardat, M.A.: Atmospheric parameters of individual components of the visual triple stellar system HIP 32475, arXiv Prepr. arXiv: 2304.03604 (2023)

22. Hussein, A.M., Abu-Alrob, E.M., Mardini, M.K., Alslaihat, M.J., Al-Wardat, M.A.: Complete analysis of the subgiant stellar system: HIP 102029. Adv. Space Res. (2023). https://doi.org/10.1016/j.asr.2023.07.045

23. Masda, S.G., Al-Wardat, M.A., Pathan, J.K.M.K.: Physical and geometrical parameters of VCBS XIII: HIP 105947. Res. Astron. Astrophys. **18**(6), 072 (2018). https://doi.org/10.1088/1674-4527/18/6/72

24. Masda, S.G., Al-Wardat, M.A., Pathan, J.M.: Stellar parameters of the two binary systems: HIP 14075 and HIP 14230. J. Astrophys. Astron. **39**(5), 58 (2018)

25. Tanineah, D.M., Hussein, A.M., Widyan, H., Al-Wardat, M.A.: Trigonometric parallax discrepancies in space telescopes measurements I: the case of the stellar binary system Hip 84976. Adv. Space Res. **71**(1), 1080–1088 (2023). https://doi.org/10.1016/j.asr.2022.09.025

26. Collaboration, G., et al.: Gaia data release 2: summary of the contents and survey properties. Astron. Astrophys. **616**, A1 (2018). https://doi.org/10.1051/0004-6361/201833051

27. Gray, R.O., et al.: Contributions to the nearby stars (NStars) project: spectroscopy of stars earlier than M0 within 40 pc–the southern sample. Astron. J. **132**(1), 161–170 (2006). https://doi.org/10.1086/504637

28. Tokovinin, A., Mason, B.D., Hartkopf, W.I., Mendez, R.A., Horch, E.P.: Speckle interferometry at soar in 2014. Astron. J. **150**(2), 50 (2015). https://doi.org/10.1088/0004-6256/150/2/50

29. Fourth Interferometric Catalog. http://www.astro.gsu.edu/wds/int4.html. Accessed 29 Mar 2023

30. Perryman, M.A.C., et al.: The HIPPARCOS Catalogue, åp, vol. 323, pp. L49–L52 (1997)

31. ESA, The Hipparcos and Tycho Catalogues (1997)

32. van Leeuwen, F.: Validation of the new Hipparcos reduction. Astron. Astrophys. **474**(2), 653–664 (2007). https://doi.org/10.1051/0004-6361:20078357

33. Tokovinin, A.: Speckle–spectroscopic studies of late-type stars. Int. Astron. Union Colloq. **135**, 573–576 (1992). https://doi.org/10.1017/S0252921100007193

34. Kurucz, R.: Solar abundance model atmospheres for 0,1,2,4,8 km/s. Sol. Abundance Model Atmospheres 0, vol. 19 (1994). https://ui.adsabs.harvard.edu/abs/1994KurCD..19.....K. Accessed 27 Mar 2023
35. Galactic DUST Reddening & Extinction. https://irsa.ipac.caltech.edu/applications/DUST/. Accessed 03 Apr 2023
36. lang: lang (1992)
37. von Steiger, R., Zurbuchen, T.H.: Solar metallicity derived from in situ solar wind composition. Astrophys. J. **816**(1), 13 (2015)
38. Girardi, L., Bressan, A., Bertelli, G., Chiosi, C.: Evolutionary tracks and isochrones for low- and intermediate-mass stars: from 0.15 to 7 $M_{\odot}$, and from $Z=0.0004$ to 0.03. Astron. Astrophys. Suppl. Ser. **141**(3), Art. no. 3 (2000). https://doi.org/10.1051/aas:2000126

The United Nations Group of Experts on the Standardization of Geographical Names (UNGEGN) and Its Relationship with the International Astronomical Union (IAU) on the Naming/Designation of Extraterrestrial Objects'

Brahim Atoui[(✉)] and Awni Khawasna

New York, USA
atoui.brahim@hotmail.fr

Abstract. In this paper, we will discuss the role of the United Nations Group of Experts on the Standardization of Geographical Names in the naming of geographical names in general and its relationship with the International Astronomical Union in the naming/designation of extraterrestrial objects. Emphasizing the need for the reactivation of cooperation between these two organizations and also the need for the establishment of close collaboration between the Arabic Geographical Names Division of the UNGEGN and the two Specialized Groups respectively the Working Group on Small Body Nomenclature (WGSBN) and the Working Group on Planetary Systems Nomenclature. (WGPSN) of the UIA in charge of naming, thus promoting a better presence and visibility of names from the Arab world in the naming of extraterrestrial objects.

Keywords: Toponymy · UNGEGN · IAU · Astronomy · standardization of geographical names · Planetary Systems Nomenclature · the naming/designation of extraterrestrial objects · romanisation system

1 Introduction

We begin our remarks by quoting this Quranic verse:

' وَعَلَّمَ آدَمَ الأَسْمَاء كُلَّهَا

' HE (Allah) taught Adam all the names. ' *Qur'an, Surah Al Baqara, verse 31.*

B. Atoui—Former Vice-Chair UNGEGN 2002–2010, Chair Task Team for Africa/UNGEGN, 2006–2023.
A. Khawasna—Chair Arab Division/UNGEGN.

H. M. K. Al Naimiy et al. (Eds.): AUASS-CONF 2023, SPPHY 420, pp. 75–82, 2025.
https://doi.org/10.1007/978-981-96-3276-3_7

As stated in this Koranic verse, the act of naming is thus, attested even before the appearance of Man (of Adam), on earth! For God taught Adam the names of all things during his stay in paradise!

Similarly, in the Bible there is mention of this act of naming attested during Man's stay in paradise; but unlike the Koran, Adam limited himself to naming only certain categories of animal species: "And the man gave names to all the cattle, to the birds of the air and to all the beasts of the field" (Bible, Genesis 2:20, Louis Segond);

While in the Koran, it is God who 'taught' (1) Adam 'all the names', in the Bible, it is 'Man' (Adam) who took charge, although he only gave names to certain categories of animals.

However, despite these differences between the Qur'an and the Bible, the act of naming is ancient and appeared even before Adam appeared on earth.

Moreover, since the earliest times, people have been interested in explaining the formation and meaning of geographical names.

In 1975, an Italian archaeological team discovered 15,000 cuneiform clay tablets in the town of Elba, 60 km from Aleppo in Syria, which were the remains of a royal archive (third millennium BC). These tablets contain 5,000 place names! This is a fine illustration of the long-standing interest in place names on the part of mankind.

If interest in place names goes back a long way, so does interest in standardising them. As far back as the time of Confucius, i.e. 500 centuries BC, the issue of standardising place names arose.

Confucius said:

"Tzu lou said: "If the Prince of Wei were waiting for you to settle public affairs with you, what would be your first concern? - To give everything its true name", replied to the Master. Really?" replied Tzeu lou. Master, you are straying far from the goal. What is the point of this rectification of names?" The Master replied: "How boorish you are! An honorable man is careful not to comment on what he does not know. If the names are not adjusted, the language is not adequate. If language is not adequate, things cannot be done. If things cannot be done right, propriety and harmony hardly flourish. If decorum and harmony do not flourish, torments and other punishments are not just. With punishments and other punishments no longer just, the people no longer know how to dance. Everything that the honorable man conceives, he can state, and stating it he can do it. The honorable man leaves nothing to chance.

In this introduction, we wanted to show how long the interest in geographical names goes back.

But what about our era?

Albert Camus said: 'Is not naming things wrongly adding to the misfortune of the world?

In addition to the importance of place names as part of a nation's cultural, historical, sociological, folkloric, and other heritage, and in order to preserve this depth, we need to ensure that these place names are standardized.

In a world without frontiers in terms of the exchange of information and communications, which with the development of information technology and communications techniques such as the Internet, Face Book, Tweeter (X) etc. encourage contact between the different countries of this world, confusion in place names, whether terrestrial or extraterrestrial, and the spelling mobility of the same place name are hardly acceptable in today's world.

This is one of the reasons why the international community, through the UN, has been setting up structures to manage and standardize toponymy at international level since 1959.

2 Presentation of Ungegn

In view of the importance of geographical names as a cultural, historical and social heritage that plays a part in the social cohesion of different nations and their role in improving international communication, of which they are an essential element, particularly for providing assistance during natural disasters, facilitating commercial, economic and cultural exchanges between different states and nations, etc., the international scientific community first, followed later by the United Nations, took an interest in this issue very early on.

From the 18th century onwards, the academic community, particularly in the West, focused mainly on establishing uniform systems for transliterating the characters of languages written in scripts other than Roman, particularly for so-called oriental languages, with a view to scientific studies. The United Nations Organisation, for its part, took an interest in geographical names with a view to standardizing their writing for the purposes of good international communication.

Fourteen years after the creation of the UN, in 1959, the Economic and Social Council (ECOSOC) recommended defining a policy for the standardization of geographical names at national and international level; to carry out this task, it recommended the creation of a United Nations Group of Experts on the Standardization of Geographical Names.

This was set up and has since met every five years as the United Nations Conference on the Standardization of Geographical Names and every two years as the Group of Experts. Its first Conference was held in 1967 in Athens.

Following reforms introduced in 2017, with new working methods and a new organisation, it now only meets in regular biennial United Nations Sessions on the standardization of geographical names.

Aims, objectives and results of the Group:

One of the fundamental principles of UNGEGN is that international standardization should be based on national standardization. To this end, it promotes the advantages of standardization among member states, in particular by emphasising the role of place names as an essential element in the preservation of cultural heritage and, above all, by encouraging states to collect place names from their respective countries, register them and standardize them at national level.

To this end, and in order to achieve concrete results, the establishment of recognised, permanent national structures for the management and standardization of geographical names in each state is strongly recommended.

As well as enabling the establishment of standardized toponymic databases and directories for both national and international use, these national structures will be called upon, particularly for countries using non-Latin characters to write their languages, to decide on the adoption and official use of a single romanisation system.

To date, UNGEGN has adopted 30 romanization systems, including Arabic characters. Thirty (30) others are still being examined.

The UNGEGN, which has over 400 members from more than 100 countries, is made up of 24 so-called geographical/linguistic divisions grouping together countries sharing one or more languages and/or belonging to the same geographical area; among these divisions is the Arab World Division, which groups together the 22 Arab states and plays a very active role within the UNGEGN.

UNGEGN also has 10 working groups responsible for monitoring technical and practical subjects and issues, including: training, digital data files and geographical directories and databases, romanisation systems, country names, terminology, publicity and funding, and toponymic guidelines.

In addition, given the specific nature of Africa, not only in terms of its linguistic particularity and diversity stemming from non-written languages, but also in terms of the small number of participants in the work of the UNGEGN, the UNGEGN has set up a Task Team for Africa, which coordinates the work of African countries in order to involve them more closely and raise their awareness of the importance of standardising geographical names.

In connection with our meeting, we can mention that among the other UNGEGN working groups that have been set up and disbanded are the 'Extraterrestrial Topographic Names' group and the 'Names of Underwater and Maritime Features' group.

3   Reminder of the Role of the Ungegn Working Group on Extraterrestrial Topographic Names

At the second UNGEGN Conference, held in London in 1972, the question of the designation of extraterrestrial topographical features was examined by the Experts present and it was recommended 'that the United Nations Group of Experts on Geographical Names study the question of drawing up an international convention on the standardization of the nomenclature of extraterrestrial features, in collaboration with the other competent international bodies'.

At the third Conference held in Athens in 1977, while recalling the provisions of the previous resolution, the Conference set out certain principles and objectives in a new resolution, this time recognising the importance and essential role of the International Astronomical Union in the naming/designation of extraterrestrial objects.

In Resolution III/23 on: Names of extraterrestrial features", the Conference, recalling Resolution 21 of the Second United Nations Conference on the Standardization of Geographical Names, notes that "recent planetary research programs have resulted in a growing need for names because of newly discovered features on the surfaces of planets and anticipates that names will be required in the future for many as yet unidentified features while recognizing on the one hand, that nations actively engaged in planetary research must meet the need for names for terrestrial and marine charts, and recognising on the other hand, that in view of the great importance of space exploration to all mankind, world-wide participation in the procedure of naming extraterrestrial details would ensure that future generations inherit a nomenclature less confused than that received from the past,'.

Furthermore, being aware that nations actively engaged in planetary exploration have developed certain procedures for the establishment of names for land and sea charts and other publications, and conscious of the role played by the International Astronomical Union in the establishment of lists of names and the application of names to details in international use, the United Nations Group of Experts on the Standardization of Geographical Names recommends, inter alia, that the United Nations Group of Experts on Geographical Names should establish close co-operation with the International Astronomical Union and other interested international bodies, and that 'nations should use their own language and writing system for the treatment of generic terms'.

In these recommendations, it is clearly stipulated that it is the UNGEGN that is responsible for the management and standardization of the names of extraterrestrial topographical details; the IAU is only a collaborator and partner in the same way as other interested bodies.

At the Fourth Conference, held in Geneva, and on the report of the Chairman of the Working Group on the Names of Extra-terrestrial Objects, which confirmed that the Working Group on Planetary System Nomenclature of the International Astronomical Union was satisfactorily carrying out the task of naming extraterrestrial details, it was recommended, in Resolution IV/13, to dissolve the UNGEGN Working Group in question and to entrust and reserve this function exclusively to the IAU.

Nevertheless, it was recommended in the same resolution that the United Nations Group of Experts on the Standardization of Geographical Names, through the intermediary of the Chairman of the disbanded Working Group on Extraterrestrial Details, should continue to liaise with the specialized Working Group or Groups of the International Astronomical Union.

This liaison, which was supposed to coordinate relations on the naming and standardization of extraterrestrial objects, has unfortunately gradually diminished over the years and no report on the subject has since been submitted to the UNGEGN's work on the subject.

The same was true of the Working Group on Names of Marine and Underwater Features, which at the same Conference limited its prerogatives to marine features only (Resolution IV/12), leaving the task of standardization to the International Hydrographic Organisation and the Intergovernmental Oceanographic Commission.

In fact, in the same way as for the names of extraterrestrial objects, the UNGEGN has not dealt with this subject since the Fourth Conference; and this Group has ipso facto been dissolved, but no resolution has been passed to support this dissolution, as was the case for the Group on extraterrestrial objects.

4 The International Astronomical Union and the Naming/Designation of Extraterrestrial Topographic Objects

Following the UNGEGN's recognition of the IAU's role and its withdrawal from this issue, the naming of extraterrestrial objects has once again become the monopoly of the International Astronomical Union, whose role has thus been strengthened and confirmed.

It should be remembered that the IAU has a great deal of experience and expertise in the field of naming extraterrestrial objects, and that as early as 1922, at its meeting in Rome, it standardized the names and abbreviations of constellations. More recently, through its two specialised groups, the Working Group on the Nomenclature of Small Bodies (WGSBN) and the Working Group on the Nomenclature of Planetary Systems. (WGPSN), the IAU has assigned and formalised object names and astronomical characteristics, as reported in its latest report for the year 2022, in which 83 new names were assigned.

In addition, these two Working Groups have drawn up guidelines, rules and procedures detailing the process to be followed for the proposal, adoption, and naming of extraterrestrial objects. This approach was updated and adopted in 2021. Among the new features introduced, mention should be made of those defining naming privileges, the officialization of names, the standardization of scripts and characters in which names must be written, etc.

The procedure for submitting and assigning names for extraterrestrial objects pursued by the International Astronomical Union can in some cases take years and years: After assigning a provisional number based on its discovery date, its discoverer is invited to propose a name that is neither commercial nor the name of animals nor that of a living person; Like the policy recommended by the UNGEGN, the attribution of a person's name is only possible after a certain period of disappearance, which is 5 to 15 years and 100 years for the IUA.

5 Conclusion

The United Nations Group of Experts on the Standardization of Geographical Names, the body responsible for the management of geographical names at international level since 1959, is responsible for coordinating the standardization of geographical names at international level; it has a certain amount of expertise and competence in this area.

The same applies to the IAU, which also has proven expertise in this field.

The names of extraterrestrial objects are important elements of the general nomenclature of geographical names.

As such, we recommend not only the reinvigoration of cooperation and coordination between the IAU and UNGEGN, but also the establishment of a strong relationship between the Arab Division of UNGEGN and the Arab Union for astronomy & space sciences.

The Arab Division of Experts on Geographical Names, which was set up when UNGEGN was founded, is made up of experts with extensive experience in the field of toponymic sciences and the naming and standardization of geographical names. It is one of the most active Divisions. It meets regularly every two years and produces many documents and directives for the benefit of the Arab countries.

The creation of a working group within the Arab Astronomical Union, to be known as the Arab Working Group for Planetary System Nomenclature (AWGPSN), would undoubtedly make a definite contribution and be a clear source of proposals for promoting the names of extraterrestrial objects in relation to the history, culture, and contribution of the Arab world to universal civilisation in general and to the space sciences in particular.

This suggested Group, if established, could set up, in partnership with the Arab Division of UNGEGN, another joint Working Group to deal with the problem of the naming of extraterrestrial objects in relation to the Arab world and make proposals to the IUA for adoption.

Furthermore, in order for the Arab contribution in the field of the naming of extraterrestrial objects to be effective, it would be necessary, in our opinion, for Arab Experts of the International Astronomical Union to join the other members of the Working Group for the Nomenclature of Planetary Systems, namely the WGPSN and the WGSBN, the groups in charge of approving the names of the bodies of the solar system.

It should be noted that no Arab expert is a member of these both important groups.

References

UNGEGN Related Documentation

United Nations Group of Experts on Geographical Names. Technical reference manual for the standardization of geographical names (PDF). United Nations, New York, pp. 44–45 (2007). ISBN 978-92-1-161500-5. Accessed 6 Apr 2011

UNGEGN - United Nations Group of Experts on Geographical Names: United Nations Conferences on the Standardization of Geographical Names Information on toponymic guidelines on the official GENUNG website

UIA Related Documentation

Naming Stars IAU. Union astronomique internationale

Acheter des étoiles et des noms d'étoiles. Union astronomique internationale

Groupe de travail de l'UAI sur les noms d'étoiles (WGSN). Union astronomique internationale

Bulletin du groupe de travail de l'UAI sur les noms d'étoiles, No. 1 (PDF). Consulté le 28 juillet 2016

Final Results of Name ExoWorlds Public Vote Released (Communiqué de presse). IAU.org. 15 décembre 2015. Archivé de l'original le 2022-11-10

Bulletin du Groupe de travail de l'AIU sur les noms d'étoiles, n° 2 (PDF). Consulté le 12 octobre 2016

Nommer les étoiles. Union astronomique internationale

Name ExoWorlds the Approved Names. Archivé de l'original le 2018-02-01. Consulté le 2016–07–28

Richard Hinckley Allen, Star-Names and Their Meanings, G.E. Stechert, New York (1899)

Robert Burnham, Jr. Burnham's Celestial Handbook, Volume 1, p. 359

Assessment of Active Space Debris Removal Methods Using the Weighted Sum Model (WSM)

Kareem Mesrega[1,2](✉) ⓘ, O. M. Shalabiea[2,3] ⓘ, Dalia Elfiky[4] ⓘ, Wesam Elmahy[2] ⓘ, and Haitham Elshimy[3] ⓘ

[1] Menofia University, Gamal Abdel Nasser Street, Shebin El-Kom 32511, Egypt
`kareemmesrega@science.menofia.edu.eg`
[2] Cairo University, Gamaa Street 1, Giza 12613, Egypt
[3] Beni-Suef University, Beni-Suef 2731070, Egypt
[4] National Authority for Remote Sensing and Space Sciences, Jozif Tito Street 23, Cairo 11769, Egypt

Abstract. With the recent growth in space exploration, the problem of space debris is becoming increasingly important. This debris could endanger active spacecraft. So, many methods are suggested to clear space debris. The best and most efficient removal method must be selected from these available options. This is our main objective. In this study, we suggest using the Weighted Sum Model (WSM) to order the chosen removal methods according to a set of deciding criteria. WSM is one of the most common Multi-Criteria Decision Analysis (MCDA) techniques. Based on how widely applicable they are, we selected five active removal methods. Thirteen decision criteria were chosen carefully. The Relative Frequency approach was used to determine the weight of these criteria. The WSM equation was used in the final stage. According to our major findings, Tethered-Deployed Nets, with a performance score of 78.9%, is comparatively the best method among the selected methods. More in-depth research on this approach is recommended for future work. To obtain more accurate results for our future research, we will use more decision criteria.

Keywords: Space Debris · Active Removal Methods · Multi-Criteria Decision Analysis · Weighted Sum Model

1 Introduction

European Space Agency (ESA) defines space debris as any man-made objects such as fragments and other related elements that are no longer functional and currently present in Earth's orbit [1]. Around the Earth in different orbits, there is a large number of space debris with different sizes and shapes.

The estimated number of space debris objects orbiting the Earth, categorized by size as follows: over 29,000 objects larger than 10 cm, more than 670,000 objects larger than 1 cm, and over 170 million objects larger than 1 mm. These objects can pose a threat to operational spacecraft and satellites [2]. It has been discovered that an increase in

H. M. K. Al Naimiy et al. (Eds.): AUASS-CONF 2023, SPPHY 420, pp. 83–97, 2025.
https://doi.org/10.1007/978-981-96-3276-3_8

uncontrolled objects in Low Earth Orbit (LEO) might cause a series of collisions and a persistent buildup of orbital debris, which would cause environmental instability. The "Kessler Syndrome" is the name for this cascading effect [3].

Several methods or systems have been suggested over the years to clear space debris. These methods could be passive removal methods like Electrodynamic tethers or active ones like Tethered-deployed nets [4]. Each of these removal methods has benefits and drawbacks. So, selecting the most effective method among these options becomes a problem that must be tackled.

An example of conducting a comparative analysis between the removal methods is found in Hakima and Emami in their study [5]. They have made a quantitative analysis using the Analytical Hierarchy Process (AHP). They evaluate different active debris removal methods for clearing large objects in LEO. The comparison was done between the net method, on-orbit laser, electro-dynamic tether, ion beam shepherd, and robotic arm. They found that net methods are the most effective overall, with on-orbit lasers and robotic arms being close contenders.

As seen in Hakima and Emami's research, they used AHP which is one of the Multi-Criteria Decision Analysis (MCDA) techniques. MCDA concentrates on problems with discrete decision spaces. In these problems, the set of decision alternatives has been predetermined to choose the best between them based on preselected decision criteria. There are many techniques used in MCDA such as the Weighted Product Model (WPM), WSM, and AHP. One of the most popular MCDA techniques used today is WSM [6].

In this study, we attempt to find the most efficient and the most suitable method for removing space debris using WSM. WSM has never been used in this field previously. Therefore, the scientific question that we are asking is, "According to WSM, which of the proposed methods is the best for removing space debris?".

In our research, we will select five active removal methods that have the highest chance for practical implementation. The decision-making process will consider thirteen relevant criteria. These decision criteria are based on the characteristics of space debris and space debris removal missions. The weight of these criteria will be calculated by using the Relative Frequency Approach. Finally, the performance scores for the chosen methods will be calculated by WSM. This will supply the decision maker with suggested options.

This paper is structured as follows: Sect. 2 describes the methodology used in this research. The results are given in Sect. 3. Section 4 represents the discussion. Section 5 provides the conclusions and suggestions for future work.

2 Methodology

In this section, we will review more details about the WSM that we will follow in this research.

2.1 Problem Definition

The problem here is choosing the most appropriate method for removing space debris.

2.2 Identification of Decision Alternative and the Decision Criteria.

The decision alternatives are the removal methods which we will choose the best of them. In this research, we selected five active removal methods. For the decision criteria, thirteen criteria represent a combination of standard space mission criteria such as the power consumed during operation, those specific to debris removal missions such as reusability, and criteria related to space debris characteristics such as the size of debris. Each criterion was described. Through this description, a conversion scale was developed for each criterion to obtain all performance values for all methods within the context of the selected criteria.

2.3 Definition of Weights

We used the Relative Frequency Approach to determine the relative weight of each criterion which helped to find out their importance.

Relative frequencies indicate the ratio or percentage of occurrences of a particular event or observation in comparison to the total number of events or observations. They are commonly employed to illustrate the frequency with which a specific category is present in a dataset [7].

So, criteria that have higher relative frequencies will be assigned higher weights because they are considered more influential in the decision.

To implement the Relative Frequency Approach, we chose 41 scientific papers that delve into these methods. We combed through these papers to identify the selected criteria. When a criterion was found discussed or mentioned in a paper, it received a score of 1; otherwise, it received a score of zero. We tallied these scores to determine how many times each criterion was mentioned or discussed within the 41 scientific papers. Consequently, if a criterion appeared frequently, it was assigned greater significance. By the relative frequencies approach, the weight of each criterion (W_m) was computed using Eq. (1) by dividing the number of mentions of mth criterion (NC) by total number of mentions of all criteria (TC) in the 41 papers:

$$W_m = \frac{NC}{TC} \tag{1}$$

2.4 Data Collection and Performance Values

Qualitative and quantitative data related to the selected criteria were collected for each method. Various sources were used to collect this data, including scientific papers, books, review papers, and websites. All gathered data is entered into the decision matrix and transformed into performance values (P) through the application of a conversion scale.

2.5 Normalization of Performance Values

To make all criteria comparable, we conduct a normalization process on the performance values for each criterion. This involves dividing the score of the performance value of nth

method in m^{th} criterion (X_{nm}) by the maximum value of mth criterion (X_m^{Max}). This type of normalization is referred to as linear normalization. So, we can obtain the normalized performance value ($\overline{X}_{nm}$) using Eq. (2):

$$\overline{X}_{nm} = \frac{X_{nm}}{X_m^{Max}} * 100 \tag{2}$$

2.6 Weighted Normalized Decision Matrix and Performance Score

We applied the WSM equation, Eq. (3), in two steps. The first step is to calculate the weighted normalized performance values by multiplying the normalized performance values of each method by the corresponding criterion weights ($W_m\overline{X}_{nm}$). W_m is the weight of mth criterion. The weighted normalized decision matrix will contain all the weighted normalized performance values. The second step is to add all weighted normalized performance values of nth method to get the performance score for n^{th} method (A_n^{WSM}) [6]:

$$A_n^{WSM} = \sum_{m=1}^{k} \left(W_m\overline{X}_{nm}\right) for\ n = 1, 2, .., l \tag{3}$$

3 Results

This section will involve a review of the chosen removal methods and decision criteria. Additionally, it will present all the results derived from both the Relative Frequency approach and the WSM.

3.1 The Identified Debris Removal Methods

Within this subsection, we will provide a concise overview of the chosen removal methods.

Ion Beam Shepherd (IBS). This method entails placing the IBS spacecraft into orbit, where it will track and meet a preselected target debris. The IBS will position itself adjacent to the target and use an ion beam to reduce the speed of the designated debris by applying a decelerating force. This force causes debris to deorbit into the atmosphere or reorbit to disposal orbit, all without the necessity of physically docking with debris [8].

Laser Systems. The laser-based technique uses a pulsed laser beam directed towards the target object, causing the object to decelerate and descend into the Earth's atmosphere. This laser system can be deployed either from a ground station (ground-based laser) or a space station (space-based laser). It serves the purpose of space debris removal. Laser systems can vaporize or ablate small debris [9].

Robotic Systems. Robotic systems are means of space debris removal. These systems are attached to the debris and subsequently propel the object into an orbit that will rapidly deorbit and degrade [10, 11]. These robotic systems come in various forms, including tentacles, single robotic arms, and multiple robotic arms [12].

Space Harpoon System. The procedure involves using a chaser spacecraft to launch a harpoon connected to a tether from a safe distance. The harpoon is required to pierce a predefined part of the debris and firmly secure itself. Subsequently, the chaser spacecraft deorbits the debris and transports it toward the upper atmosphere and ultimate destruction [13].

Tethered-Deployed Nets. This system functions as a capture mechanism using a flexible net in conjunction with a spacecraft. The flexible net is tethered to the spacecraft using an extended rope, and mechanical mechanisms are employed to cinch the net. Then, the spacecraft pulls the net with the debris down into the atmosphere [12].

3.2   The Identified Decision Criteria

Table 1 contains the description and the conversion scale of each criterion. The conversion scale replaced the qualitative data with quantitative data and scores to help us in applying WSM.

Table 1. The description and the conversion scale of each criterion.

The criterion	Performance values
Applied orbit	- The method can be used in one orbit (LEO or geostationary orbit (GEO)) $= 1$ - It can be used in both orbits (LEO or GEO) $= 2$
Policy and legal concerns	- The method may be used as a space weapon $= 1$ - It is not used as a space weapon $= 2$
Flight proven. A space mission was launched to test the method or to test the technology on which this method is based	- There is no space mission was launched $= 1$ - A space mission was launched $= 2$
The ability of the method to deal with different shapes of debris	- The method don't have this ability $= 1$ - It has this ability $= 2$
Size of debris which the method can deal with. The space debris has been divided into three categories according to size: Small debris (S) (< 1mm), Medium debris (M) (10 cm to 1 mm), and Large debris (L) (> 10cm) [14]	- The method can deal with one category whether small, medium, or large $= 1$ - It can handle two categories $= 2$ - It can handle three categories $= 3$

(continued)

Table 1. (*continued*)

The criterion	Performance values
Reusability. The ability of the method to remove more than one piece of debris through one system	- The method can only remove one piece of debris with the same system $= 1$ (one for one) - The method can remove more than one piece of debris with the same system $= 2$ (one for more)
Weight of the method	- The method weighs 4000 kg or more $= 1$ - It weighs from 3000 kg to 4000 kg $= 2$ - It weighs 2000 kg to 3000 kg $= 3$ - It weighs 1000 kg to 2000 kg $= 4$ - It weighs less than 1000 kg $= 5$
Ability to deal with tumbling debris	- The method cannot deal with tumbling debris $= 1$ - It cannot deal with high tumbling debris only $= 2$ - It can deal with and remove tumbling debris $= 3$
If the method needs docking or a close approximation	- The method needs a docking mechanism $= 1$ - It needs only a close approximation to the debris $= 2$ - It does not need any docking or a close approximation $= 3$
Technology readiness level (TRL)	- The method will be evaluated according to its TRL based on the National Aeronautics and Space Administration (NASA) classification [15]
Risk of Collision of the method with the other debris. This risk depends mainly on the cross-sectional area of the spacecraft. The risk of collision is classified into three stages, high, medium, and low risk of collision	- The method has a high risk of collision $= 1$ - It has a medium risk $= 2$ - It has a low risk $= 3$
Contamination of the surrounding environment	- The method can pollute the surrounding environment $= 1$ - It isn't polluting the surrounding environment $= 2$
Power used during the operation of the method	- The used power is within the limits of megawatts $= 1$ - The used power is within kilowatts $= 2$ - The power used is within the limits of watts $= 3$ - There is no power used $= 4$

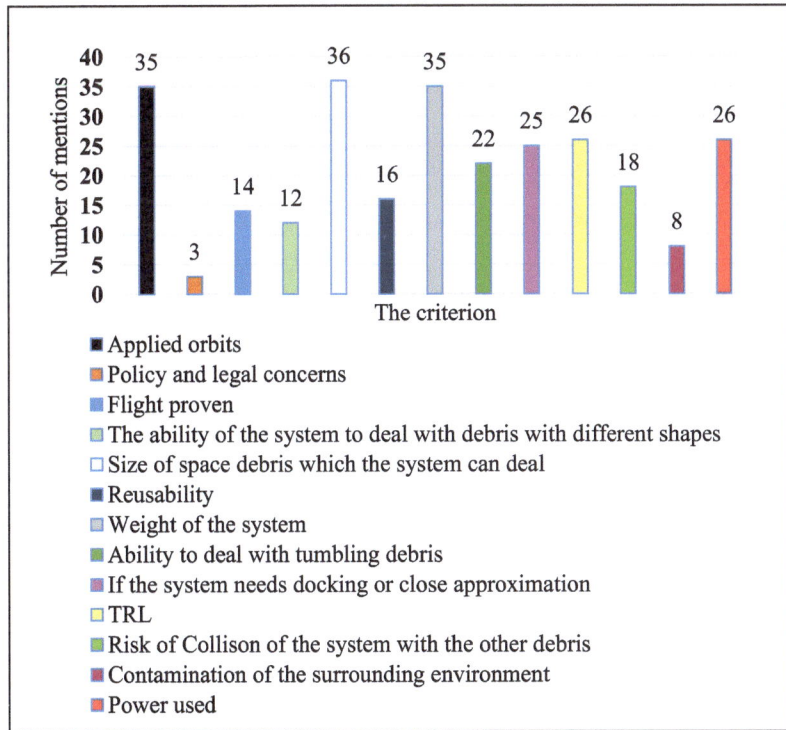

Fig. 1. Number of times each criterion was mentioned in the 41 papers. (Relative frequency histogram).

3.3 Weights of the Decision Criteria

The number of times each criterion was mentioned or discussed in the 41 papers was collected to calculate the weight. Figure 1 is a graph illustrating the frequency of mentions for each criterion within the 41 papers.

Using Eq. (1), the weight for each criterion was determined. It was observed that the criterion "size of debris which the method can deal with" holds the utmost significance, with 36 mentions in the 41 papers and the highest weight of 0.130. Conversely, "policy and legal concerns" is the least significant criterion, with only 3 mentions in the 41 papers and the lowest weight of 0.011. The order of the criteria and their respective weights are detailed in Table 2.

Based on the findings in Table 2, it becomes evident that a method capable of effectively addressing a broad range of debris sizes, functioning in multiple orbits, and being lightweight holds a significant advantage over alternative methods.

3.4 Decision Matrices

All the gathered data and the corresponding performance values have been presented in Table 3 and Table 4. Within the decision matrices, the criteria are organized in descending order based on their relative weights, from the most important to the least important one.

Table 2. The weight of the thirteen criteria.

The criterion	Weight of each criterion
1-Size of debris which the method can deal with	0.130
2-Applied orbit	0.127
3-Weight of the method	0.127
4-Power used during the operation of the method	0.094
5-TRL	0.094
6-If the method needs docking or a close approximation	0.091
7-Ability to deal with tumbling debris	0.080
8-Risk of Collision of the method with the other debris	0.065
9-Reusability	0.058
10-Flight proven	0.051
11- The ability of the method to deal with different shapes of debris	0.043
12-Contamination of the surrounding environment	0.029
13-Policy and legal concerns	0.011

Table 3. Decision matrix for active removal methods (IBS, Laser systems, and Robotic systems).

Criterion	Method					
	IBS	P	Laser Systems (ground and space-based)	P	Robotic Systems	P
1	L. debris [18]	1	L., M., and S. debris [9, 16, 17]	3	L. debris [18]	1
2	LEO, GEO [19, 20]	2	LEO, GEO [21, 22]	2	LEO, GEO[23, 24]	2
3 (Chaser dry mass)	About 500 kg for large debris [20, 25]	5	About 2300 kg (for space-based) [26]	3	About 700 kg for 1.5 arm length [26]	5
4	Up to 15 KW for large debris [20]	2	Some MW [27]	1	Handers of W [26]	3
5	3 [26]	3	3 [26]	3	7 [28]	7
6	Needs close approximation [19]	2	No docking or close approximation [29]	3	Needs docking [4, 30]	1
7	Can deal [20]	3	Can deal [31]	3	Cannot deal [4, 28]	1
8	Low risk [26]	3	Low risk for space-based	3	Medium risk	2
9	One for more [19]	2	One for more [22]	2	One for more [24]	2
10	No	1	No	1	ETS-VII [4]	2
11	Can deal [25]	2	Can deal [32]	2	Can't deal [33]	1
12	Yes [20]	1	Yes	1	Yes	1
13	Yes [17]	1	Yes [17]	1	Yes [11]	1

Table 4. Decision matrix for active removal methods (Space harpoon system and Tethered-deployed nets).

Criterion	Method			
	Space Harpoon System	P	Tethered-Deployed Nets	P
1	L. debris [17]	1	L. and M. debris [34–36]	2
2	LEO, GEO [26]	2	LEO, GEO [12, 37, 38]	2
3 (Chaser dry mass)	About 150 kg [26]	5	About 1300 kg [26]	4
4	Up to 20 W [39]	3	About power 12 KW [26]	2
5	9 [39]	9	9	9
6	Needs close approximation [13, 39]	2	Needs close approximation [34]	2
7	Can deal (problem with high tumbling rate) [13]	2	Can deal [38]	3
8	Medium risk	2	Medium risk	2
9	One for one [17]	1	One for one [36]	1
10	RemoveDebris [40]	2	RemoveDebris [40]	2
11	Can deal [41]	2	Can deal [36]	2
12	Yes	1	Yes	1
13	No	2	No [11]	2

3.5 Normalized Decision Matrix

By using Eq. (2) on the decision matrices for the active removal methods, we can generate the normalized decision matrix. This matrix includes the normalized performance values. Table 5 illustrates the Normalized decision matrix. Within this matrix, the criteria are organized in descending order based on their relative weights.

Table 5. Normalized decision matrix for active removal methods.

Criterion	Method				
	IBS	Laser Systems (ground and space-based)	Robotic Systems	Space Harpoon System	Tethered-Deployed Nets
1	33.3%	100%	33.3%	33.3%	66.6%
2	100%	100%	100%	100%	100%
3	100%	60%	100%	100%	80%

(*continued*)

Table 5. (*continued*)

Criterion	Method				
	IBS	Laser Systems (ground and space-based)	Robotic Systems	Space Harpoon System	Tethered-Deployed Nets
4	50%	25%	75%	75%	50%
5	33.3%	33.3%	77.7%	100%	100%
6	66.6%	100%	33.3%	66.6%	66.6%
7	100%	100%	33.3%	66.6%	100%
8	100%	100%	66.6%	66.6%	66.6%
9	100%	100%	100%	50%	50%
10	50%	50%	100%	100%	100%
11	100%	100%	50%	100%	100%
12	50%	50%	50%	50%	50%
13	50%	50%	50%	100%	100%

3.6 Weighted Normalized Decision Matrix and Performance Scores

By applying Eq. (3), we obtain a weighted normalized decision matrix, from which the performance scores for all methods are computed. Table 6 illustrates the weighted normalized decision matrix, while Fig. 2 depicts the performance scores for each removal method.

Table 6. Weighted normalized decision matrix for active removal method.

Criterion	Method				
	IBS (%)	Laser Systems (ground and space-based) (%)	Robotic Systems (%)	Space Harpoon System (%)	Tethered-Deployed Nets (%)
1 (W = 0.130)	4.329	13	4.329	4.329	8.658
2 (W = 0.127)	12.7	12.7	12.7	12.7	12.7
3 (W = 0.127)	12.7	7.62	12.7	12.7	10.16
4 (W = 0.094)	4.7	2.35	7.05	7.05	4.7

(*continued*)

Table 6. (*continued*)

Criterion	Method				
	IBS (%)	Laser Systems (ground and space-based) (%)	Robotic Systems (%)	Space Harpoon System (%)	Tethered-Deployed Nets (%)
5 (W = 0.094)	3.13	3.13	7.30	9.4	9.4
6 (W = 0.091)	6.06	9.1	3.03	6.06	6.06
7 (W = 0.080)	8	8	2.664	5.328	8
8 (W = 0.065)	6.5	6.5	4.329	4.329	4.329
9 (W = 0.058)	5.8	5.8	5.8	2.9	2.9
10 (W = 0.051)	2.55	2.55	5.1	5.1	5.1
11 (W = 0.043)	4.3	4.3	2.15	4.3	4.3
12 (W = 0.029)	1.45	1.45	1.45	1.45	1.45
13 (W = 0.011)	0.55	0.55	0.55	1.1	1.1

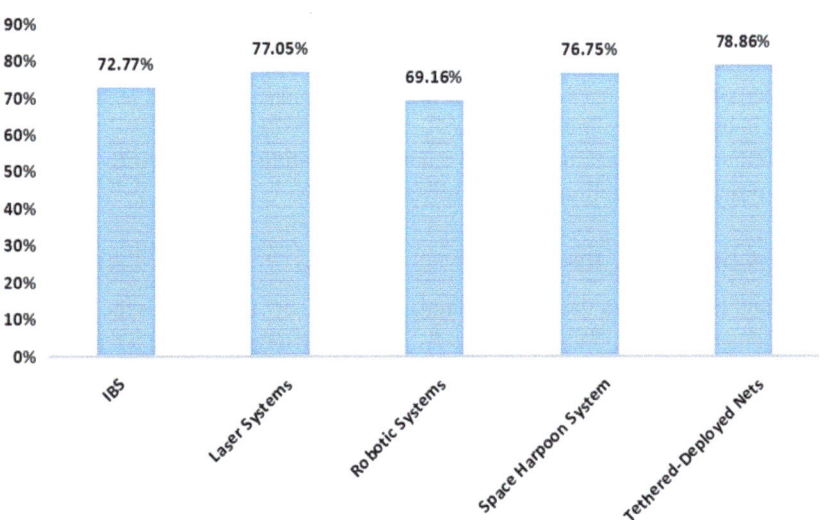

Fig. 2. Assessment of active removal methods.

4 Discussion

Through this section, we will discuss the importance of the results we obtained and the factors that can change the results. We will also explain how these results can be improved to get accurate ones.

4.1 Method ranking

Through ranking the selected methods, we offer valuable information that can assist in the selection of the most suitable method for space debris removal.

Following the application of Eq. (3) to derive the performance score for each method, these methods can be sorted in order of their performance score. The method with the highest performance score is typically regarded as the favored choice.

The research determined that the most effective method is Tether-Deployed Nets, 78.9%. However, a significant drawback is its inability to deal with small debris. Following closely behind is the laser systems, having a performance score of 77.0%, yet it necessitates a higher power supply and has policy and legal concerns. Subsequently, the space harpoon system achieved a performance score of 76.7%.

It's important to note that each method has its own set of pros and cons. Hence, the methods identified as the best are relatively most suitable when compared to the other methods.

Several factors within this study can influence the outcome and the ranking of the removal methods. Firstly, the selection of papers, which serves as the basis for calculating the weight of each criterion, is a critical factor. Altering these papers could potentially lead to changes in the weighting of each criterion. Secondly, modifying the decision criteria used in the evaluation is another influential factor. We made diligent efforts to ensure that the assessment of each method was as reliable as possible.

Enhancing the quality of these results can be achieved by expanding the sample size, which would enable more accurate weight calculations. Additionally, the possibility exists to introduce additional decision criteria and provide a more comprehensive study of the interconnections between these criteria. Such measures would lead to increased precision and confidence in the results.

In summary, we can affirm that we have successfully responded to the scientific question posed before initiating this research.

4.2 Verification of the Results

Hakima and Emami used the AHP method, as previously discussed in the introduction, whereas we utilized the WSM in our study. The first two methods are identical between our study and theirs, with the first method being net methods and the second being on-orbit lasers [5].

5 Conclusions and Future Work

Space debris has a variety of negative effects, including the malfunction or partial loss of operating satellites. Addressing this problem demands extensive efforts from all organizations participating in the space industry. Amidst numerous proposed methods for space debris removal, we selected five active removal methods for examination in this study. These methods were ranked using the WSM, leading to the identification of Tether-Deployed Nets as the top-performing method with a performance score of 78.9%. A significant advantage of tethered deployed nets is their versatility, as they can function across various orbits, handle debris of different shapes, and even address tumbling debris. This research can contribute to the process of decision-making and can provide recommendations for further testing of the methods that have achieved high-performance scores.

In future work, we will work to increase the number of scientific papers used in the sample. Also, increasing the decision criteria will help in judging the methods and make the results more confident and realistic. The use of other MCDA techniques and comparing the results of these techniques with each other will be essential.

References

1. ESA: FAQ: Frequently asked questions. https://www.esa.int/Space_Safety/Space_Debris/FAQ_Frequently_asked_questions. Accessed Apr 2021
2. ESA: How many space debris objects are currently in orbit?. https://www.esa.int/Space_Safety/Clean_Space/How_many_space_debris_objects_are_currently_in_orbit. Accessed 2021
3. Kessler, D.J., Cour-Palais, B.G.: Collision frequency of artificial satellites: the creation of a debris belt. J. Geophys. Res. Space Phys. **83**(A6), 2637–2646 (1978)
4. Zhao, P., Liu, J., Wu, C.: Survey on research and development of on-orbit active debris removal methods. Sci. China Technol. Sci. **63**(11), 2188–2210 (2020)
5. Hakima, H., Emami, M.R.: Assessment of active methods for removal of LEO debris. Acta Astronaut. **144**, 225–243 (2018)
6. Guzman, L.M.: Multi-criteria decision Making Methods: A Compartive Study. In: Applied optimization. Evangelos Triantaphyllou, 288 p. . Kluwer Academic Publishers, Wiley, Hoboken (2001)
7. Heumann, C., Shalabh, M.S.: Introduction to Statistics and Data Analysis. Springer, Cham (2016)
8. Ahedo, E., et al.: Space debris removal with an ion beam shepherd satellite: target-plasma interaction. In: 47th AIAA/ASME/SAE/ASEE Joint Propulsion Conference and Exhibit (2011)
9. Choi, S.H., Pappa, R.S.: Assessment study of small space debris removal by laser satellites. Recent Patents Space Technol. **2**(2), 116–122 (2012)
10. Pelton, J.N., Ailor, W.: Space Debris and Other Threats from Outer Space. Springer, Cham (2013)
11. Pelton, J.N.: Technological approaches to debris removal or mitigation. In: Space Debris and Other Threats from Outer Space, pp. 35–44 (2013)
12. Ru, M., et al.: Capture dynamics and control of a flexible net for space debris removal. Aerospace **9**(6), 299 (2022)
13. Dudziak, R., Tuttle, S., Barraclough, S.: Harpoon technology development for the active removal of space debris. Adv. Space Res. **56**(3), 509–527 (2015)
14. Chunlai, L., et al.: Chemical classification of space debris. Acta Geologica Sinica-Engl. Ed. **78**(5), 1090–1093 (2004)
15. NASA: Technology Readiness Level. https://www.nasa.gov/directorates/heo/scan/engineering/technology/technology_readiness_level. Accessed 28 Oct 2012
16. Klinkrad, H., et al.: Effects of debris mitigation measures on environment projections. In: Space Debris: Models and Risk Analysis, pp. 165–198 (2006)

17. Chatterjee, J., Pelton, J.N., Allahdadi, F.: Active orbital debris removal and the sustainability of space. In: Handbook of Cosmic Hazards and Planetary Defense, pp. 921–940. Springer, Cham (2015)
18. Dubanchet, V., et al.: Modeling and control of a space robot for active debris removal. CEAS Space J. **7**(2), 203–218 (2015)
19. Bombardelli, C., Pelaez, J.: Ion beam shepherd for contactless space debris removal. J. Guid. Control. Dyn. **34**(3), 916–920 (2011)
20. Bombardelli, C.: Ion beam technology for space debris mitigation (2017)
21. Thind, M.K., Lokesh, C.: Removal of space debris using laser. Adv. Aerosp. Sci. Appl. **3**(2), 107–112 (2013)
22. Phipps, C.R., et al.: A Laser Optical System to Remove Low Earth Orbit Space Debris. Lawrence Livermore National Lab. (LLNL), Livermore, CA (USA) (2013)
23. Jankovic, M., et al.: Robotic system for active debris removal: requirements, state-of-the-art and concept architecture of the rendezvous and capture (RVC) control system. In: 5th CEAS Air and Space Conference Proceedings. CEAS, Delft, the Netherlands (2015)
24. Pelton, J.N.: New technological approaches to orbital debris remediation. In: New Solutions for the Space Debris Problem, pp. 53–68 (2015)
25. Kitamura, S., Hayakawa, Y., Kawamoto, S.: A reorbiter for GEO large space debris using ion beam irradiation. In: 32nd Electric Propulsion Conference (International). IEPC-2011-087 (2011)
26. Michal, M.: Aktivní technologie řešení kosmického odpadu, České vysoké učení technické v Praze. Vypočetní a informační centrum (2021)
27. Khanolkar, N.P., et al.: Advanced space debris removable technique and proposed laser ablation technique: a review. In: 2017 International Conference on Infocom Technologies and Unmanned Systems (Trends and Future Directions) (ICTUS). IEEE (2017)
28. Jankovic, M., Kirchner, F.: Trajectory generation method for robotic free-floating capture of a non-cooperative, tumbling target. In: Stardust Final Conference: Advances in Asteroids and Space Debris Engineering and Science. Springer, Cham (2018)
29. Shen, S., Jin, X., Hao, C.: Cleaning space debris with a space-based laser system. Chinese J. Aeronaut. **27**(4), 805–811 (2014)
30. Nishida, S.-I., et al.: Lightweight robot arm for capturing large space debris. J. Electr. Eng. **6**(5), 271–280 (2018)
31. Phipps, C.: Clearing space debris with lasers. SPIE Newsroom **20** (2012)
32. Pelton, J.N.: Examining the case for active orbital debris removal. In: New Solutions for the Space Debris Problem, pp. 39–51 (2015)
33. Vyas, S., Jankovic, M., Kirchner, F.: Momentum based classification for robotic active debris removal. J. Space Saf. Eng. **9**(4), 649–655 (2022)
34. Benvenuto, R., Lavagna, M.: Flexible capture devices for medium to large debris active removal: simulations results to drive experiments. In: 12th Symposium on Advanced Space Technologies in Automation and Robotics. ASTRA Noordwijk, The Netherlands (2013)
35. Forshaw, J.L., et al.: The active space debris removal mission RemoveDebris. Part 1: From concept to launch. Acta Astronautica **168**, 293–309 (2020)
36. Lavagna, M., et al: Debris removal mechanism based on tethered nets. In: Robotics and Automation in Space (iSAIRAS 2012) International Symposium on Artificial Intelligence, Robotics and Automation in Space (iSAIRAS 2012) (2012)
37. Mark, C.P., Kamath, S.: Review of active space debris removal methods. Space Policy **47**, 194–206 (2019)
38. Stadny, K., Hovell, K., Brewster, L.: Space debris removal with sub-tethered net: a feasibility study and preliminary design. In: Proceedings of 8th European Conference on Space Debris (Virtual), Darmstadt, Germany (2021)

39. Reed, J., Barraclough, S.: Development of harpoon system for capturing space debris. In: 6th European Conference on Space Debris (2013)
40. AIRBUS: RemoveDEBRIS, Testing technology to clear out space junk. https://www.airbus.com/en/products-services/space/in-space-infrastructure/removedebris. Accessed 2023
41. Zhao, W., et al.: A simulation and an experimental study of space harpoon low-velocity impact, anchored debris. Materials **15**(14), 5041 (2022)

Thermal Control and Analysis of a 3U Nanosatellite with Deployed Panels

Amir Ashraf⬤ and Mostafa Mohamed(✉) ⬤

Faculty of Engineering, Suez University, Suez, Egypt
Amir.ASSe@eng.suezuni.edu.eg, mostafa3302269@gmail.com

Abstract. The success of a satellite mission depends on a lot of factors. The satellite thermal control system (TCS) is one of the most important subsystems that ensures that the satellite's components remain within an acceptable temperature range. Thermal control for satellites includes monitoring the energy that flows into and out of the spacecraft. This monitoring helps to prevent overheating or freezing of critical components. To execute satellite thermal modeling, the mathematical model must be solved while accounting for both the heat rejected out of the internal electrical components of the satellite subsystems and the effect of the different forms of external fluxes. This paper focuses on changing the design of a 3U CubeSat by adding a deployment mechanism. And using passive and active thermal control components guarantees that the satellite stays within the temperature range that allows for satellite operation. At an altitude of 421 km, the spacecraft is in the Low Earth Orbit. SINDA FLUINT and Thermal Desktop Software were used to analyze the satellite thermal data to make sure that the temperatures of the satellite's electrical equipment are kept within ranges that are suitable for regular operation. The process was validated, and the results were checked by performing hand calculations and computer simulation to improve our thermal analysis's reliability. The results of the modified design have achieved a better temperature range and better solar panel exposure area to the sun. It is imperative to underline that this work is integral to building a 3U Cube Satellite as a graduation project.

Keywords: 3U · CubeSat · Thermal Analysis · Deployed Panels · Thermal desktop

1 Introduction

Satellite missions have revolutionized our ability to observe and communicate across the globe and explore the depths of space. Especially the CubeSats [10] and small satellites which provide not only easier design, manufacturing, and testing but also lower cost of production Which contributes to facilitating the conduct of tests on new systems, as well as achieving benefits at a cheaper price. The success of these missions demands proper planning and execution, with satellite thermal control emerging as a critical component of this equation. Maintaining the temperature of satellite components within a predetermined and ideal range is the major goal of satellite thermal control systems [5].

© The Author(s) 2025
H. M. K. Al Naimiy et al. (Eds.): AUASS-CONF 2023, SPPHY 420, pp. 98–111, 2025.
https://doi.org/10.1007/978-981-96-3276-3_9

In the context of this paper, we will focus on modifying the satellite design by adding a deployment system to the four solar panels to improve the overall functionality and efficiency of the satellite and achieve the optimum exposure area of the solar panels. Then manipulate the passive thermal control components see [8, 15] such as tapes, paints, etc., and the heaters (active thermal control components), and make sure that all the changes caused by this modification are handled.

2 Satellite Energy Balance

The transient thermal energy balance equation was solved using the thermal analysis software tool. This required accounting for different types of sun fluxes, including solar flux, albedo, and Earth infrared (IR) radiation. Figure 1 [9] lists the various sun fluxes and energy balance terms.

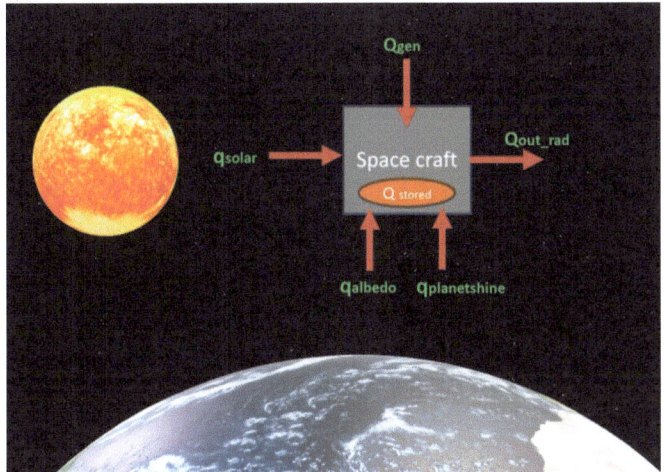

Fig. 1. Satellite energy balance [9].

3 Governing Equations

Equation 1, which states the Thermal Balance Equation, is solved by the software, which also provides definitions for all terms. To keep the inside and outside of the satellite's thermal equilibrium [5].

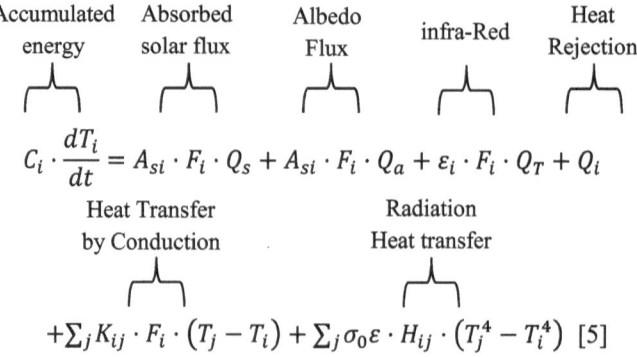

$$C_i \cdot \frac{dT_i}{dt} = A_{si} \cdot F_i \cdot Q_s + A_{si} \cdot F_i \cdot Q_a + \varepsilon_i \cdot F_i \cdot Q_T + Q_i$$

$$+ \sum_j K_{ij} \cdot F_i \cdot \left(T_j - T_i \right) + \sum_j \sigma_0 \varepsilon \cdot H_{ij} \cdot \left(T_j^4 - T_i^4 \right) \ [5]$$

Where Q_i is the node I of the rejected heat out of the satellite body in Watt/m^2, Q_a is the albedo (energy reflected from the earth) in Watt/m^2, Q_s is the solar flux emitted from the sun on the satellite body in Watt/m^2, Q_T is the infra-red energy emitted from the earth to the satellite body in Watt/m^2, F_i represents the node I of the area in m^2, K_{ij} is the thermal conductivity in Watt/K, σ_0 is the Stefan-Boltzmann's constant which equals 5.67×10^{-8} W/m^2/K^4, C_i is the node I of the Thermal capacitance in J/K, ε $_{Hij}$ represents the radiative heat transfer coefficient in m^2, A_{si} is the node I of the factor of absorptivity, Ti and Tj represent the i and j nodes of temperature [5].

4 Hand Calculations

Satellite temperatures can be estimated manually to confirm the thermal model results. The purpose of this section is to estimate the steady-state temperature of the highest hot case [2].

4.1 Solar Flux

$\dot{q}_{solar} = \alpha A S \cos\varphi$, $\varphi = 0$ [1]

where $\dot{q}_{solar}$ represents the solar flux in Watt, S is the solar constant in Watt/m^2, A is the surface area in m^2, α is the absorptivity, and φ is the incident angle which represents the angle between the vector of the solar radiation and the vector perpendicular to the specified surface. The total area of the satellite is 2840 mm^2.

$2840 = 4\pi r^2$ $r = 15$

Projected area $= \pi r^2 = 710$ mm^2

Note: The same calculations were performed for $A_{SolarCells} = 808$, $A_{Panels} = 512$

Note: A_{Panels} is the area of deployed panels that are not covered with solar cells.

$\dot{q}_{solar\,SolarCells} = 0.637 \times 202 \times 10^{-4} \times 1400 = 18W$

$\dot{q}_{solar\,Panels} = 0.248 \times 128 \times 10^{-4} \times 1400 = 4.5W$

$\dot{q}_{solar\,Structure} = 0.16 \times 710 \times 10^{-4} \times 1400 = 16W$

$\dot{q}_{solar} = 18 + 4.5 + 16 = 38.5W$ [2]

4.2 Earth Albedo

$\dot{q}_{albedo} = \alpha ASA_f F_{Earth \rightarrow Surface}, A_f = 0.3$ [12]

$F_{Earth \rightarrow Surface} = \frac{1}{2}\left[1 - \sqrt{1 - \frac{1}{h^2}}\right], h = \frac{R_{earth} + altitude}{R_{earth}}$

$R_{earth} = 6378$ km, $h = 421$ km

$F_{Earth \rightarrow Surface} = 0.327$

$\dot{q}_{Albedo_{SolarCells}} = 0.637 \times 808 \times 10^{-4} \times 1400 \times 0.3 \times 0.327 = 7.1W$

$\dot{q}_{Albedo_{Panels}} = 0.248 \times 512 \times 10^{-4} \times 1400 \times 0.3 \times 0.327 = 1.75W$

$\dot{q}_{Albedo_{Structure}} = 0.16 \times 2840 \times 10^{-4} \times 1400 \times 0.3 \times 0.327 = 6.25W$

$\dot{q}_{albedo} = 7.1 + 1.75 + 6.25 = 15.1W$ [21]

4.3 Earth IR

$\dot{q}_{IR\ temperature} = \sigma \varepsilon A F_{Earth \rightarrow surface} T_E^4$ [12]

$\dot{q}_{Albedo_{SolarCells}} = 5.67 \times 10^{-8} \times 0.9 \times 808 \times 10^{-4} \times 0.327 \times 255^4 = 5.71W$

$\dot{q}_{Albedo_{Panels}} = 5.67 \times 10^{-8} \times 0.924 \times 512 \times 10^{-4} \times 0.327 \times 255^4 = 3.71W$

$\dot{q}_{Albedo_{Structure}} = 5.67 \times 10^{-8} \times 0.03 \times 2840 \times 10^{-4} \times 0.327 \times 255^4 = 0.7W$

$$\dot{q}_{IR} = 5.71 + 3.71 + 1.11 = 10.1W$$

4.4 Steady State Energy Balance

$\dot{q}_{in} + \dot{q}_{generated} = \dot{q}_{out}$ [21].

$\dot{q}_{in} = $ Total *Solarflux* + Total *Albedo* + Total *Infra_Red*.

$\dot{q}_{in} = 38.5 + 15.1 + 10.53 = 63.7W$

$\dot{q}_{generated} = 5.77W$

$\dot{q}_{out} = 63.7 + 6 = 69.7W$

4.5 Satellite Heat Rejection

$\dot{q}_{rejected} = \sum_{i=1}^{n} A_i \varepsilon_i \sigma T_i^4$ [2]

$69.7 = 5.67 \times 10^{-8} \times 10^{-4} \times T(808 \times 0.9 + 512 \times 0.924 + 2840 \times 0.03)$

$T = 313K = 40C$

5 Numerical Simulation

5.1 Validation

The mathematical model and numerical methods are used to do the thermal analysis to check the maximum and minimum temperature limits that can be achieved during ten revolutions in satellite orbit. The initial construction (the design with body-mounted solar panels) was obtained from ISISPACE as an off-the-shelf option as in [22, 23]. The thermal analysis cases and their orbital parameters are shown in Table 1 for the initial design. The results for the hot case (normal operational situation along the hottest possible orbit) and the survival case (coldest possibility with the minimum number of subsystems in operation during the coldest possible situation) were validated compared to the results provided by Isaac Foster [2] (see Tables 2 and 3). Also, the Satellite heat loads are shown in Table 3. For the components heat loads (see Table 4) (Figs. 2 and 3).

Table 1. Initial design thermal analysis cases [2].

Case	Altitude (Km)	Beta Angle (Deg)	Solar Flux (W/m^2)	Albedo (-)	Earth IR (W/m^2)
Operation Case (Hot)	421	75.1	1414	0.3	265
Operation Case (Cold)	421	0	1317	0.15	227
Transmission Case	421	75.1	1414	0.3	265
Survival Case	421	0	1317	0.15	227

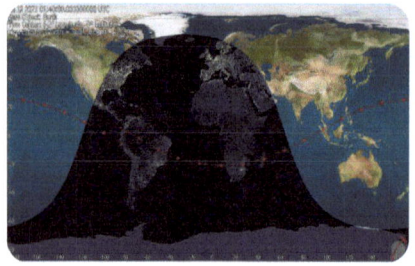

Fig. 2. 2D map of the Cold orbit showing the shadow zone (*Results of FreeFlyer simulation*).

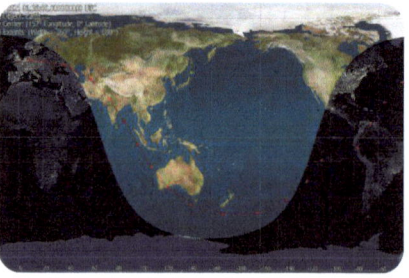

Fig. 3. 2D map of the Hot orbit showing the shadow zone (*Results of FreeFlyer simulation*).

5.2 Modified Model

The original design was modified by adding a deployment mechanism to the solar panels including the hinges as in [6] and wire cutting mechanism which is required to deploy the panels see [13]. The passive and active thermal control components were modified to be more suitable for the new design and a 4-W heater was added to the antenna to cover the heat loss due to the huge surface area of the deployed panels (see Table 5) [3]. The two models (initial and modified) are shown in Fig. 4.

The mathematical calculations and the thermal analysis are done on five cases with the characteristics shown in Table 6. The satellite orbits at 0 and 75.1 Beta angles are shown in Figs. 5, 6 and 7. Also, the thermophysical and orbital properties were obtained and applied [1, 2]. Figures 6 and 7 show the difference between the two different modes of the satellite attitude during the operation case (hot only) pointing and shooting cases. Note that the 0 Beta angle was chosen for the cold cases due to the longer in-shadow periods and the 75.1 orbit was chosen for the hot cases because of the shorter in-shadow periods [2]. The 2D maps of the orbits were calculated and drawn on 2021/07/12 using FreeFlyer software (see Figs. 2, 3) and the hot orbit showed 10.4 min in the Earth's penumbral and 8.1 min in the Earth's umbral. Otherwise, the cold orbit showed 44.89 min

Table 2. Operation Case (Hot) validation comparison.

Component	Operation Limits [3, 11]		Reference Operation Case (Hot) [2]		Reperformed Operation Case (Hot)	
	Tmin (C)	Tmax (C)	Tmin (C)	Tmax (C)	Tmin (C)	Tmax (C)
On Board Computer	−25	65	34	44	34	44
Structure	−40	80	33	35	33	35
ADCS	−40	70	34	48	34	43
Transceiver	−20	60	33	39	35	43
Antenna	−20	60	33	35	33	35
Solar Panels	−40	125	33	35	33	35
Payload	−10	50	36	37	33	34
Electrical Power Subsystem	−20	70	34	35	34	35

Table 3. Survival Case validation comparison.

Component	Operation limits [3, 11]		Reference Survival Case [2]		Reperformed Survival Case	
	Tmin (C)	Tmax (C)	Tmin (C)	Tmax (C)	Tmin (C)	Tmax (C)
On Board Computer	−25	65	−11	9	−11	8
Structure	−40	80	−13	5	−13	5
ADCS	−40	70	−11	5	−11	5
Transceiver	−20	60	−11	5	−11	5
Antenna	−20	60	−13	4	−13	5
Solar Panels	−40	125	−13	5	−13	5
Payload	−10	50	−6	4	−5	4
Electrical Power Subsystem	−20	70	−6	5	−6	6

in the Earth's penumbral and 44.49 min in the Earth's umbral. Also, the wire-cutting mechanism is shown in Fig. 8 (Fig. 9).

Table 4. Satellite Generated heat loads from different subsystems [3, 11].

Component	Operation Case (Hot), [W]	Operation Case (Cold), [W]	Transmission Case [W]	Survival Case [W]
On Board Computer	0.4	0.4	0.4	0
ADCS	1.2	1.2	1.2	0.18
Transceiver	0.48	0.48	4	0
Antenna	0.04	0.04	0.04	0
Payload	2	0	0	0
Electrical Power Subsystem	0.13	0.13	0.13	0.01
Total	4.25	2.25	5.77	0.59

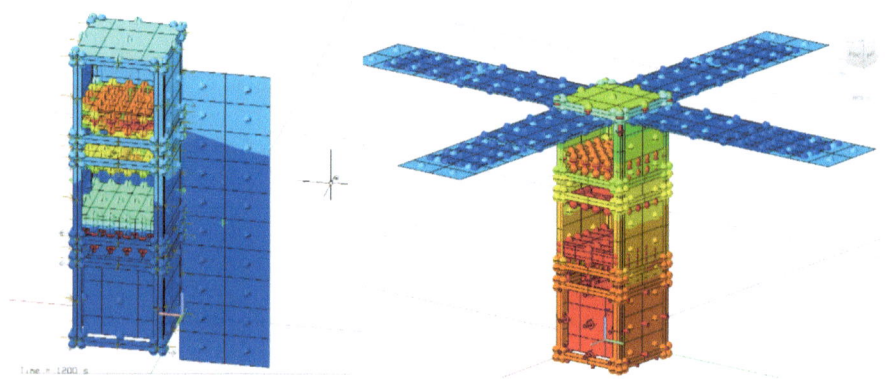

Fig. 4. Initial (left) [2] and modified (right) thermal model on the Thermal Desktop.

Table 5. Material and Location of Passive and Active Thermal Control [3, 29].

Type	Location	Material & Heater Power
Coating	Deployed panels	White epoxy paint
Tape	Electrical Power Subsystem	850-3M aluminized Mylar
Radiator	Panels 1,3	Teflon Aluminized
Heater 1	EPS	2.5 W
Heater 2	Payload	2 W
Heater 3	Antenna	4 W

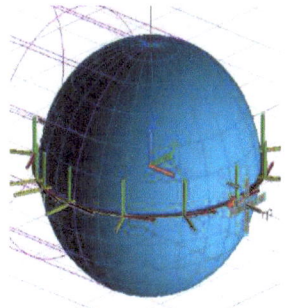

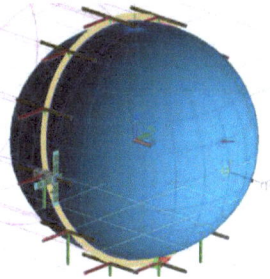

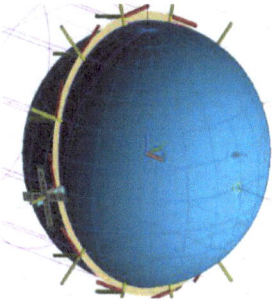

Fig. 6. (Middle) shows the
Orbit at 75.1° Beta angle
Pointing Case (Hot) Orbit.

Fig. 5. (Left) shows the
Cold orbit at 0° Beta angle.

Fig. 7. (Right) shows the
Orbit at 75.1° Beta angle
Shooting Case (Hot) Orbit [2].

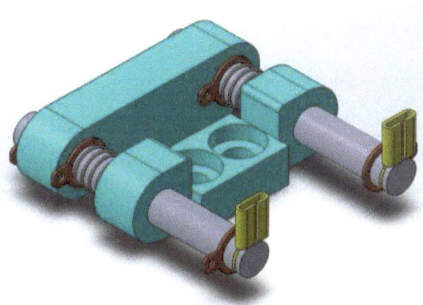

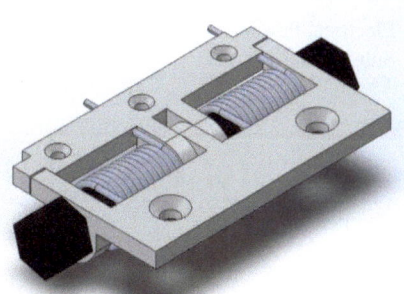

Fig. 8. Burn wire mechanism design (CAD
Model Design).

Fig. 9. Assembled hinge design (CAD
Model Design).

6 Results

The thermal model was built, and the thermal analysis was performed on the five cases
to indicate the maximum and minimum temperature limits for each subsystem of the
satellite and compared with their ultimate operating limits, the Results are shown in
Tables 6, 7 and 8. The results should be within the operating limits with a reasonable
margin to ensure the safe operation of the components.

From the obtained results the different extreme values of temperature are averaged
together, then the average temperature would be:

$$T_{Average} =$$

$$\frac{46 + 40 + 48 + 40 + 46 + 39 + 43 + 38 + 37 + 36.5 + 41 + 37 + 43 + 40.5 + 44 + 42}{16} \; C$$

$$= 41.3$$

This analysis results closely resemble the results of the computed hand calculation.
Which demonstrates the reliability of hand calculations when creating early forecasts.

Table 6. Survival case & operation case (cold) results.

Component	Operation limits [3, 11]		Survival Case		Operation Case (cold)	
	Tmin (C)	Tmax (C)	Tmin (C)	Tmax (C)	Tmin (C)	Tmax (C)
On Board Computer	−25	65	−14	8	−10	16
Structure	−40	80	−22	14	−20	14
ADCS	−40	70	−10	8	−6	14
Transceiver	−20	60	−14	8	−10	16
Antenna	−20	60	−16	14	−14	16
Solar Panels	−40	125	−30	20	−30	20
Payload	−10	50	−4	5	−1	8
Electrical Power Subsystem	−20	70	−4	6	−2	10

Table 7. Transmission case component results.

Component	Operation limits [3, 11]		Hot Case Transmit	
	Tmin (C)	Tmax (C)	Tmin (C)	Tmax (C)
On Board Computer	−25	65	16	46
Structure	−40	80	3	20
ADCS	−40	70	14	26
Transceiver	−20	60	18	46
Antenna	−20	60	4	10
Solar Panels	−40	125	−6	5
Payload	−10	50	15	16.5
Electrical Power Subsystem	−20	70	15	22.5

Also, to ensure the safe operation of the satellite we should keep an eye on the Temperature Profiles of the Transmitter (the most critical component in the satellite) as its temperature range is very critical and may approach the maximum operating limits due to its location and high heat generation range. So that the Temperature Profiles of the Transmitter are shown for the two most extreme cases of Hot transmission and pointing in Figs. 10 and 11.

Power consumption-wise, the total heater power consumption for the initial design was 177419 J during the 10 revolutions [2]. Otherwise, the total heater power consumption for the modified design was 151568 J during the 10 revolutions. The heater on/off &

Table 8. Pointing & Shooting operation case (hot) Results.

Component	Operation limits [3, 11]		Pointing operation case (hot)		Shooting operation case (hot)	
	Tmin (C)	Tmax (C)	Tmin (C)	Tmax (C)	Tmin (C)	Tmax (C)
On Board Computer	−25	65	40	48	−1	10
Structure	−40	80	37	40.5	−10	7
ADCS	−40	70	39	46	2	10
Transceiver	−20	60	40	46	−1	9
Antenna	−20	60	42	44	−10	−4
Solar Panels	−40	125	41	43	−18	−7
Payload	−10	50	36.5	37	6	7
Electrical Power Subsystem	−20	70	38	43	−10	8

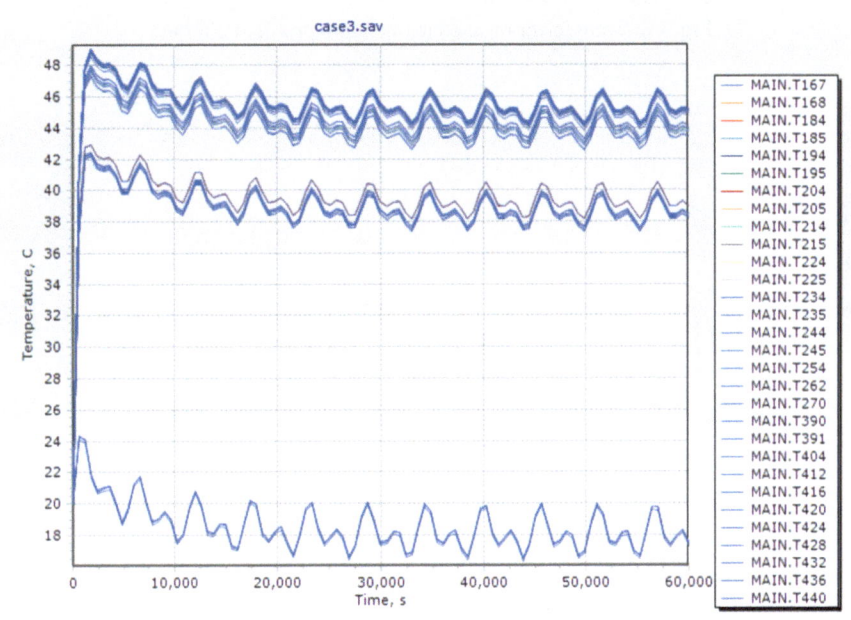

Fig. 10. Transceiver resulted temperature profile (Hot Transmit).

temperature profile for 10 revolutions (cold survival case) is shown in Fig. 12 (Figs. 13 and 14).

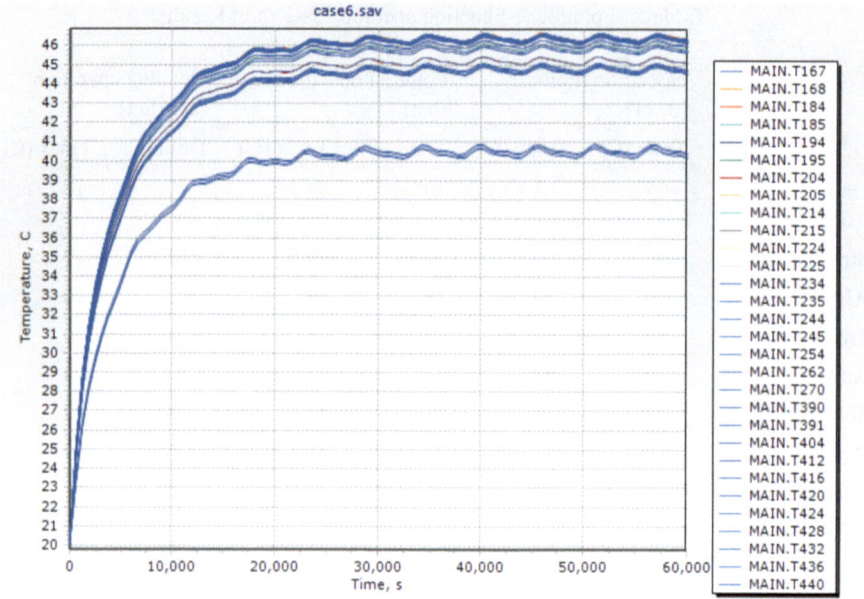

Fig. 11. Transceiver resulted temperature profile (Pointing).

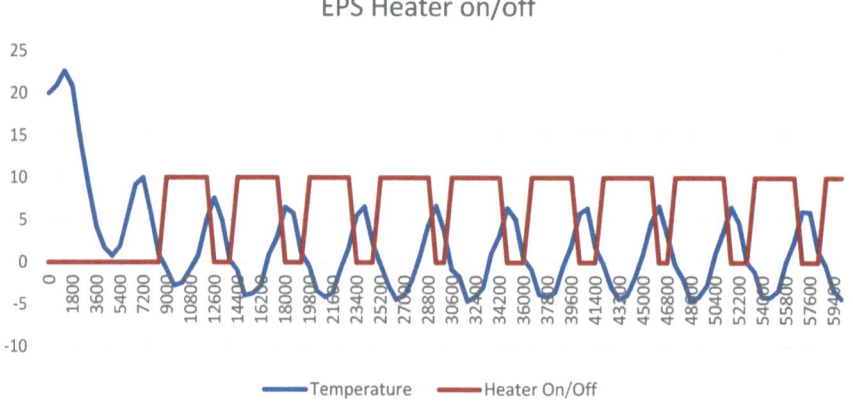

Fig. 12. EPS Heater on/off & EPS temperature profile for 10 revolutions (survival case).

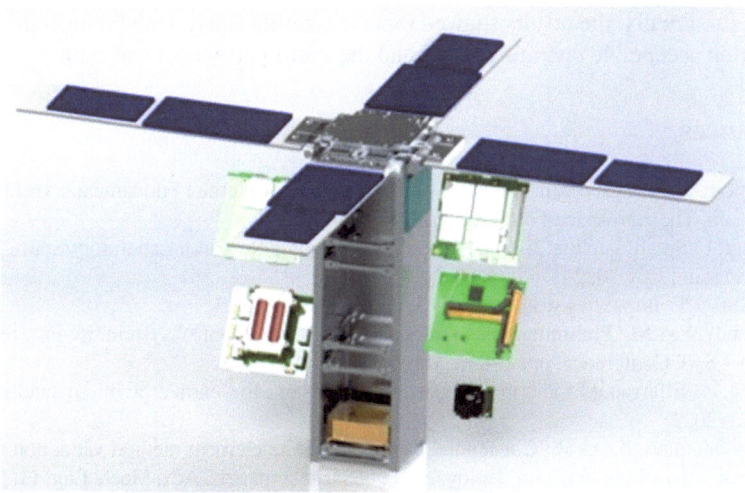

Fig. 13. Satellite design (Cad Model).

Fig. 14. Satellite First Prototype Model.

7 Conclusion

Thermal analysis of 3U nanosatellites in the LEO was provided to compare the initial and modified designs. First, the thermal analysis process was validated by taking the maximum and minimum satellite subsystems' temperatures into consideration. Second, the design was modified by adding a deployment mechanism to the four solar panels, including hinges and a wire-cutting mechanism. The thermal control was performed by modifying the passive and active thermal control elements to correct any changes caused by the modifications. Then the thermal analysis was done using Thermal Desktop and SINDA FLUINT as the software tools on the modified model to make sure that the satellite subsystems would remain within acceptable operating limits. Third, the hand calculations were done to increase the reliability of the calculations and were found to be very useful. Finally, the results showed that the satellite subsystems' temperature range falls within acceptable operating limits and the Prototype Model was built.

References

1. Gilmore, D.G.: Spacecraft Thermal Control Handbook Volume I Fundamental Technologies, 2nd edn. The Aerospace Press, El Segundo, CA (2002)
2. Foster, I.: Small Satellite Thermal Modeling, Air Force Research Laboratory/Space Vehicles Directorate, July 2022
3. ISISSPACE. https://www.isispace.nl. Accessed 18 Nov 2023
4. Elgendy, Y.A.M.: Preliminary design for satellite thermal control system. In: Proceedings of 12th ASAT Conference, pp. 29–31, May 2007
5. Farag, A., Elfarran, M.: Thermal design and analysis of a low earth orbit micro-satellite. JAET **41**(2) (2022)
6. Solís-Santomé, A., et al.: Conceptual design and finite element method validation of a new type of self-locking hinge for deployable CubeSat solar panels. Adv. Mech. Eng. **11**(1) (2019)
7. Ziegler, L.A.: Configuration, manufacture, "assembly, and integration of a university microsatellite", Master's thesis, University of Missouri-Rolla (2007)
8. Elhefnawy, A., et al.: Passive thermal control design and analysis of a university-class satellite. J. Therm. Analy. Calorimetry (2021)
9. Rickman, S.L.: Introduction to On-Orbit Thermal Environments. In: Thermal and Fluids Analysis (2014)
10. Chandrashekar, S.: Thermal analysis and control of MIST CubeSat. Ph.D. dissertation (2017)
11. ISISPACE: iEPS Electrical Power System. https://www.isispace.nl/product/ieps-electrical-power-system. Accessed 18 Nov 2023
12. Czernik, S.: Design of the Thermal Control System for Compass-1. Diploma thesis, University of Applied Sciences Aachen, Germany (2004)
13. Choi, J., et al.: Design, fabrication and test of qualification model of wire thermal cutting based non-explosive separation device for a small satellite. Proc. Inst. Mech. Eng., Part G: J. Aerosp. Eng. **229**(4), 612–620 (2015)
14. Opromolla, R., et al.: A new star tracker concept for satellite attitude determination based on a multi-purpose panoramic camera (2017)
15. Yang, L., et al.: Quasi-all-passive thermal control system design and on-orbit validation of Luojia 1–01 satellite (2019)
16. Corpino, S., et al.: Thermal design and analysis of a nanosatellite in low Earth orbit. Acta Astronaut. **115**, 247–261 (2015)
17. Kreith, F.: Principles of Heat Transfer, 3rd edn. Intext Educational Publishers, New York (1976)
18. Cullimore, B.A.: Computer code SINDA '85/FLUINT System Improved Numerical Differencing Analyzer and Fluid Integrator, Version 2.3. Martin Marietta (2010)
19. Versteeg, C., Cotten, D.L.: Preliminary Thermal Analysis of Small Satellites (2018)
20. Farag, A.: Thermal investigation of spacecraft optical electronic observation system. Int. J. Curr. Adv. Res. 23897–23902 (2021)
21. Boushon, K.E.: Thermal Analysis and Control of Small Satellites in Low Earth Orbit. Missouri University of Science and Technology, Rolla (2018)
22. ISISPACE: 3-Unit CubeSat Structure. ISISSPACE. https://www.isispace.nl/product/3-unit-cubesat-structure/. Accessed 18 Nov 2023
23. ISISPACE, VHF uplink/UHF downlink Full Duplex Transceiver. ISISSPACE. https://www.isispace.nl/product/isis-uhf-downlink-vhf-uplink-full-duplex-transceiver/. Accessed 18 Nov 2023

24. Wesley, J.: Thermal analysis of low layer density multilayer insulation test results. In: AIP Conference Proceedings, pp. 1434–1519 (2012)
25. Mavromatidis, L.E., et al.: Numerical insulation. Build. Environ. **49**, 227–237 (2012)
26. Binder, J.: Planning space missions with FreeFlyer, vol. 43, no. 1, pp. 24–25 (2005)
27. Lapidus, L., Pinder, G.F.: Numerical solution of partial differential equations in science and engineering. Wiley, New York (1982)
28. Boushon, K.E.: Thermal analysis and control of small satellites (2018)
29. National Aeronautics and Space Administration: Small Spacecraft Technology State of the Art. Mission Design Division, Ames Research Center, Moffett Field, California, NASA/TP-2015–216648/REV1, December 2015

Advancements in Astronomy and Space Sciences in Jordan: Contributions from Experts and Astrophysical Institutions

A. A. Abushattal[1]([✉]), Ala'a A. A. Azzam[2,3], Mashhoor A. Al-Wardat[4], Hatem Widyan[5], Mohammad Mardini[6], Ali Taani[7], and Mohammed Talafha[8]

[1] Department of Physics, Al-Hussein Bin Talal University, P.O. Box 20, Ma'an 71111, Jordan
ahmad.abushattal@ahu.edu.jo
[2] Department of Physics, The University of Jordan, Queen Rania Street, Amman, Jordan
[3] Research Department, AstroJo Institute, Wasfi Al-Tal Street, Amman, Jordan
[4] Department of Applied Physics and Astronomy, College of Sciences, and Sharjah Academy for Astronomy, Space Sciences and Technology, University of Sharjah, 27272 Sharjah, United Arab Emirates
[5] Department of Physics, Al al-Bayt University, Mafraq 25113, Jordan
[6] Department of Physics, Zarqa University, Zarqa 13110, Jordan
[7] Physics Department, Faculty of Science, Al Balqa Applied University, Salt 19117, Jordan
[8] Department of Space Physics and Space Technology, Wigner Research Centre for Physics, Budapest, Hungary

Abstract. Humanity has always been fascinated by astronomy and space sciences. Two distinct groups of Jordanians show interest in this subject: specialists and amateurs. Experts, often academicians and researchers, immerse themselves in the scientific mysteries of the universe, contributing to global knowledge. Their tools, telescopes, and methodologies are advanced, and they often collaborate with international institutions and agencies. Meanwhile, amateurs contribute significantly to popularising astronomy by driving passion and curiosity. Sometimes they make significant discoveries while stargazing, tracking meteors, or performing basic observational activities. The enthusiasm and grassroots approach of these individuals are essential for fostering a love for the cosmos and disseminating knowledge among the younger generation, even without the advanced tools of specialists. This synergy has led to a vibrant astronomical community in Jordan, where both groups complement each other's efforts to understand and appreciate the universe's vastness. Through specialized lenses such as research focuses and published academic studies, this seminal study explores the evolution of astronomy in Jordan. Space science and astronomy experts from both within and outside the country are examined in this work. Furthermore, this study highlights the key role played in various facets of this development, including observational procedures, recreational astronomical endeavors, community involvement in these activities, and a critical discussion of the challenges facing astronomy in Jordan.

Keywords: Astronomy and Space Sciences · Jordan · Specialists · Amateurs

© The Author(s) 2025
H. M. K. Al Naimiy et al. (Eds.): AUASS-CONF 2023, SPPHY 420, pp. 112–130, 2025.
https://doi.org/10.1007/978-981-96-3276-3_10

1 Introduction

The fascination with astronomy and space sciences has been a part of humanity's innate curiosity about the cosmos throughout history. In the heart of the Middle East, Jordan stands as a testament to this enduring interest, with two distinct groups leading the charge: specialists and amateurs. Jordanian scientists, often affiliated with academic or research groups, study the intricacies of the universe in depth. In addition to their local contributions, they also have a global impact. Their collaborations with renowned international institutions and agencies push the boundaries of our understanding of the universe with advanced telescopes and cutting-edge methodologies. In contrast, amateurs are key to popularising astronomy in Jordan because of their sheer passion and curiosity. They are no less significant for their grassroots approach, which lacks the sophisticated tools that specialists possess. In addition to satisfying their celestial curiosity, they can ignite the same in others through activities such as stargazing and meteor tracking. It is their dedication and keen observational skills that led to their discoveries, sometimes groundbreaking. Moreover, they help instill a love of the cosmos in young people. As a result of the harmonious coexistence of these two groups, Jordan has developed a vibrant astronomical community. As a result of their combined efforts, each complementing the other, we now understand and appreciate the vastness of the universe. Using specialized lenses, such as research focuses and academic publications, this seminal study examines the evolution of astronomy in Jordan. Experts from Jordan and abroad provide insights into space science and astronomy. It also emphasizes the multifaceted development of the field, which encompasses observational techniques, recreational astronomical pursuits, and community participation. As well as discussing the challenges facing the Jordanian astronomical community, it also provides solutions for future discussions.

2 Historical Background

Astronomy in Jordan has deep historical roots, tracing back to the ancient Nabateans who once observed the heavens from their desert fortresses (Alzoubi, 2016; Belmonte, González–García, & Polcaro, 2013). The clear desert skies of Jordan provide excellent conditions for stargazing and celestial observations (Al-Naimy & Konsul, 2004; Weaver, 2011). In recent years, the Kingdom has taken significant strides in promoting astronomy as both a science and a means of eco-tourism. The Jordanian Astronomical Society, established in the late 20th century, plays a pivotal role in raising public awareness in the field of space science and astronomy. On the other hand, AstroJo Institute has focused on astronomical research and education in the country since it was established in 2017. Moreover, Jordan hosts the astronomical observatory on Umm Al Dami, a testament to its commitment to the field. For enthusiasts and tourists, places like the Wadi Rum desert, with its minimal light pollution, offer spectacular views of the Milky Way and other celestial wonders. Jordan's blend of ancient traditions and modern advancements make it a unique destination for those interested in the mysteries of the universe (Eichhorn, Accomazzi, Grant, Kurtz, & Murray, 2003; Henneken et al., 2009). Jordan's commitment to the study of celestial bodies and phenomena goes beyond its historical significance and modern-day observatories. Education plays a crucial role in its astronomical journey. Numerous schools and universities within the country have integrated astronomy into their curriculum, ensuring that future generations remain connected to the cosmos. The Jordanian people are also showing a growing interest in the astronomical events. Solar and lunar eclipses, meteor showers, and other significant celestial happenings often turn into communal viewing events. Local organizations and astronomy clubs, such as AstroJo Institute, and The Jordanian Astronomical Society frequently organise these gatherings, offering telescopes and expert guidance to attendees. The government and private sectors in Jordan also recognise the potential of astronomical tourism. By developing infrastructure like star parks and observatories in prime locations, they aim to attract both local and international tourists. These efforts not only bolster the nation's economy but also position Jordan as a focal point in the middle eastern astronomical community. International collaborations further enhance Jordan's stature in the field. The country is involved in various projects and partnerships with global astronomical institutions, paving the way for exchange programs, joint research, and shared knowledge (Hussein et al., 2022; A Taani, Abushattal, Khasawneh, Almusleh, & Al-Wardat, 2020; Ali Taani, Abushattal, & Mardini, 2019). Lastly, Jordanian folklore and traditional stories, many rooted in the nomadic Bedouin culture, contain numerous references to stars, planets, and other celestial phenomena. These tales, passed down through generations, highlight the timeless bond between the Jordanian people and the night sky. With its rich history, active scientific community, and profound cultural ties to the cosmos, Jordan continues to be a beacon of astronomical knowledge and appreciation in the region.

3   SAO/NASA Astrophysics Data System (ADS)

The SAO/NASA Astrophysics Data System is an online database of more than 15 million astronomy and physics papers. It is managed by the Smithsonian Astrophysical Observatory (A. A. Abushattal, Docobo, & Campo, 2019; M. Kurtz, 1994; M. J. Kurtz et al., 1993). Astronomers save hundreds of millions of dollars every year by using the ADS search engine. Global trends in astronomical research can be analyzed using ADS usage statistics. It is a fact that there are a greater number of astronomers per capita in a country when the gross domestic product (GDP) is higher per person and the amount of research astronomers conduct is greater per capita when the GDP is higher. So, in a country, the amount of research done is proportional to the square of its GDP divided by the population (J. Docobo, Balega, Campo, & Abushattal, 2018; Ali Taani, Karino, et al., 2019). A system of electronic indexing was developed by astronomers when they realized they could index astronomical research papers on the Internet. Abstracts of journal papers were suggested to be included in a database in 1987. SIMBAD and ADS databases were integrated in 1991. As proof of 1988, the ADS Abstract Service launched in 1988 and became generally available in April 1993. Transatlantic scientific databases were accessed simultaneously for the first time over the Internet (A. A. Abushattal et al., 2019; M. Kurtz, 1994; M. J. Kurtz et al., 1993).

ADS works with almost all astronomical journals, which provide abstracts to the Astrophysical Journal's online edition, which began publishing in 1995. Over eight million documents are now available on the service. ADS Labs launched Streamlined Search in 2011 and ADS 2.0 in 2013. In 2015, ADS-beta released Bumblebee, a new search engine codenamed Bumblebee (J Docobo, Y Balega, et al., 2018; JA Docobo,

PP Campo, & AA Abushattal, 2018). Dynamic page loading is supported by client-side micro-services API. Twelve mirror sites on five continents distribute ADS worldwide, with updates triggered from a central location. There is a bibliographic record for each paper in the database, which contains details of the journal in which it was published, as well as various other details. Originally only astronomical references were in the ADS, but now also includes physics, astronomy, and preprints from arXiv. The most advanced is the astronomy database. ADS runs CentOS 5.4 Linux at the Harvard & Smithsonian Center for Astrophysics in Cambridge, Massachusetts.

- **Indexing**

Nearly two hundred journal sources submit abstracts or tables of contents to ADS, and ADS creates one bibliographic reference based on the most accurate information. Searches are performed using a database of author names maintained by ADS. Author names are one of the more difficult to convert to standard Surname, Initial format. It is easy to extract references from electronic articles or to convert a reference database into citations for scanned articles.

- **Search engine**

Assuming the user is well-versed in astronomy and capable of interpreting the results, the ADS search engine is tailored for searching astronomical abstracts. An "inverted file" is generated by gathering synonyms and simplifying search terms. Author names are indexed by surname and initial and a list of variations is used to account for possible spelling variations. Following pre-processing, the database is queried for synonyms and specifically astronomical synonyms related to the revised search term. ADS returns papers in languages other than English by searching for English language synonyms of foreign search terms.

4 Astrophysical Scientists and Experts

Researchers classified under specialized researchers in this section are those with at least one published research in the field of astronomy in the SAO/NASA Astrophysics Data System (ADS) (A. A. Abushattal et al., 2019; A. A. M. Abushattal, 2017). ADS is a platform that the Smithsonian Astrophysical Observatory (SAO) operates under a NASA grant and gives access to astronomical data for researchers. In Jordan, the field of astronomy and space sciences is enriched by the contributions of dedicated specialists and institutions. These specialists exemplify Jordan's commitment to advancing astronomy and space sciences through their remarkable contributions to research and science communication. Additionally, Jordan boasts educational and research institutions that offer courses and training in these fields, providing a fertile ground for the growth of aspiring astronomers and space scientists (Al-Tawalbeh et al., 2021; Al-Wardat, Docobo, Abushattal, & Campo, 2017; Alameryeen, Abushattal, & Kraishan, 2022; Algnamat, Abushattal, Kraishan, & Alnaimat, 2022). These efforts often involve collaborations with international institutions, agencies, and observatories, reinforcing Jordan's position in the global astronomical community.

Prof. Mashhoor Al-Wardat is a highly accomplished astrophysicist with extensive experience in research and teaching. His work has contributed significantly to our understanding of celestial objects and their properties, particularly in the field of binary stars and multiple stellar systems. He has made significant contributions to the study of celestial objects and their properties. Al-Wardat's educational background includes a Ph.D. in High-Resolution Imaging from the Special Astrophysical Observatory in Russia, an M.Sc. in Institute of Astronomy and Space Sciences from Al al-Bayt University in Jordan, and a B.Sc. in Physics from Yarmouk University in Jordan.

Throughout his career, Al-Wardat has held various academic positions, including serving as a Professor of Applied Physics and Astronomy at the University of Sharjah in the United Arab Emirates since 2019. He has also been a professor et al. al-Bayt University and served as the Dean of the Faculty of Science et al.-Hussein Bin Talal University in Jordan. Al-Wardat's research has been published in numerous journals and covers a wide range of topics in astrophysics, including the identification of new meteor showers in the UAE, the study of multiple stellar systems, and the observation of variable stars and asteroids using small telescopes in the UAE. He is an active member of several international astronomical organizations, including the International Astronomical Union and the Arab Union of Astronomy and Space Sciences.

Prof. Hatem Widyan is a professor of theoretical astroparticle physics. He is interested in the field of phase transition in the early Universe as well as theoretical astrophysics. After obtaining his bachelor's degree in physics from Yarmouk University in Jordan, he joined the master's program in physics at Aligarh Muslim University in India. From Delhi University he earned his Ph.D. Widyan started his teaching career at Jordan University of Science and Technology. After serving there for two years he moved to Al-Hussein bin Talal University. Currently, he is a professor et al. al-Bayt University in Jordan. During his academic career, he earned good experience in teaching as well as in research. He supervised and co-supervised more than ten M.Sc. Students. Widyan's research has been published in numerous journals and covers topics in phase transitions and binary systems in astrophysics. Widyan served as the Dean of Scientific research et al.-Hussein Bin Talal University and the dean of the Faculty of Aviation et al. al-Bayt University.

Dr. Ali Taani is an accomplished Associate Professor of Astrophysics and Space Science et al. Balqa Applied University in Jordan, boasting a rich and diverse professional journey that has taken him across the globe. He has made significant contributions to various scientific organizations, including the International Astronomical Union (IAU) and the Committee on Space Research (COSPAR). With a Ph.D. in astrophysics and space science from the University of the Chinese Academy of Sciences and a prestigious Chinese Academy of Sciences President's International Fellowship Initiative in 2018, he conducted groundbreaking research as a visiting scientist in China. Dr. Taani's research interests encompass a wide array of topics in astronomy and space science, including the study of cyclotron lines in magnetized neutron star systems, the distribution of double neutron star systems and their connection to supernova explosions, the formation and evolution of binary stellar systems, and the investigation of the oldest metal-poor stars in our Galaxy, providing crucial insights into cosmic evolution and the aftermath of the Big Bang.

In his scholarly pursuits, Dr. Taani focuses on unraveling the complexities of accretion processes in magnetized neutron star systems by analyzing cyclotron lines in High-Mass X-ray Binary systems (HMXBs). Additionally, he investigates the distribution and origins of double neutron star systems, shedding light on their cosmic lineage and connections to supernovae (J Docobo, Y Balega, et al., 2018). His research also delves into the formation and evolution of binary stellar systems, considering factors such as stellar evolution, magnetic fields, and orbital periods (Ali Taani, Karino, et al., 2019; Ali Taani,

Khasawneh, Mardini, Abushattal, & Al-Wardat, 2020). Furthermore, Dr. Taani leads a dedicated team in the exploration of metal-poor stars, the oldest celestial bodies in our Galaxy, which offer valuable insights into the early stages of cosmic evolution and the conditions following the Big Bang.

Awni Moh'd Khasawneh who is currently the Secretary General of the Arab Union for Astronomy &Space Sciences has an extensive curriculum vitae that reflects a remarkable career that spans a wide range of academic and professional experiences. With a background in astrophysics, civil engineering, and geodetic survey, Dr. Khasawneh has earned a Ph.D. in Astrophysics and Radio Astronomy from the Armenian National Academy of Sciences and held teaching positions at various universities in Jordan. His academic journey includes a Higher Diploma in Geodetic Survey from the United Kingdom and a bachelor's degree in civil engineering from Pakistan, numerous courses in Geographical information systems and remote sensing, photogrammetry, demonstrating his commitment to continuous learning and interdisciplinary expertise.

Dr. Khasawneh's contributions go beyond academia, as he has taken on leadership roles in numerous organizations related to astronomy, geospatial sciences, and space technology. Notably, he has served as the General Director of the Jordanian Geographical Center, Director of the Military Surveying directory, Chairman of the Jordanian Astronomical Society, Chairman of the Jordanian Geographical Society and Director General of the Regional Center for Space Science & Technology Education for Western Asia (UN-RCSSTEWA) additional to that he is still the Head of the Arab Division of Geographical Name Experts, one of the divisions affiliated with the United Nations, Dr. Al-Khasawneh is also a member of the selection committee for applicants to membership in the International Astronomical Union… His involvement in these organizations has been instrumental in promoting astronomical awareness, geographic and genome research, and space science education particularly he is currently the execution director of the Astronomical Office for development in the Arab region and beyond.

Furthermore, Dr. Khasawneh has an impressive publication research, with research papers and books covering topics ranging from astronomy to geographical names and Geographical Information System Remote sensing. His dedication to scientific research, supervision of numerous Master's theses, and being the supervisor and editor for specialized scientific journals underscore his commitment to advancing knowledge and education in various fields. Awni Moh'd Khasawneh's career exemplifies a multifaceted approach to science, education, and leadership that has left a lasting impact on the academic and scientific communities in Jordan and the broader Arab region.

Even now, when his name is mentioned before the scientific community, he has become associated with the founder of modern astronomy in Jordan in particular and the Arab region in general.

Dr. Ala'a A. A. Azzam graduated from University College London in 2013 with a PhD in atomic and molecular physics for space science applications. She received a scholarship from the University of Jordan to obtain this degree and returned to work at the University of Jordan after obtaining her doctorate in the Physics Department as an assistant professor. Since her return to Jordan in 2013, Dr. Al-Azzam has implemented many activities in the field of space sciences and astronomy at the level of the Physics Department, at the university level, and at the community level, intending to increase public awareness of the importance of this field. Also, Dr Azzam was working to help young people interested in this field to specialize in it. To help achieve these goals on a large scale, Dr. Azzam established the AstroJo Institute in 2017.

Dr. Azzam has much specialized research in calculating the spectra of molecules important for life detection in the atmospheres of exoplanets, such as sulfur hydroxide, aluminum monochloride, and carbon dioxide. Also, Dr Azzam is a member of the ExoMol project (see exomol.com).

Dr. Mohammad Mardini: His work is characterized by its depth and diversity, contributing significantly to our understanding of the universe's mysteries. Dr. Mohammad Mardini is a luminary in the realm of Galactic Archaeology, renowned for his unwavering commitment to unraveling the mysteries of our universe. His journey began with a bachelor's degree in physics from Al-Hussain Bin Talal University, where he developed a profound fascination for Black Holes. Driven by an insatiable thirst for knowledge, he pursued a Master's degree in Physics from Yarmouk University, where he delved into advanced physics courses, pioneering innovative techniques to analyze the nucleosynthesis and reaction rates of fluorine 19 in the Sun. Dr. Mardini's insatiable curiosity led him to obtain a Ph.D. in Galactic Archaeology from the University of Chinese Academy of Sciences, where his groundbreaking research focused on deciphering the formation and evolution of the Milky Way. By meticulously analyzing extensive observational data using state-of-the-art telescopes and cutting-edge data analysis techniques, he made groundbreaking discoveries that shed light on the Milky Way's formation history over billions of years (Rah, Yatman, Taani, Abushattal, & Mardini, 2024).

Dr. Mardini's exceptional dedication and contributions to the field have earned him numerous accolades, including recognition as an excellent Ph.D. student for his discovery of six of the most metal-poor stars in the third data release of LAMOST, revolutionizing the search for these ancient celestial objects in large spectroscopic surveys. Beyond his academic pursuits, Dr. Mardini has developed a valuable GitHub repository called The-ORIENT, designed to facilitate the sharing of code implementations related to trajectories in time-varying gravitational potentials, derived from Milky Way analogs selected from the Illustris-TNG simulation. Dr. Mardini invites the scientific community to explore and contribute to this repository, emphasizing the importance of collaboration in advancing collective scientific knowledge. Outside the realm of academia, Dr. Mardini is a passionate advocate for public outreach and science communication. He has delivered captivating talks at international conferences, captivating audiences with his enthusiasm for Galactic Archaeology and its profound implications for our understanding of the universe. Dr. Mardini's remarkable journey is a testament to his relentless pursuit of knowledge and his dedication to inspiring and educating others about the wonders of the cosmos.

Dr. Mohammad Talafha is a dedicated researcher with a specialization in Solar Physics and MHD Theory, with a keen interest in astrophysical simulations. His academic journey began at Yarmouk University in Jordan, where he earned both a B.Sc. And an M.Sc. in physics, focusing on the abundance of chemical elements in the Sun. During his master's program, Dr. Talafha developed an open-source package for calculating and analyzing chain reactions occurring in the solar interior. His pursuit of knowledge then led him to Budapest, where he completed his Ph.D. in astronomy at Eötvös Loránd University in 2022. During his doctoral research, Dr. Talafha focused on magneto-hydrodynamic theory (MHD) and its significance in predicting solar cycles and understanding space weather, which has wide-ranging implications for daily life and various applications. He investigated the effects of varying parameters and different types of non-linearities in the equations describing the evolution of the Sun's large-scale magnetic field. His work aimed to identify the signatures in statistical properties that correspond to specific non-linearities or parameter combinations, shedding light on the long-term behavior of the solar dynamo. Dr. Talafha's research has potential applications in predicting space weather events and understanding solar cycle variations. Currently, Dr. Talitha serves as a postdoctoral researcher at WIGNER RCP in Budapest, where he focuses on parameter investigations for MHD simulations of the inner-heliosphere magnetic field. He takes advantage of the unique opportunities presented by the NASA Parker Solar Probe and ESA Solar Orbiter missions to study the inner heliosphere with unprecedented proximity to the Sun. Dr. Talafha's commitment to advancing our understanding of solar physics and MHD theory is evident through his comprehensive list of publications and contributions to the field.

Dr. Ahmad Abushattal is an eminent figure in the field of astrophysics in Jordan, where his research on galactic dynamics and binary star systems has greatly advanced our understanding. Dr. Abushattal has earned a Ph.D. from the University of Santiago de Compostela and has been recognized for his groundbreaking research during his post-doctoral fellowship. This thesis has been hailed for its clarity and depth as it presents new models of spectroscopic binaries. In addition to his academic excellence, Dr. Abushattal is also known for his commitment to enriching the intellectual landscape of the physics department, fostering critical thinking, and stimulating scientific curiosity in students et al.-Hussein Bin Talal University in Maan, Jordan.

In addition to celestial research, Abushattal has made a significant contribution to renewable energy research in the field of solar cell technology. He contributes to the development of energy sources that are more efficient and sustainable through his research on solar cell operation and experiments with new materials (A. Abushattal,

Al-Wardat, Taani, Khassawneh, & Al-Naimiy, 2019; A. Abushattal, Alrawashdeh, & Kraishan, 2022; Boukortt et al., 2022). Moreover, this work is vital to Jordan's national plan for enhancing renewable energy capacities and reducing fossil fuel imports (A. A. Abushattal, Al-Wardat, et al., 2024; A. A. Abushattal, Loureiro, & Boukortt, 2024; Yaylacı et al., 2022). In addition to his expertise in astrophysics, Dr. Abushattal is keenly interested in applying these principles to real-world challenges, creating a synergy between astrophysics and modern technology. Furthermore, by publishing articles in peer-reviewed journals and presenting at international conferences, Dr. Abushattal is an active member of the scientific community. He maintains his status as a thought leader in his field by serving as a review editor for esteemed scientific publications. As a researcher with a passion for research, Dr. Abushattal's skills in securing funding demonstrate that the scientific community believes in the transformative power of his work.

5 Data Analysis for Jordanian Researchers

The SAO/NASA Astrophysics Data System (ADS) is a powerful tool for researchers in the field of astrophysics, offering comprehensive data analysis capabilities. It allows researchers to delve into various aspects of their work by providing valuable metrics such as the number of papers authored, the number of reads those papers have received, and the citation count for each article. These metrics are crucial for assessing the impact and reach of a researcher's work within the scientific community(J Docobo, Y Balega, et al., 2018). Additionally, ADS provides insights into the Author Network, which includes author occurrences, paper citations, and paper downloads. This author network feature helps researchers identify their connections within the academic community, explore collaborative opportunities, and gain a deeper understanding of the influence of their work in the astrophysics field. With ADS, researchers in Jordan and beyond can access vital data analysis tools to enhance their scientific endeavors and contribute to the advancement of astrophysics knowledge (J. A. Docobo, Griffin, Campo, & Abushattal, 2017).

Figure 1 Shows the number of scientific articles published during the period of (1995–2023) with a total of 230 papers. Figure 2 describes the citations number of citations during the period of 1999 with 10 citations to 1806 for 2023 with a total number of 5216 citations with 395 self-citations for these 26 years with an average of 23 citations per article. Furthermore, Fig. 3 shows 20908 reads for all the articles published from 1997 to 2023, with an average of 91 reads for each article. Despite the low number of astrophysics and space science researchers in Jordan, these statistics indicate the global interest in these subjects. In light of these numbers, this specialized group in astronomy and space sciences pursues important research topics. As a result, they continue to be a key point of reference for numerous researchers worldwide due to their commitment to staying up to date with scientific advancements.

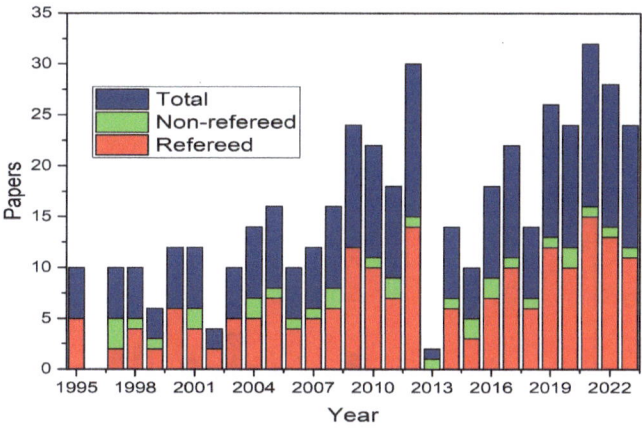

Fig. 1. Numbers of Papers vs Year

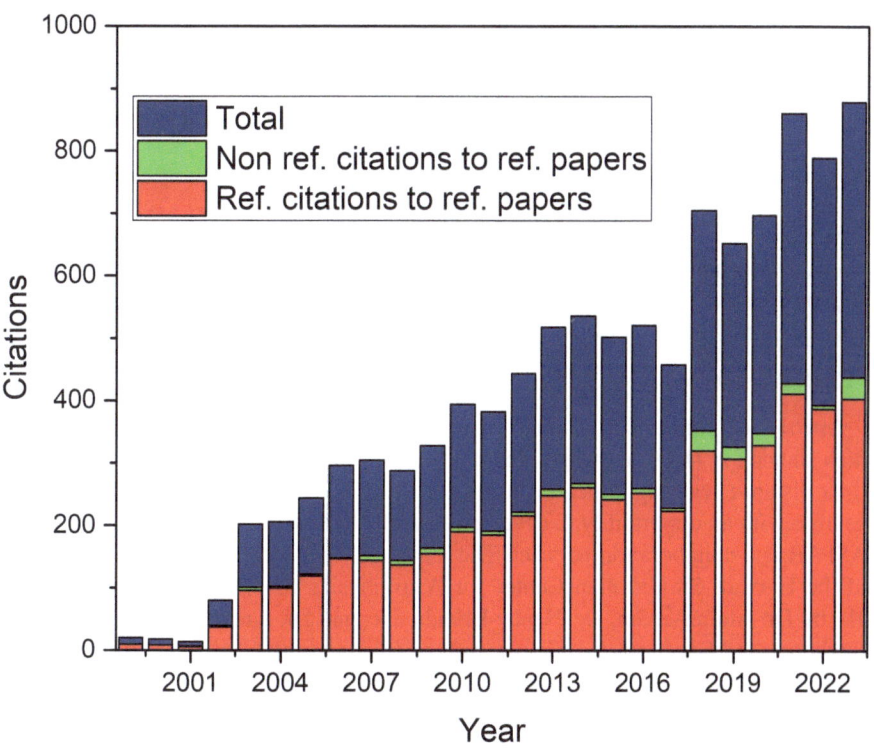

Fig. 2. Number of Citations vs Year

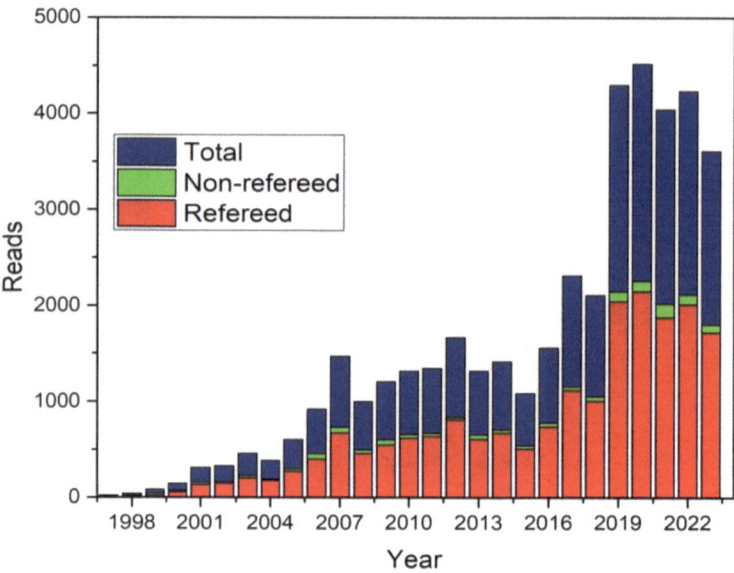

Fig. 3. Numbers of reads vs Year.

6 Astrophysical Institutions

AstroJo is a space sciences institute based in Jordan – Amman. It was founded by Dr. Alaa A. A. Azzam in 2017, as part of her vision to enrich space sciences in the Middle East, specifically Jordan, by experts. Dr Azzam is currently an assistant professor at the physics department at the University of Jordan (see astrojo. space).

- **Research projects:**

AstroJo hosts three main research groups. The first research group is called CASPAR, which stands for The Cambridge-Amman Seminar Project on Astrophysical Research. It is a distance learning educational initiative to provide promising young Jordanian scientists with an opportunity to develop research skills in astronomy and astrophysics. CASPAR is a collaboration between The University of Jordan (JU), AstroJo Institute, and the Harvard-Smithsonian Center for Astrophysics (CfA) that started in January 2017 by Dr. Ala'a Azzam, Dr. Nancy Brickhouse, and Dr. Andrew Szentgyorgyi. Drs. Brickhouse and Szentgyorgyi are research scientists at the CfA. CASPAR provides a special opportunity for an emerging generation of Jordanian scientists.

CASPAR seminars focus on exoplanet science, exobiology, exoplanet habitability, and the impact of cool host star activity on exobiotic systems. The CASPAR curriculum consists of a mix of lectures by Azzam, Brickhouse, and Szentgyorgyi, followed by a series of student-led seminars. The final phase of the seminar consists of individual research investigations into the habitability of exoplanets. Fifteen students participated in this program and were trained, five of them had the chance to make a night's visit to the Farid & Moussa Raphael observatory (FMRO) at Notre Dame University Louaize (NDU) in Lebanon. This observatory is a teaching, research, and outreach facility located at the main campus of the university. The five students were trained to observe exoplanets by the transit method by Dr. Roger Hajjar.

The second research group is part of the ExoMol project (see exomol.com) led by the Professor Jonathan Tennyson at University College London [23,24,25]. The project aims to review and expand the spectral line lists of molecules that are considered Bio-signatures. Such as AlCl [2], carbon dioxide isotopologues $16O13C16O$ [3], $16O12C18O$ [4], and $16O13C18O$ [5]. Hydrogen Cyanide (HCN) is also one of the molecules of interest [6]. This research group started in 2018, and twenty-five students have participated in this program since its establishment.

The third research group is part of the ORBYTS program based at University College London (see orbyts.org). ORBYTS is an education program in which secondary school students work on original research projects under the supervision of PhD students, Post-Docs, and other young scientists. Eleven Jordanian students, whose ages ranged between thirteen and sixteen years, participated in the first session held in AstroJo of this program. All these students live in Amman, and their schools belong to the private and public sectors. Around 36% of them are girls.

The research topic for the first session of this program was to calculate the energy levels of carbon dioxide $16O12C16O$ from empirical rovibrational transitions. Also, AstroJo Institute hosts training programs in space sciences, such as the Data Intensive Science (DIS) program, which started in Feb 2020 and ended in March 2023. This program was funded by the Khalidi-Newton Fund, organized, and hosted by AstroJo, and sponsored by University College London. The DIS program aimed to teach programming and machine learning to the young people of Jordan, with applications in astrophysics set in mind. Twenty-nine participants were able to complete this program, with the girls' participation rate reaching more than 60 percent see this website https://www.ucl.ac.uk/data-intensive-science-industry/training-outreach/newton-khalidi-fund-project.

- **Basic training program:** AstroJo hosts a space science basic training program that people ages eleven and older can join. It is a six-month program consisting of one lecture per week, during which theoretical lectures and practical workshops are carried out. This program aims to provide young people with the knowledge and tools to join research groups hosted by AstroJo. Over fifty participants have completed this program to date. Public activities: AstroJo Institute is interested in the cultural aspect of society in the field of space science and astronomy. Therefore, a group of programs was designed that is concerned with promoting the culture of space science and astronomy for the local community and working to correct all. Common mistakes in this field in society [19–22]. These programs are as follows: - Portable planetarium The portable planetarium is one of the services provided by the AstroJo Institute, which helps. Achieve the educational and cultural goals of the Institute. The planetarium can be transported to any region of Jordan. The dome can be placed inside the school campus. Its system relies on the use of modern educational panoramic (360°) display systems, and the program contains a comprehensive library of educational programs that cover and extend the teaching curricula. The goals of this program can be summarized as follows:

1. Enabling students and teachers in all regions of the country to attend planetarium dome shows without the need to leave their schools and endure the hardships of the journey to reach the dome.
2. Enabling attendees to learn various topics in space science with the help of researchers specialized in space science.
3. Enabling schools to conduct specialized educational activities in space science for all students in the school in conjunction with the normal course of the teaching process in the school and without the need to cancel an entire school day for students.
4. Removing many common fallacies and misunderstandings of many topics related to planet Earth, our solar system, and the universe.
5. Spreading awareness among people of the possibility of participating in space science research with simple tools. More than sixty thousand students distributed in different regions of Jordan have been able to watch the portable planetarium shows so far.

- **Teachers Training:** The AstroJo Institute offers a training program for teachers in the field of space science and astronomy. This program serves teachers working in all education sectors in Jordan, such as the government, private, and UNRWA sectors. It is a one-week program, during which intensive theoretical lectures are implemented in addition to workshops, where different tools and materials are used to achieve the program objectives. One of the most important goals of this program is to train teachers whose school curricula and which they teach include topics in the field of space science and astronomy. In addition to training teachers who implement extracurricular activities for their students in the field of space science. So far, more than seventy teachers distributed in different regions of Jordan benefited from this program. - Students and Teachers Competition This competition is held annually in October in conjunction with World Space Week, and the goal of the competition is to spread the culture of space science and astronomy in Jordanian society through schools. Every year from 2021, AstroJo organizes a large competition, focusing on space sciences (A. Abushattal, Kraishan, & Alshamaseen, 2022; J Docobo, P Campo, & A Abushattal, 2018). The competition is aimed at school students and tests their knowledge of astronomy. The competition also involves teachers who help the students learn about astronomy. Students are ranked based on the number of questions they can answer, and how quickly they can answer on the day of the competition, while teachers are ranked by the activities in space sciences, they hold with their students at their schools during the second semester. The grand prize in this competition is a visit to institutions specialized in the field of space science and astronomy in one of the countries. Egypt was selected for the 2022 competition, the United Arab Emirates for the 2023 competition, and Morocco for the competition which going to be held in 2024. A total of more than two hundred students have participated in the competition so far and during the three editions that have been implemented so far. -Stargazing trips AstroJo organizes monthly trips to the space camp near Qasr al-Kharana which is a desert castle located about 60 km east of Amman. The trips are available to the public, accompanied by trained stargazers who help everyone identify the night sky. Many stargazing programs are also implemented in schoolyards in many regions of

Jordan. All these activities are implemented using a six-inch telescope and another eight-inch telescope which are usually attached to DSLR cameras attached to data shows. More than five thousand people participated in the stargazing trips that have been carried out so far.

7   Conclusion

- Jordan's fascination with astronomy and space sciences has deep historical roots and continues to thrive today, thanks to the dedication and contributions of both specialists and amateurs. Specialists like Dr. Al-Wardat, Dr. Taani, Dr. Khasawneh, Dr. Azzam, Dr. Mardini, Dr. Talafha, and Dr. Abushattal have made significant strides in advancing our understanding of the universe through their research and scientific endeavors. Institutions like AstroJo have played a pivotal role in promoting space sciences through research projects, training programs, and public activities.
- The harmonious coexistence of specialists and amateurs has created a vibrant astronomical community in Jordan. Amateurs, driven by their passion and curiosity, contribute to popularizing astronomy and inspire the younger generation. They often make significant discoveries while stargazing and participating in observational activities.
- Jordan's commitment to space sciences extends beyond its borders through international collaborations, making it a focal point in the Middle Eastern astronomical community. The country's rich history, active scientific community, and cultural ties to the cosmos make it a unique destination for those interested in the mysteries of the universe.
- Despite the challenges facing astronomy in Jordan, such as light pollution and limited resources, the dedication of specialists and amateurs, along with the support of institutions and the government, ensures a bright future for the field. The synergy between these two groups continues to foster a love for the cosmos and disseminate knowledge, ultimately enriching our understanding of the vastness of the universe.

8   Recommendations and Future Work

By implementing these recommendations and focusing on future work in these areas, Jordan can continue to nurture its vibrant astronomical community, inspire the next generation of scientists, and contribute meaningfully to our understanding of the cosmos.

- **Enhancing Public Outreach**: To further promote astronomy and space sciences in Jordan, it is recommended to expand public outreach efforts. Organize more stargazing events, workshops, and lectures aimed at engaging and educating the general public. Consider partnerships with schools and local community centers to reach a broader audience.
- **Youth Engagement:** Encourage more young people to get involved in astronomy and space sciences by offering educational programs and competitions. Establish astronomy clubs in schools and universities, providing students with hands-on experiences and opportunities for research projects.

- **Research Collaborations:** Continue and expand international research collaborations with renowned astronomical institutions. Foster partnerships that enable knowledge exchange, joint research projects, and access to advanced telescopes and technologies.
- **Infrastructure Development:** Invest in the development of state-of-the-art observatories and research facilities to support advanced research in space sciences. Seek funding from governmental and private sources to enhance the country's research capabilities.
- **Teacher Training:** Continue and expand the teacher training program to ensure that educators are well-equipped to teach astronomy and space sciences effectively. This will help in spreading knowledge and interest among students.
- **Astronomy Tourism:** Promote astronomy tourism as an eco-tourism initiative. Develop more star parks and observatories in prime locations to attract both local and international tourists interested in stargazing and celestial observations.
- **Data Sharing and Open Access:** Encourage researchers and institutions to share their astronomical data and findings openly. Promote the use of open-access journals and data repositories to facilitate collaboration and knowledge dissemination.
- **Educational Resources:** Develop and distribute educational materials and resources related to astronomy and space sciences for schools and the general public. This includes books, digital content, and interactive tools to make learning more accessible.
- **Space Science Curriculum Integration:** Collaborate with educational institutions to integrate space science and astronomy into the regular curriculum. Ensure that students have access to quality education in these fields from an early age.
- **Funding Opportunities:** Advocate for increased funding for space science research and educational programs. Seek partnerships with governmental agencies and private companies interested in supporting scientific endeavors.
- **Long-Term Planning:** Develop a long-term strategic plan for the advancement of astronomy and space sciences in Jordan. Set clear goals and milestones to measure progress over time.

References

Abushattal, A., Al-Wardat, M., Taani, A., Khassawneh, A., Al-Naimiy, H.: Extrasolar Planets in Binary Systems (Statistical Analysis). Paper presented at the Journal of Physics: Conference Series (2019)

Abushattal, A., Alrawashdeh, A., Kraishan, A.: Astroinformatics: the importance of mining astronomical data in binary stars catalogues. Commun. BAO **69**(2), 251–255 (2022)

Abushattal, A., Kraishan, A., Alshamaseen, O.: The exoplanets catalogues and archives: an astrostatistical analysis. Commun. BAO **69**(2), 235–241 (2022)

Abushattal, A.A., et al.: The 24 Aqr triple system: a closer look at its unique high-eccentricity hierarchical architecture. Adv. Space Res. **73**(1), 1170–1184 (2024)

Abushattal, A.A., Docobo, J.A., Campo, P.P.: The most probable 3D orbit for spectroscopic binaries. Astron. J. **159**(1), 28 (2019)

Abushattal, A.A., Loureiro, A.G., Boukortt, N.E.I.: Ultra-high concentration vertical homo-multijunction solar cells for cubesats and terrestrial applications. Micromachines **15**(2), 204 (2024)

Abushattal, A.A.M.: The modeling of the physical and dynamical properties of spectroscopic binaries with an orbit. Universidade de Santiago de Compostela (2017)

Al-Naimy, H., Konsul, K.: Basic space sciences in Jordan. Paper presented at the Developing Basic Space Science World-Wide: A Decade of UN/ESA Workshops (2004)

Al-Tawalbeh, Y.M., et al.: Precise masses, ages, and orbital parameters of the binary systems HIP 11352, HIP 70973, and HIP 72479. Astrophys. Bull. **76**, 71–83 (2021)

Al-Wardat, M., Docobo, J., Abushattal, A., Campo, P.: Physical and geometrical parameters of CVBS. XII. FIN 350 (HIP 64838). Astrophys. Bull. **72**, 24–34 (2017)

Alameryeen, H., Abushattal, A., Kraishan, A.: The physical parameters, stability, and habitability of some double-lined spectroscopic binaries. Commun. BAO **69**(2), 242–250 (2022)

Algnamat, B., Abushattal, A., Kraishan, A., Alnaimat, M.: The precise individual masses and theoretical stability and habitability of some single-lined spectroscopic binaries. Commun. BAO **69**(2), 223–230 (2022)

Alzoubi, M.: The Nabataean timing system. Acta Orientalia Academiae Scientiarum Hungaricae **69**(3), 301–309 (2016)

Belmonte, J.A., González-García, A.C., Polcaro, A.: Light and Shadows over Petra: astronomy and landscape in Nabataean lands. Nexus Netw. J. **15**, 487–501 (2013)

Boukortt, N.E.I., et al.: Electrical and Optical Investigation of 2T–Perovskite/u-CIGS Tandem Solar Cells With~ 30% Efficiency. IEEE Trans. Electron Devices **69**(7), 3798–3806 (2022)

Docobo, J., Balega, Y., Campo, P., Abushattal, A.: Double Stars Inf. In: Circ (2018)

Docobo, J., Campo, P., Abushattal, A.: IAU commiss. Double Stars **169**(1) (2018)

Docobo, J.A., Griffin, R.F., Campo, P.P., Abushattal, A.A.: Precise orbital elements, masses and parallax of the spectroscopic–interferometric binary HD 26441. Mon. Not. R. Astron. Soc. **469**(1), 1096–1100 (2017)

Eichhorn, G., Accomazzi, A., Grant, C.S., Kurtz, M.J., Murray, S.S.: Access to the astronomical literature through the NASA astrophysics data system from developing countries. Paper presented at the Library and Information Services in Astronomy IV (LISA IV) (2003)

Henneken, E., et al.: The SAO/NASA astrophysics data system: a gateway to the planetary sciences literature. Paper presented at the 40th Annual Lunar and Planetary Science Conference (2009)

Hussein, A.M., et al.: Atmospheric and fundamental parameters of eight nearby Multiple stars. Astron. J. **163**(4), 182 (2022)

Kurtz, M.: The future of memory: archiving astronomical information. Paper presented at the Symposium-International Astronomical Union (1994)

Kurtz, M.J., et al.: Intelligent text retrieval in the NASA astrophysics data system. Paper presented at the Astronomical Data Analysis Software and Systems II (1993)

Rah, M., Yatman, M., Taani, A., Abushattal, A.A., Mardini, M.K.: Unraveling the Origins and Development of the Galactic Disk through Metal-Poor Stars. arXiv preprint arXiv:2402.07045 (2024)

Taani, A., Abushattal, A., Khasawneh, A., Almusleh, N., Al-Wardat, M.: Jordan journal of physics. Jordan J. Phys. **13**(3), 243–251 (2020)

Taani, A., Abushattal, A., Mardini, M.K.: The regular dynamics through the finite-time Lyapunov exponent distributions in 3D Hamiltonian systems. Astron. Nachr. **340**(9–10), 847–851 (2019)

Taani, A., et al.: On the wind accretion model of GX 301-2. Paper presented at the Journal of Physics: Conference Series (2019)

Taani, A., Khasawneh, A., Mardini, M., Abushattal, A., Al-Wardat, M.: Probability Distribution of Magnetic Field Strengths through the Cyclotron Lines in High-Mass X-ray Binaries. arXiv preprint arXiv:2002.03011 (2020)

Weaver, D.: Celestial ecotourism: new horizons in nature-based tourism. J. Ecotour. **10**(1), 38–45 (2011)

Yaylacı, E.U., Öner, E., Yaylacı, M., Özdemir, M.E., Abushattal, A., Birinci, A.: Application of artificial neural networks in the analysis of the continuous contact problem. Struct. Eng. Mech. Int. J. **84**(1), 35–48 (2022)

The Central Distance Distribution of Hα and γ-ray Burst Solar Flares

Ramy Mawad[(✉)] [iD]

Astronomy and Meteorology Department, Faculty of Science, Al-Azhar University, Nasr City , Cairo 11488, Egypt
ramy@azhar.edu.eg

Abstract. The angular distance of the solar flares to the projective point of the center of the solar disk on the solar spherical surface has been studied by the heliographical or helioprojective coordinates, during the periods 1975–2021 for GOES events and 2002–2021 for RHESSI events, hereafter "distance." It gives a specific distribution curvature. It has also been noted that when using the number of solar flare events in each satellite, GOES or RHESSI, or even using the sum of the flux (class) or importance parameter, it obtains the same result, which is that the shape of the distribution curve remains in its shape without any significant change. In addition, it has been shown that the distribution curve contains a specific number of peaks. These peaks have a specific distance from the center of the solar disk that is very similar to the projection of the solar interior layers on the solar disk. For this reason, the names of these four main peaks have been given as follows: (1) the core circle (0–15°): it is a projection of the solar core onto the solar disk, (2) radiative ring (15–45°), and (3) the convection ring (45–55°). The limb ring is 80–90°. This result makes us wonder why the number of events in the middle of the solar disk is few, and also small at the solar limb, while many in the other parts in the solar disk. This suggests that we need to understand the sun better than before, and it also suggests that solar flares are connected to each other through the solar interior layers, the extent of which may reach the convection zone or perhaps beyond that, or the opacity of the convection zone may be less than the currently estimated value.

Keywords: The Sun · Solar Flare · Solar Disk · Solar layers · Photosphere · Solar Interior layers

1 Introduction

The sun is a gaseous body, considered spherical in shape, according to the photosphere layer that is visible through various observation devices and naked eye (i.e., visible light), which is also called the surface of the sun. The surface extends to the chromosphere layer, where most solar phenomena occur [1]. Many phenomena have been observed on the solar surface [2, 3], the most famous of which are sunspots, which played an important role in determining the rotational axis of the sun and the solar equator [4]. Solar physicists have been able to determine coordinates on the surface of the sun using

© The Author(s) 2025
H. M. K. Al Naimiy et al. (Eds.): AUASS-CONF 2023, SPPHY 420, pp. 131–138, 2025.
https://doi.org/10.1007/978-981-96-3276-3_11

different coordinate systems, such as heliographic and heliocentric coordinate systems [5, 6]. As a result, they have been able to monitor the sun and its events that appear on it, and understand many aspects of solar dynamics [7–17].

Solar flares are violent and significant solar events that appear on the surface of the sun, especially in the chromosphere layer. They are intense explosions that occur as a result of the release of tremendous energy due to the reconnection of magnetic field lines in active regions. Solar flares emit radiation across all wavelengths. Their earliest observations were in visible light, where they could be detected in the H-alpha band [18]. Through various space missions, they have been observed solar flares in X-ray and γ-ray bands [19].

The distribution of solar flares with respect to latitude has been previously studied [8–13]. Previous studies have found that the main occurrence latitude range is in the range 13°–15° [8]. In addition, there is homogeneity between the northern and southern hemispheres, which varies with solar activity. Additionally, it has been observed that there is no specific distribution of solar flares with respect to longitude, as the Sun rotates on its axis and solar flares are momentary events that can last for minutes or even hours [13–17]. A study [9] attempted to investigate this distribution by combining latitude and longitude. It founds that active regions giving rise to solar flares occur at specific latitudinal belt. This eruptive belt is varying according to solar activity.

The latitude and longitude has been combined by other method [1], called "angular distance", hereafter "distance". This solar flare distance is measured between solar flares and the center of the solar disk angularly. It accords on the coordinate system that used, which is represented by the point where equator and Zero-longitude (i.e., the observer's direction) intersect (i.e., center of solar disk). This study found that the curvature shape, of the solar flare distance distribution, does not change with a change in the type of coordinate systems used, which expresses the direction of different observers. Nor even with a change in the cycle of solar activity. Nor with a change in the observed wavelength in the X-ray range, whether hard- or soft- x-ray. Even the same distribution curve appears when using the number of events at the same distance or when using the total amount of flux or importance. Which indicates the importance of studying this distance distribution. This distribution concludes that solar flares do not occur at the center of the solar disk, while a few appear in the middle of the solar disk. At the distance region (15°-30°), the number of solar flares is maximized. The number of solar flares then decreases as it moves away from the center of the solar disk towards the solar limb, where the number of solar flares is minimized.

This previous study focused on the X-ray solar flares. But the current study will focus on applying and examine distance distribution for solar flares which are detected in Hα and by γ-ray burst monitor.

2   Data Sources and Distance Determination

The Hα solar flare data is obtained from National Geophysical Data Center (NGDC) – by national oceanic and atmospheric administration (NOAA), during the period 1955–2010, from URL (https://www.ngdc.noaa.gov/stp/space-weather/solar-data/solar-features/solar-flares/h-alpha/events/). This catalog recorded the solar flare coordinates by heliographical coordinates system. Therefore, we can use the flowing equations to determining the flare distance [1]:

$$D = arccos[cos(\lambda)cos(\beta)] \tag{1}$$

where D is the flare's distance between the projection point of the center of the solar disk on the spherical surface and the solar flare position by any coordinates system such as heliographical or heliocentric coordinates. λ and β are the flare's latitude and longitude, respectively.

In addition, the Fermi γ-ray Burst Monitor (GBM) solar flares have been used. This data source is available on the Fermi satellite web page at the following URL: (http://hesperia.gsfc.nasa.gov/fermi/gbm/qlook/fermi_gbm_flare_list.txt). The catalog of the GBM solar flare events records the coordinates by Cartesian-projective coordinates (x, y). Therefore, it needs to be converted into projective coordinates by the following formula, driven by [1].

$$\lambda = arcsin\left[\frac{x}{1 - y^2}\right] \tag{2}$$

$$\beta = arcsin\left[y\right] \tag{3}$$

In addition, the Fermi γ-ray Burst Monitor (GBM) solar flares have been used, during the period 2008–2023. This data source is available on the Fermi satellite web page at the URL: (http://hesperia.gsfc.nasa.gov/fermi/gbm/qlook/fermi_gbm_flare_list.txt). The catalog of the GBM solar flare events records the coordinates by Cartesian-projective coordinates (x, y). Therefore, it needs to be converted into projective coordinates by the following formula, driven by [1].

The solar flare events that occur at every angular distance (integer value with steps 1°) in the range 0–90° can be counted, then plot that distribution.

The solar activity cycle C can be calculated using the following equation:

$$C = -157.43 + 0.090253Y \tag{4}$$

where Y is the year of occurred solar flare event.

3 Results

The distance between solar flare position and the projection point of the center of the solar disk on the solar spherical surface, which is represents the observer's direction (i.e., Zero-longitude) was estimated during the period 1955–2010 in the Hα, and during the period 2008–2023 for events detected in the γ-ray burst monitor.

Despite the distance for solar flares in both bands being based on a different coordinate system, which represents a different perspective of the Sun from the observer's angle, the distance distribution curve gave approximately the same distribution curvature shape as shown in Fig. 1 and 2. This result is consistent with a previous study conducted on X-rays [1].

With applying Eq. 4, solar flare events can be categorized according to the solar activity cycle. It has been found that each solar cycle has a peak in the distribution curve, whose position slightly varies from one cycle to another depending on the strength of the cycle. It also appears that the stronger the amplitude of the solar activity cycle, the higher the range of the peak point, depending on strength of the solar activity. Thus, the peak point in the distribution curve rises accordingly. Additionally, there seem to be smaller peaks following the main peak, with approximately five peaks, which is evident in Fig. 3.

Main peak of the distance distribution curve is at about distance 25°–30° from center of solar disk (i.e., the core circle). It is compatible to previous result by [1].

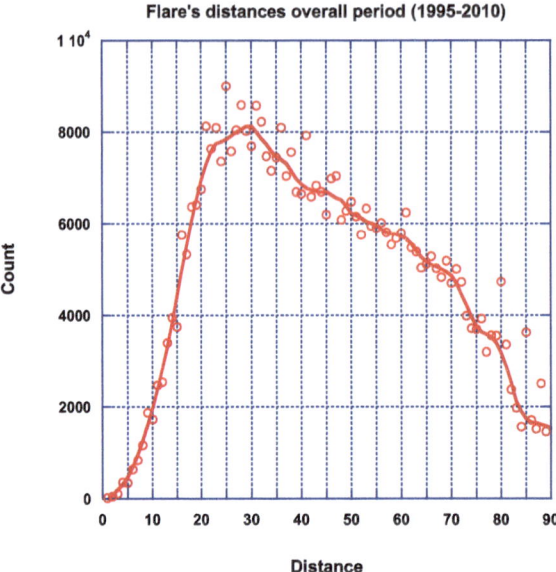

Fig. 1. The distance distribution of the Hα solar flares, during the period 1995–2010, for all events.

4 Discussions

The distribution curve of distance shows that it remains the same during different solar activity cycles and different observations depending on the wave-length bands. It is suggested that one of the reasons for the alignment of the distribution curve in Hα with X-rays is that initially, the location of X-ray solar flares observed in the Geostationary Operational Environmental Satellites (GOES) catalog was dependent on those measurements from the Hα catalog until 1997. However, after that, a different measurement of the solar flare location was adopted. This is according to GOES reports. Additionally, the GBM catalog includes solar flares in the X-ray band as well.

Nevertheless, the same solar flare observed by different satellites or tele-scopes gives each of them a slightly different coordinate. Comparing this with the persistence

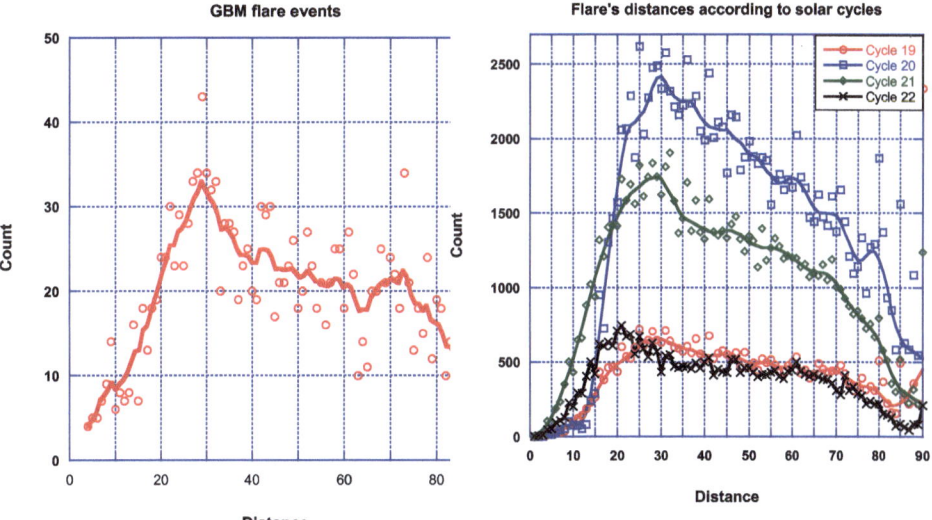

Fig. 2. The distance distribution of the γ-ray burst monitor solar flares, during the period 2008–2023.

Fig. 3. The distance distribution of the Hα solar flares, during the period 1995–2010, according to solar cycle.

of the distribution curve regardless of the observer's direction, it suggests that the actual position of solar flares cannot be measured from a single solar image, which is represented to a one observer's direction. This measured coordinates from image directly are the projection point of onto the solar spherical surface the actual position. To determine the actual coordinate, it requires at least three simultaneous observations from different viewing angles, each one being 120° apart from the others, as an example.

It is similar to coronal mass ejections (CMEs), where the initial observations from the Large Angle and Spectrometric Coronagraph (LASCO) telescope on the Solar and Helio-spheric Observatory satellite (SOHO) did not accurately represent the actual direction of motion of the detected CMEs in the image. The direction in the image is a projection of the desired actual motion direction. To determine the actual direction, it requires at least two additional satellites to observe the Sun in space from a 3-dimensional perspective. These two satellites are the Solar Terrestrial Relations Observatory (STEREO) A and B [20].

The implication of this result is that most solar flares occur in association with active regions, containing sunspot groups. However, some solar flares may originate from inactive regions. Previous studies have shown that sunspots (which often accompany solar flares) are present at latitudes 30–40° north or south. However, at the end of the solar activity cycle, the average latitudes become 5–10° north and south, forming the butterfly diagram. This law is called the Spörer law [21].

Solar flares are the same, as the images do not determine the height above the surface, thus introducing an error in estimating the position of solar flares. Methods that rely on estimating the longitude and latitude of solar flares from solar image assume that the flare occurred precisely at the surface, exactly at a height equals to the radius of the Sun from its center. Certainly, this introduces an error in estimating the heliographical or heliocentric coordinates of the solar flare. Therefore, if the solar flare itself is observed in different satellites, it will provide different positions.

Therefore, it is necessary to observe solar flares from different satellites that are separated from each other around the Sun by an angle of approximately 120°.

5 Conclusions

The central angular distances of solar flares, known as "distance", were studied using available observations from the Hα catalog from 1955 to 2010 and the γ-ray burst monitor catalog from 2008 to 2023. It was found that these distance observations for solar flares follow the same distribution curve as previously studied [1]. Despite using different wavelengths than those studied in the previous research, specifically X-rays, the same result was obtained.

Different coordinate systems were also applied, including heliographic and helioprojective coordinates. Both yielded the same result.

This distribution was studied for each solar activity cycle, and it was found that the distribution curve of central distances gives a consistent result.

All of this indicates that the determination of the position of a solar flare varies with the observer direction. Therefore, the position of observed solar flares in solar images represents a projection on the surface of the solar sphere from the true position of the solar flare. Estimation method of true position of solar flares needs to be developed to determine this true position.

Hence, if the same solar flare is observed by different satellites, it will yield different positions due to the slight angular differences between observers' directions. Therefore, this study proposes launching additional satellites (at least two) to simultaneously observe the Sun from different directional angle, with a separation of approximately 120° in central solar angles.

Acknowledgements. The author thanks the National Geophysical Data Center (NGDC) – by national oceanic and atmospheric administration (NOAA) and the Fermi γ-ray Burst Monitor (GBM) for providing the data of solar flares that helped to complete this study.

In addition, the author would also like to thank and appreciate the 14th Arabic Conference of the Arab Union for Astronomy and Space Sciences and the University of Sharjah in supporting the attendance of the conference and publication of this research paper. The author also extends his thanks to both Prof. Hamid M K Al Naimiy (Chancellor of the University of Sharjah, President of the Arab Union for Astronomy and Space Science and Conference Chair), and Prof. Mashhoor Al Wardat (Coordinator of the conference and Professor at University of Sharjah).

References

1. Mawad, R.: The study of angular distance distribution to the solar flares during different solar cycles. AAPPS Bulletin (2024). https://doi.org/10.1007/s43673-023-00102-6
2. Mawad, R., Shaltout, M., Ewaida, M., Yousef, M., Yousef, S.: Filaments disappearances in relation to solar flares during the solar cycle 23. Adv. Space Res. **55**(2), 696–704 (2015). https://doi.org/10.1016/j.asr.2014.11.003

3. Mawad, R., Mosalam Shaltout, M., Yousef, S., Yousef, M.E.: Filaments disappearance in relation to coronal mass ejections during the solar cycle 23. Adv. Space Res. **55**(2), 688–695 (2015). https://doi.org/10.1016/j.asr.2014.11.002

4. Carrington, R.C.: Observations of the spots on the sun, pp. 221–244. Williams and Norgate, London (1863)

5. Thompson, W.T.: Coordinate systems for solar image data. A&A, **449**(2), 791–803 (2006). https://doi.org/10.1051/0004-6361:20054262

6. Ulrich, R.K., Boyden, J.E.: Carrington Coordinates and Solar Maps. Sol. Phys. **235**(1–2), 17–29 (2006). https://doi.org/10.1007/s11207-006-0041-5

7. Gnevyshev, M.N.: On the 11-years cycle of solar activity. Sol. Phys. **1**, 107–120 (1967). https://doi.org/10.1007/BF00150306

8. Abdel-Sattar, W., Mawad, R., Moussas, X.: Study of solar flares' latitudinal distribution during the solar period 2002–2017: GOES and RHESSI data comparison. Adv. Space Res. **62**(9), 2701–2707 (2018). https://doi.org/10.1016/j.asr.2018.07.024

9. Mawad, R., Abdel-Sattar, W.: The eruptive latitude of the solar flares during the Carrington rotations (CR1986-CR2195). Astrophys. Space Sci. **364**(197), 2701–2707 (2019). https://doi.org/10.1007/s10509-019-3683-0

10. Mawad, R., Moussas, X.: Sympathetic solar flare: characteristics and homogeneities. Astrophys. Space Sci. **367**, 107 (2022). https://doi.org/10.1007/s10509-022-04145-3

11. Shrivastava, P.K., Singh, N.: Latitudinal distribution of solar flares and their association with coronal mass ejections. Chinese J. Astron. Astrophys. **2**, 198–2025 (2005)

12. Pandey, K.K., Yellaiah, G., Hiremath, K.M.: Latitudinal distribution of soft X-ray flares and disparity in butterfly diagram. Astrophys. Space Sci. **356**(2), 215–224 (2015). https://doi.org/10.1007/s10509-014-2148-8

13. Zharkova, V.V., Zharkov, S.I.: Latitudinal and longitudinal distributions of sunspots and solar flare occurrence in the cycle 23 from the solar feature catalogues. In: Marsch, E., Tsinganos, K., Marsden, R., Conroy, L. (eds.) Proceedings of the Second Solar Orbiter Workshop. ESA-SP 641. European Space Agency}, Noordwijk (2007). ISBN 92–9291–205–2. http://adsabs.harvard.edu/abs/2007ESASP.641E..90Z

14. Loumou, K., Hannah, I.G., Hudson, H.S.: The association of the Hale sector boundary with RHESSI solar flares and active longitudes. Astron. Astrophys. **618**(A9), 12 (2018). https://doi.org/10.1051/0004-6361/201731050

15. Cliver, E.W., Mekhaldi, F., Muscheler, R.: Solar longitude distribution of high-energy proton flares: fluences and spectra. Astrophys. J. Lett. **900**(1), id.L11 (2020). https://doi.org/10.3847/2041-8213/abad44

16. Li, H., Feng, H., Liu, Y., Tian, Z., Huang, J., Miao, Y.: A longitudinally asymmetrical kink oscillation of coronal loop caused by a diagonally placed flare below the loop system. Astrophys. J., 881(111), 2, 6 (2019). https://doi.org/10.3847/1538-4357/ab2bf7

17. Jetsu, L ; Pohjolainen, S ; Pelt, J ; Tuominen, I.: Longitudinal distribution of major solar flares, Astrophysics and Astronomy, 9th Cambridge Workshop on Cool Stars, Stellar Systems and the Sun, Cambridge, UK, 1995 Report number NORDITA-95–76-A (1995). 3 p.

18. Tandberg-Hanssen, E., Emslie, A.G.: The Physics of Solar Flares. Cambridge University Press (1988)

19. Aschwanden, M.J.: Irradiance observations of the 1–8 Å solar soft x-ray flux from goes. Sol. Phys. J. **152**(9), 53–59 (1994). https://doi.org/10.1007/BF0147318

20. Egorov, Y.I., Fainshtein, V.G.: A simple technique for identifying the propagation direction of CMEs in 3D space. Sol. Phys. **296**, 161 (2021). https://doi.org/10.1007/s11207-021-019 04-3

21. Spörer, G.F.W.: Ueber die periodicität er sonnenflecken seit dem jahre 1618: vornehmlich in bezug auf die heliographische breite derselben, und nachweis einer erheblichen störung dieser periodicität während eines langen zeitraumes, vol. 53. Blochmann, Germany (1889)

Update of the Solar Radiation Atlas for the Arab Republic of Egypt 2023

U. Ali Rahoma$^{(\boxtimes)}$, A. H. Hassan, Samy A. Khalil, and Ashraf S. Khamees

National Research Institute of Astronomy and Geophysics. NRIAG, Cairo, Egypt
UsamaAliRahoma@Yahoo.com

Abstract. The Solar Radiation Atlas of the Arab Republic of Egypt is funded and supported by the Institute's management and is integrated into the 2020/2030 Strategic Plan.. The Arab Republic of Egypt has one of the greatest levels of solar radiation in the world, with an average of 3,000 h of sunshine per year in most sections of the country. In the long run, solar energy is regarded as the most well-known energy source. Based on the various maps and tables in this atlas, which represent the most important elements of solar radiation in its various components, as well as the atmospheric elements that directly affect the amount of solar radiation received, such as temperature, humidity, and wind speed, and by including height above sea level, as well as flat, open areas with easy and flat terrain. The best area to set up solar energy projects is located between latitude 22–26° N and longitude 28–30° E, after entering the most appropriate height above sea level within the range of 500 m and flat, open areas with no seismic activity. The mean values of global horizontal solar radiation (GHI) is 6.18 ± 1.5 KWh/m^2/d, while the global solar radiation on the inclined by the latitude angle of the place (GTI) is 6.4 ± 0.82 Kwh/m^2/d.

Keywords: Global solar radiation · Diffuse solar radiation · Direct solar radiation · UV Solar radiation

1 Introduction

If the world is looking for new sources of energy because of their importance in technological growth [1], then the time has come to find out where the Solar Queen's golden square is on Earth. We know from past solar energy research how important southern Egypt is because it receives the most sun radiation in the globe [2]. Several past studies have been undertaken throughout Egypt, including: Egypt has a high potential for renewable energy (RE), which includes solar, wind, and biomass energy. Renewable energy technologies (RET) and systems have varying needs for research and development, demonstration, and market development assistance.

Solar-terrestrial physics began in the early twentieth century at Egypt's Helwan observatory. From February 1914 to December 1923, the levels of solar radiation at normal incidence were measured at Helwan observatory for air masses 1 and 2 [3]. Another study obtained hourly rates of total solar radiation at normal incidence for clear

© The Author(s) 2025

H. M. K. Al Naimiy et al. (Eds.): AUASS-CONF 2023, SPPHY 420, pp. 139–159, 2025.

https://doi.org/10.1007/978-981-96-3276-3_12

sky conditions in Helwan observatory (NRIAG) from 1922 to 1927 [3], and from 1935 to 1954 at Sant Catrine mountain in Sinai under the supervision of professor Abbot, the director of the Smithsonian Institution's Astrophysical Laboratory at the time [4].

Search for an antique global atlas: Nepal solar energy atlas based on regional adaptations of the solargis model [5, 6], the European solar radiation atlas [7], the global solar atlas [8], and surface solar irradiance in Egypt for energy generation [9].

Solar radiation atlas for the Arab Republic of Egypt; this research is one of many produced by the NRIAG, one of the national research institutes and centres connected with the Egyptian Ministry of Higher Education and Scientific Research. The atlas of solar radiation for the Arab Republic of Egypt is funded and supported by the Institute's management and is included in the Arab Republic of Egypt's strategic plan 2020/2030 (energy axis). NRIAG has various research and stations in Egypt, as well as publications dating back to than 120 years. Field observations were carried out by the institute's scientists beginning in the nineteenth century to measure the solar constant radiation from 1900 [3], Egyptian solar radiation atlas was, 1981 [10], 1991 [11], "Solar energy distribution over Egypt using cloudiness from meteosat images" [12], "Energy and Power Engineering" [13], prediction of diffuse solar energy on horizontal at several specified locations. A thorough examination of solar energy components utilising several models on horizontal and inclined surfaces for various climate zones [14], (ERA-5) Verification of Solar Energy Measurements and Its Impact on North African Electricity Costs [15], Performance Evaluation and Statistical Analysis of Solar Energy Modelling [16]. The foundation of solar radiation and its prediction [17, 18]. Solar radiation in Egypt - Arab Republic of Egypt [19], Climate and its impact on standard evapotranspiration in Egypt [20].

Establishing ground stations to measure solar radiation in its various components requires financial capabilities, including expensive equipment and a high cost of maintenance and periodic calibration. That is why empirical models (EMs) are widely used to predict the amount of solar radiation based on available meteorological data.

Therefore, basic data can be collected through measurement, modeling, or a combination of the two.

In practice, measurements of all solar radiation components cannot be used exclusively for the following reasons: 1) there are no long periods of measured data in most regions around the world; 2) Even if they exist, they often contain gaps that need to be filled by modeling; and 3) performing quality measurements is much more expensive than operating models [21–23].

The five countries of north Africa have diverse socioeconomic and energy circumstances. With an annual solar irradiation intensity ranging from 1800 to 2600 kWh/m²/d, the region with the highest solar energy intensity is largely located between 10 and 35° N. As a result, north Africa has a lot of solar energy resources. The majority of the countries in the region, including Morocco, Algeria, Libya, and Egypt, have significant potential to develop solar power infrastructure [24, 25].

Increased usage of renewable energy will increase energy security in the region, lower long-term energy prices, and have a number of positive economic repercussions. The nations in the region have set goals to increase the share of renewable energy in their electrical mix. Egypt plans to use 42% renewable energy for power by 2035 [15].

Energy intensity has remained practically constant over the last thirty years, indicating that there are enormous opportunities to encourage increased energy efficiency in all industries. Energy efficiency projects have faced a number of problems, ranging from a lack of institutional capacity to the inability to create the requisite markets to attract the necessary investment [6–8]. The ERA5 generation reanalysis dataset from the European Centre for Medium-Range Weather Forecasts (ECMWF) has an hourly temporal resolution and a geographical resolution of 31 km^2. This collection contains records dating back to 1950 [8].

The solar radiation atlas aims to create modern maps and tables on the various components of solar radiation falling on various regions of the country and to choose the most suitable areas for establishing solar energy plants, which contributes to making the most of the solar radiation falling on Egypt.

2 Data Cllections and Methodology

2.1 Solar Radiation Compounds [17]

- DNI is an acronym for normal irradiance, which represents the quantity of light that strikes the surface perpendicularly.
- GHI is an acronym for global horizontal irradiance, which measures the total quantity of shortwave radiation received from above by a surface that is horizontal (parallel) to the ground.
- Diffused Horizontal Irradiance (DHI) = Global Horizontal Irradiance (GHI) - Direct Normal Irradiance (DNI)x [cos (solar zenith angle)]
- GTI is an abbreviation for Global slanted Irradiance, which is irradiation that falls on a slanted surface. A tilted surface, like a horizontal surface parallel to the ground, receives a small quantity of ground-reflected radiation (REF). GTI is an approximation for calculating the energy yield of fixed installed tilting as the PV panels.
- Go is The intensity of solar irradiation directly outside the earth's atmosphere on a horizontal surface is almost constant as the average of $1,366 \pm 0.066$ W/m^2 (Solar Constant).
- Kt (GHI/Go) is the clearness index is a measure of the clearness of the atmosphere. It is the fraction of the solar radiation that is transmitted through the atmosphere to strike the surface of the Earth. It is a dimensionless number between 0 and 1, defined as the surface radiation divided by the extraterrestrial radiation
- Kd (DIF/GHI) is the ratio of diffuse solar irradiation (DIF) to total (global) irradiation on the terrestrial surface (GHI) is called cloudiness index or diffuse fraction (Kd).

2.2 Area of Study

Egypt's climate is affected by several factors, the most important of which are geographical location, surface features, the general system of pressure, air depressions, and bodies of water. All of this helped to divide Egypt into several distinct climatic regions. Egypt is located in the dry tropical region, except for the northern edges, which fall within the warm temperate region, which enjoys a climate similar to the Mediterranean climate region, which is characterized by heat and dryness in the summer months and mildness in the winter with rainfall [19]. A few grow on the coast Topography and geography of Egypt which is located in the northeastern corner of Africa, from latitude of 22 to 31 °N and longitude of 24 to 36 °E. It is bordered to the east by the red sea, to the west by Libya, to the north by the mediterranean sea, and to the south by Sudan. The total land area is 997,688 Km^2 and includes five main geographical regions: the Nile valley (Upper and Lower Egypt), the Nile delta, the eastern desert, sinai, and the western desert [20]. In addition, it is considered ideal because of its stable climatic rates in terms of temperature, humidity, and wind speed, which have a direct impact on the intensity of solar radiation [21, 22]. Total horizontal solar radiation increases in Egypt from north to south, and its value ranged 5–7 $KWh/m^2/d$, the north coast and more than 7 $KWh/m^2/d$ in the far south of Egypt, while the average number of days of sunshine reaches about 300 days annually. These values are among the highest in the world, as the annual average number of hours of sunshine per day ranges from 9 h in the north. to 11 h in the south, which means increased investment opportunities in the field of various solar energy applications in Egypt [23]. A combination of solar radiation data from the institute's meteorological stations, and satellites is used [24–26].

2.3 Data Collection and Processing

Database; the findings aim to address needs of policymakers; validation of solar radiation atlas and solar resource database, contained will depend on integrating available data, whether ground-based observations, satellites, or simulation programs for development and improvement in the 2023 atlas. The data was verified through verification and calibration and was published in peer-reviewed scientific journals [27–36].

2.3.1- Ground measured data are made to compare with the data of the National Institute for Astronomical and Geophysical Research (NRIAG), which is the reliable source for daily calibration and monitoring on a regular basis (every minute).

2.3.2- Modeling is done using specialized programs such as (Golden software Grapher 12) - Surfer 10 - Excel 2016.

2.3.3- The necessary curves are drawn as stated in the published and mentioned research and the research that has been published in the Atlas of Solar Radiation for Egypt. More than one source of solar radiation data was used from satellites and others, such as SODA HelioClim (Database of Daily Solar Irradiance v4.0 (derived from satellite data) [37].

2.3.4. ECMWF(European center for medium range weather forecast International data sites were used to issue many global atlases in the field of solar radiation, including: This data was compared with the ground stations of NRIAG at Helwan, Suez, and Matrouh stations, and also compared with previous work from previously published atlases until we reach the highest correlation coefficient and the lowest error in mapping. The program is developed through the ERA-5 system to develop data until we reach a correlation coefficient of no less than 95%. ERA-5 Land; It is a reanalysis dataset that provides a consistent view of the evolution of land variables over several decades with improved comparative resolution. And also, produced by rerunning the land component of the ECMWF ERA5 climate analysis. The reanalysis combines model data with observations from around the world into a complete and globally consistent data set using the laws of physics. Atlas maps: Grads (Grid analysis and display system) was used to draw maps of solar radiation [38, 39].

3 Results and Disscusion

Solar radiation measurement stations are chosen to cover the various regions of climate change in Egypt [19, 20]. Air temperature distributions are one of the inputs to solar radiation simulation models. Table 1 shows the annual and monthly averages of the air temperature (°C) (1973–2020) for the regions that have solar radiation stations, noting that the air temperature does not change except by a small percentage, not exceeding one degree, during the last four centuries. It is clear from this table that the maximum temperature value of 33.56°C was in Kharga and the lowest value of 23.95°C was in Matrouh, and this indicates relative climatic stability in temperature changes over Egypt. Note that the temperature mentioned is the average during the entire day (over 24 h). Relative humidity (R.H, %) is considered one of the most important influences on solar radiation falling on the ground and is one of the inputs in simulation programs. The monthly and annual average number of hours of sunshine in the north is 6–11 h/day, while in southern Egypt it rises to 9–12.4 h/day. Wind speed (m/s) is considered an important influence because of its connection to the components of the atmosphere, such as dust and water vapor, on solar radiation falling on the Earth, and it is one of the inputs in simulation programs [19, 20].

3.1 Annual Distribution of Total Horizontal Solar Radiation (GHI) Over Egypt 1980–2020.

Figure 1 shows yearly global solar radiation horizontal (GHI) over Matruh 1980–2020, For example as one of 17 station under test; this table show the yearly distribution of monthly averages, which range from the highest values of ranged 7–8 KWh/m^2/d at south and the lowest values of 4.5–5.5 KWh/m^2/d at the north. The available values show that the highest monthly average of total horizontal radiation (GHI) over Egypt is in July, as the average amount of total solar radiation is about 7.9 KWh/m^2/d, and this month represents the highest average in all parts of Egypt during the whole year. December represents the lowest month of the year in the amount of total horizontal solar radiation falling on Egypt, as the average amount of radiation during this month is 3.22 KWh/m^2/d.

Table 2 shows the highest average values of total horizontal solar radiation (GHI) over Egypt are in Aswan with a value of 6.34 KWh/m^2/d, followed by Asyut with 6.18 KWh/m^2/d. We conclude from this, the highest values of solar radiation in Egypt are those located between latitude 24 -27° N [6].

Figure 2 shows, yearly values of the global horizontal solar radiation (GHI) over Egypt dependent on the distribution of latitute and longitude map, with aveage 6.18 ± 1.5 KWh/m^2/d. and also shows the most suitable places for establishing new and renewable energy projects in Egypt.

Figure 3 and Table 3: shows the global solar radiation tilted by the latitude place, GTI, SD (KWh/m^2/d), the annual average is 6.4 ± 0.82 (KWh/m^2/d) in the period 1980–2020, and the highest values are located in the southeast of Egypt from 23–25 °N and

32–34°E. It is one of the best places of all, which has the highest values of GTI are 6.8 ± 0.4 and 6.7 ± 0.5 (KWh/m²/d) in Aswan and Asyut respectively.

Table 1. Annual averages of Temp., RH, WS, and Sun shine duration stations in Egypt in the period from 1973–2020 [19, 20].

Station	Sunshine (h)	WS (m/s) س(m/9)	RH (%)	Temp (°C)
Matruh	8.70	4.60	70.30	24.00
Alex	8.70	4.00	68.50	24.70
Damietta	8.80	2.30	74.00	24.40
Port Said	8.90	5.90	69.90	24.00
Arish	8.90	2.30	70.80	24.30
Tanta	9.20	2.40	67.70	26.90
Cairo	9.20	2.20	67.20	28.50
Menia	9.90	2.90	53.00	29.70
Asyut	10.10	4.10	49.10	28.00
Aswan	10.50	4.70	26.40	33.50
Fayum	9.70	2.70	57.70	29.70
Bahria	10.00	2.30	43.90	29.40
Kharga	10.40	4.10	38.20	33.60
Hurghada	10.00	5.50	49.60	28.10
Helwan	9.70	3.90	54.40	29.10
Suez	9.80	3.60	49.70	28.50
Owainat	10.40	4.60	30.10	31.30

3.2 The Best Suitable Places for Establishing Solar Energy Projects

Choosing the best suitable places for establishing solar energy projects, taking into account the elevation above sea level, seismic activity, and geological surface formations.

Figure 4 shows the level of elevation above sea level (m), and Fig. 5 shows the distribution of seismic activity over Egypt. In addition to the information contained in from Figs. 4 and 5 and Table 3; it appears that the average values of global solar radiation on the plane inclined by the latitude place is GTI = 6.4 ± 0.82 KWh/m²/d.

Figure 6 shows the best location in which all the features are available, between latitude 22–26 N° and longitude 28–30 E°. The best use of the GTI system is which gives a greater value of intenisty and less in standard deviation than the average values of GHI especilly in the case of photovoltaic systems and thermal systems as the solar heaters (Fig. 6).

Table 2. Monthly of GHI (Average, Max., Min., and SD) (KWh/m^2/d) at Egypt 1980–2020.

Months	Average	Max	Min	SD
Jan.	3.75	4.6	2.9	0.7
Feb.	5.25	6	4.5	0.7
Mar.	6.1	7.1	5.1	0.6
Apr.	7.1	7.7	6.5	0.4
May.	7.95	8.4	7.5	0.2
Jun.	8.1	8.4	7.8	0.2
Jul.	8.25	8.8	7.7	0.2
Aug.	**8.00**	**8.6**	**7.4**	**0.2**
Sep.	6.95	7.1	6.8	0.3
Oct.	5.5	6.4	4.6	0.5
Nov.	4.25	5.1	3.4	0.2
Dec.	3.7	4.8	2.6	0.7

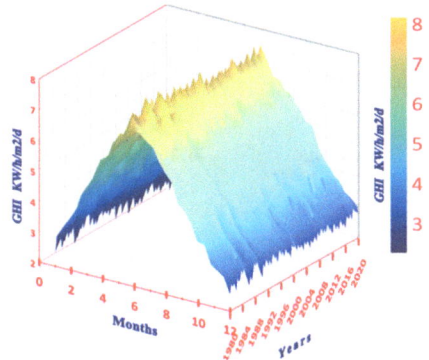

Fig. 1. Yealy global solar horizontal radiation (GHI) over Matruh 1980–2020.

Fig. 2. Average distribution of global solar radiation horizontal (GHI) over Egypt 1980–2020.

4 Annual Distribution of Direct Solar Radiation (DNI) Over Egypt

Table 4 and Fig. 7; shows the monthly and annual mean variations of DNI which is 7.5 ± 1.00 (KWh/m^2/d) over Egypt 1980–2020.

5 Annual Distribution of Diffuse Solar Radiation (DIF) Using Latitude and Longitude Coordinates Over Egypt 1980–2020

Scattered solar radiation originates primarily in the Earth's atmosphere as a result of the collision of direct solar radiation with the layers of the atmosphere components, which is a mixture of gases surrounding the Earth and attracted to it by the Earth's gravity. Natural

Table 3. Global solar radiation tilted by the latitude place GTI, SD (Kwh/m2/d).

GTI: Global Tilted Solar Radiation (kWh/m²/d)	Station	Average	SD
	Matruh	5.9	1.0
	Alex	6.0	1.1
	Damietta	6.2	1.0
	Port Said	6.4	0.9
	Arish	6.2	0.9
	Tanta	6.2	0.9
	Cairo	6.4	0.8
	Menia	**6.5**	**1.0**
	Asyut	6.7	0.5
	Aswan	6.8	0.4
	Fayum	6.5	0.8
	Bahria Oasis	6.5	0.8
	Kharga	6.6	0.7
	Hurghada	6.6	0.5
	Helwan	6.4	1.0
	Suez	6.3	0.9
	Owienat	6.6	0.7
	Average	**6.4**	**0.82**

pollutants (dust, sand, volcanic output…) and industrial pollutants such as fumes and gases from human activity waste such as factory outputs, car exhausts, etc.

Table 5 shows that the annual average of diffuse solar radiation for all sites DIF is 1.95 $\pm$ 0.60 KWh/m²/d and the highest diffuse values in all sites are in June and the lowest are in January. DIF is considered the one of strongest indicator of the good transparency of the atmosphere and its almost of freedom from air pollutants and suspended particles. Increasing atmospheric turbidity leads to an increase in diffuse values, and these values in general give an ideal picture of Egypt's location in terms of climatic stability, which directly affects solar radiation values at the general level.

Figure 8 shows that the annual average level of dispersion, diffuse fraction, Kd (DIF/GHI) depends on the ratio of dispersed solar radiation to total solar radiation. The results showed that the dispersion level values ranged between the highest value of 45% and the lowest value of 32%.

Table 5 and Fig. 8: shows the diffuse solar radiation DIF (KWh/m²/d) with longitude and latitude coordinates during the period 1980–2020. The difference between the highest and lowest values of scattered solar radiation DIF is 0.27 KWh/m²/d. Which means that the amount of natural and industrial pollutants is the main factor in increasing the values of diffuse solar radiation, which is relatively small in all cities of Egypt, which

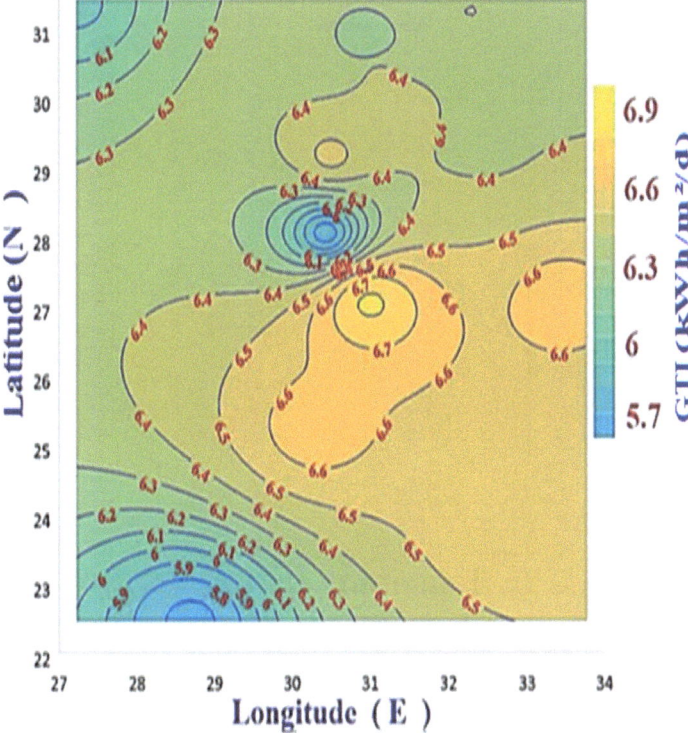

Fig. 3. Distribution of annual average values of GTI (Kwh/m2/d).

means relative stability to create appropriate conditions for establishing solar energy projects.

6   Clearness index Kt (GHI/Go)

Table 6: represents the annual average of the clearness index Kt for the 17 most important sites in Egypt, well distributed covering all climates by 0.575 ± 0.0485 which is considered one of the highest global rates, especially the western desert region, specifically the Aswan, Bahria, and Kharga regions (0.66), which are considered among the most stable regions in terms of receiving the amount of solar radiation over Egypt. It is clear from Table 6, the annual average of Kd (DIF/GHI) by latitude and longitude places is 0.341 ± 0.02, which is considered one of the most stable in the atmosphere and free from natural and industrial pollutants in terms of its effect on the receipt of solar radiation in Egypt.

6.1   Distribution of Ultraviolet Solar Radiation (UV) Over Egypt 1980–2020

Table 7 and Fig. 9, shows monthly mean values of UV (250-450nm) at Egypt 1980–2020 is 0.45 ± 0.05 KWh/m^2/d. It is lowest possible in the winter months and highest possible in the summer months.

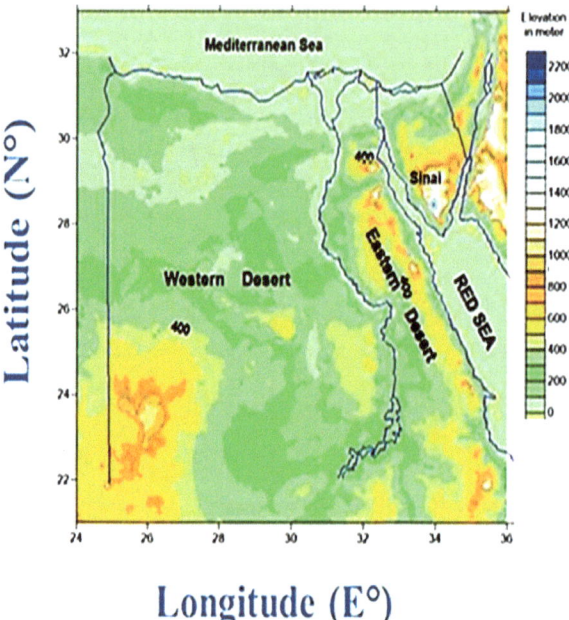

Fig. 4. Shows the height above sea level in Egypt

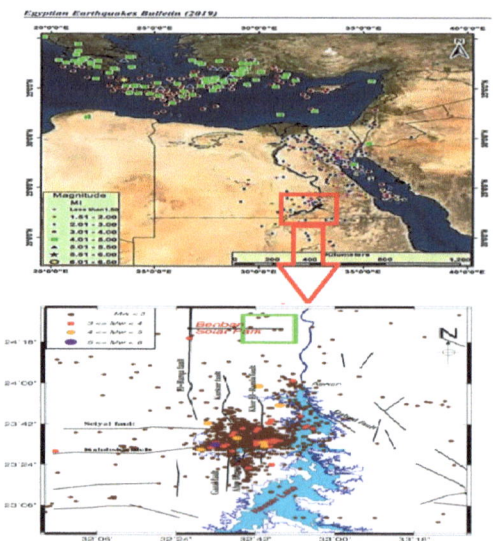

Fig. 5. Distribution of seismic activity over Egypt.

Fig. 6. The best proposed location for establishing solar energy projects in Egypt

Table 4. Monthly mean values of DNI (KWh/m^2/d) at Egypt 1980–2020.

Months	Average	Max	Min	SD
Jan	6.0	7.9	4	0.97
Feb	6.7	8.8	4.5	1.07
Mar	6.9	8.8	5	0.94
Apr	7.8	9.8	5.7	1.03
May	8.4	10.3	6.5	0.95
Jun	8.8	10.1	7.5	0.65
Jul	8.6	9.9	7.3	0.65
Aug	8.7	9.6	7.7	0.65
Sep	8.0	9.2	6.8	0.6
Oct	7.1	9.1	5	0.9
Nov	6.7	8.9	4.5	1.03
Dec	6.3	8.4	4.2	1.02
mean	7.5	9.23	5.73	1.00

6.2 Measurements of Ultraviolet Solar Radiation Index (UVI) Over Egypt 2011 – 2019

Figure 10, Shows the monthly averages of solar radiation values of ultraviolet rays (UVI) for Helwan station during the period from 2011 to 2019. It is observed from this figure that the highest values of ultraviolet rays are during the summer season and range between 7–9 and the lowest values during the winter season and range between 2–3.

Table 5. Annual average values of DIF (KWh/m²/d) at Egypt 1980–2020.

Station	Average	SD
Matruh	1.86	0.63
Alex	1.78	0.61
Damietta	1.83	0.57
Port Said	1.82	0.64
Arish	1.96	0.62
Tanta	1.84	0.56
Cairo	1.85	0.59
Menia	1.93	0.63
Asyut	2.19	0.61
Aswan	2.24	0.57
Fayum	1.89	0.55
Bahria	2.15	0.59
Kharga	2.14	0.62
Hurghada	1.82	0.64
Helwan	1.84	0.63
Suez	1.81	0.58
Owainat	2.26	0.61
Mean	1.954	0.603

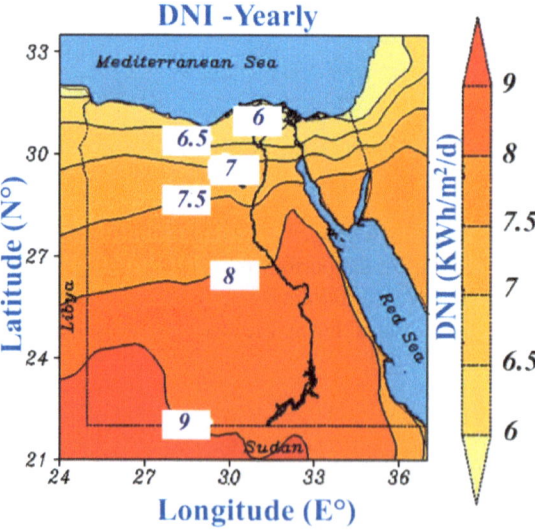

Fig. 7. Distribution of annual values for DNI (KWh/m²/d) over Egypt 1980–2020.

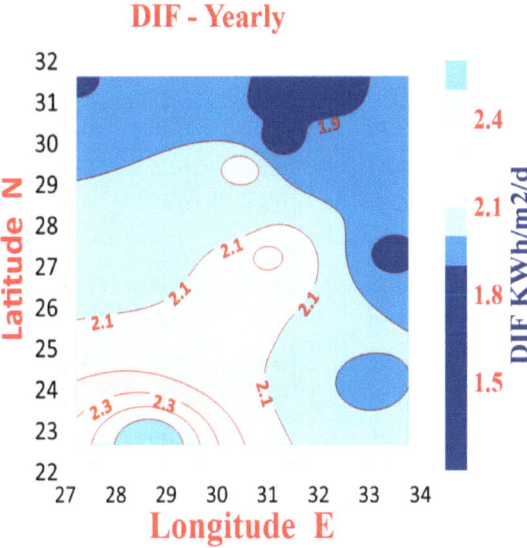

Fig. 8. Distributing of annual values of DIF (KWh/m^2/d) by the latitude and longitude degree over Egypt 1980–2020.

Table 6. Average annual distribution of Kd and Kt for 17 most important sites at Egypt 1980–2020

Station	Kd	Kt
Matruh	0.344	0.545
Alex	0.324	0.546
Dommiat	0.331	0.561
Por Said	0.326	0.543
Arish	0.373	0.509
Tanta	0.330	0.529
Cairo	0.344	0.573
Menia	0.317	0.596
Asyut	0.357	0.633
Aswan	0.353	0.670
Fayum	0.339	0.571
Bahria Oasis	0.347	0.629
Kharga	0.346	0.664
Hurghada	0.307	0.517
Helwan	0.341	0.543
Suez	0.326	0.568
Owienat	0.386	0.581
Minimum	0.307	0.509
Maximum	0.386	0.670
Range	0.079	0.161
Mean	0.341	0.575
S.D	0.020	0.048

Table 7. Average monthly distribution of the UV KWh/m^2/d at Egypt 1980–2020

Month	Average	Max	Min	SD
Jan	0.30	0.43	0.16	0.08
Feb	0.41	0.52	0.25	0.07
Mar	0.54	0.63	0.42	0.06
Apr	0.53	0.59	0.42	0.05
May	0.55	0.62	0.48	0.03
Jun	0.59	0.65	0.48	0.05
Jul	0.55	0.6	0.48	0.03
Aug	0.55	0.61	0.46	0.04
Sep	0.48	0.55	0.39	0.04
Oct	0.34	0.44	0.21	0.01
Nov	0.33	0.42	0.28	0.08
Dec	0.27	0.39	0.21	0.08
Mean	0.453	0.537	0.353	0.052

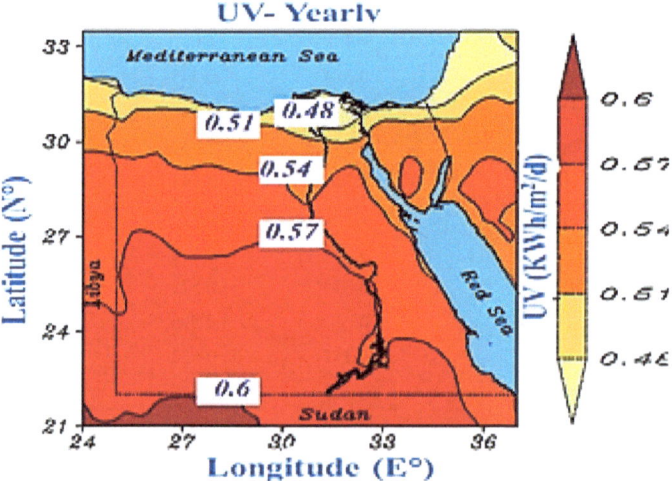

Fig. 9. Average annual distribution of the UV (KWh/m^2/d) over Egypt 1980–2020

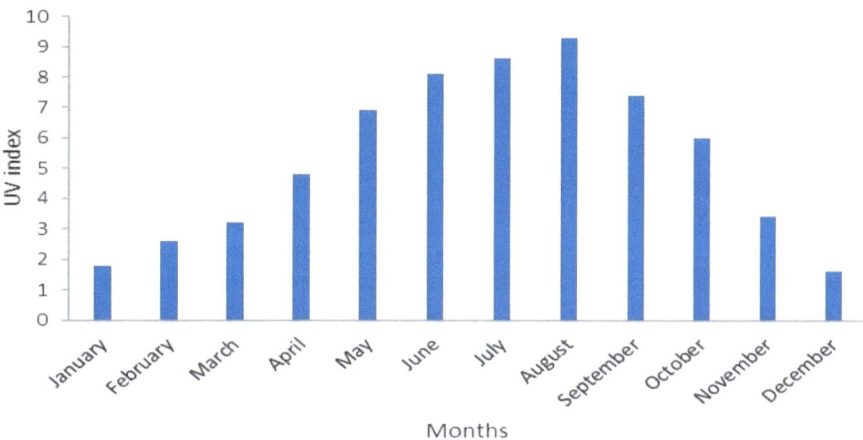

Fig. 10. Monthly mean values of UVI for the main radiation center at Helwan 2011–2019.

7   Conclusion

Egypt is located in the northeastern corner of Africa, from latitude 22 to 31° N and longitude 24 to 36° E. The data used in this atlas (about 40 years, mostly in the period 1980–2020), is a mixture of ground data and relevant satellite measurements, and the most widely used model as EARA5, which mainly includes and depedent on the most important climate variables: air temperatures (T,°C), relative climaty (R.H, %), wind speed (m/s), and sunshine durations (h/d).

The most important results extracted from this work:

- The annual difference in average air temperatures between the highest and lowest values is within ten degrees Celsius over Egypt (ranged 24 h), where the maximum temperature value is 33.56 °C at Kharga, and the lowest value is 23.95 °C in Matrouh. This indicates relative climate stability. The monthly and annual average of relative humidity (R.H, %) shows that the maximum value is 74% in Damietta and the lowest value is 26% in Aswan. The average wind speed (m/s) indicates that its maximum is 5.5 m/s in Hurghada, 5.9 (m/s) in Port Said, and its lowest in Cairo is 2.2 m/s. Therefore, areas with low wind speeds are suitable for setting up solar energy systems (solar cells, etc.), and areas with high wind speeds are suitable for establishing stations to generate electricity from wind energy. The monthly and annual average sunshine durations (hour/day) in the north is 6–11 h/d, while in southern Egypt it rises to 9–12.4 h/d.
- The mean values of global horizontal solar radiation (GHI) is 6.18 ± 1.5 KWh/m^2/d, while the global solar radiation on the inclined by the latitude angle of the place (GTI) is 6.4 ± 0.82 KWh/m^2/d, which represents one of the most important solar radiation values that can be used in practical applications for solar energy uses, which gives its highest value at Aswan, 6.8 ± 0.4 KWh/m^2/d. The incidence of solar radiation at a fixed and inclined angle to the horizon with the latitude of the place gives an increase in the received solar energy and a decrease in the standard deviation from the horizontal.

- The mean values of diffuse solar radiation (DIF) is 1.95 ± 0.6 KWh/m^2/d, mean that the amount of natural and industrial pollutants is the main factor in increasing the values of diffuse solar radiation, which are relatively few in all cities of Egypt, which means relative stability in order to create the appropriate conditions for establishing solar energy projects.
- The annual average of the clearness index Kt (GHI/Go) for the 17 most important sites in Egypt is 0.575 ± 0.0485, while the annual average of Kd (DIF/GHI) by latitude and longitude coordinate is 0.341 ± 0.02. The western desert region, specifically; Aswan, Bahria, and Kharga regions (Kt $= 0.66$), are considered among the most stable regions in terms of receiving the amount of solar radiation over Egypt.
- The average values of direct solar radiation (DNI) is 7.5 ± 1.0 KWh/m^2/d.
- The annual averages of ultraviolet solar radiation (UV) is 0.6 ± 0.11 KWh/m^2/d.
- The ultraviolet solar radiation index (UVI) for Helwan station during the period from 2011 to 2019 gives the highest values in summer and ranged of 7–9 and the lowest values during the winter season and ranged of 2–3.
- IRA-5 is considered a good model for predicting solar radiation values in Egypt, and this is clearly evident in the very small difference between IRA-5 values and ground measurements (especially in open areas away from cities).
- One of the basic outcomes of updating this atlas is to determine the most suitable flat area for longitude and latitude and the highest intensity of solar radiation suitable for establishing new and renewable energy projects and their applications in Egypt, especially solar cell farms, which are located between latitude 22–26 °N and longitude 28–30 °E, after entering the most appropriate height above sea level within the range of 500 m and flat, open areas with no seismic activity.

Acknowledgments. This atlas is the result of a series of great efforts exerted over an extended period of time, beginning in 1981, which was initiated by Prof. Dr. Muslim Shaltut, until I reached this atlas in 2023, and those who did this work are from the same school of this great science in solar radiation in our Institute (NRIAG).

Appendex

The Validation of Data:
(i) Solar radiation data; validation and quality control

Validation and quality control of radiological data are as important as the measurement procedures themselves, because data can only be guaranteed after data accuracy and completeness are guaranteed. Various methods can be identified to verify the quality of measured solar radiation data, including any quality assurance monitoring of the data archive (accuracy, completeness, etc.). Controls that address data validation (evaluating data accuracy by identifying "measurement errors," equipment malfunctions and inhomogeneities.

Where n is the number of data, $K_{measured}$ (K_{meas}) is the observed data from ground stations, K_{Era5} is the ERA-5 data, x expresses the variable values, and x_a is the average of the total value over the period. Table 8 shown, Statistical evaluation of monthly mean GHI (KWh/m^2) from Era5 compared to the measured data. Table 9, shown The descriptive statistical for solar radiation (KWh/m^2/d) from Era5 and the measured data as a selected five stations. From Tables 8 and 9, it is clear that the region with the best homogeneity in the data is the Aswan and Kharga regions, and this appears through the very small difference between the mean and the median values, as well as the lowest values of different type of error as MBE, RMSE and MAPE. ERA-5 is considered a very good model for predicting solar radiation values in Egypt, and this is clearly evident in the very small difference between ERA-5 values and ground measurements (especially in open desert areas such as Aswan and Kharga). While it is less efficient in areas with high population density, such as Cairo.

(ii) Sample data is available from the solargis database

Model solar and meteorological data are available for all regions of the Earth between 60°N and 45°S. detailed information is available in the technical report. Below is a summary of the data that can be downloaded from the atlas. Where:

$$MBE = \frac{1}{n} * \sum_{i=1}^{n}(k_{Meas} - k_{Era5})$$

$$RMSE = \sqrt{\sum_{i=1}^{n}(k_{Meas} - k_{Era5})^2 * \frac{1}{n}}$$

$$MAPE = \frac{\sum_{i=1}^{n}(|k_{Meas} - k_{Era5}|)/k_{Meas}}{n}$$

$$SD = \frac{\sum_{i=1}^{n}(x - x_a)}{n}$$

$$R^2 = 1 - \frac{\sum_{i=1}^{n}(k_{Meas} - k_{Era5})^2}{\sum_{i=1}^{n}(k_{Meas} - \overline{k_{Meas}})^2}$$

Table 8. Statistical evaluation of monthly mean GHI (KWh/m^2/d) from Era5 compared to the measured data.

Stations	R^2	MBE	RMSE	MAPE
Cairo	0.988	0.77	0.88	0.16
Asyut	0.992	0.44	0.66	0.08
Aswan	0.958	−0.02	0.13	−0.001
Kharga	0.988	0.23	0.48	0.04

Table 9. The descriptive statistical for GHI ($KWh/m^2/d$) from Era5 and the measured data at the available five stations

		Asyut	Aswan	Cairo	Kharga
Max	measured	7.9	8.0	7.5	8.0
	Era5	8.3	7.9	8.1	8.1
Min	measured	3.6	4.4	2.9	4.1
	Era5	4.0	4.4	3.3	4.4
Range	measured	4.3	3.6	4.6	3.9
	Era5	4.3	3.5	4.8	3.7
Mean	measured	6.0	6.5	5.2	6.3
	Era5	6.4	6.5	6.0	6.5
Median	measured	6.2	6.7	5.4	6.5
	Era5	6.7	6.8	6.2	6.8
SD	measured	1.6	1.2	1.7	1.4
	Era5	1.5	1.2	1.8	1.3

References

1. Sengupta, M., Habte, A., Gueymard, C., Wilbert, S., Renné, D., Stoffel, T.: Best practices handbook for the collection and use of solar resource data for solar energy applications (2017). https://www.nrel.gov/publications
2. Sianturi, Y., Marjuki and Sartika, K.: Evaluation of ERA5 and MERRA2 Reanalyses to Estimate Solar Irradiance Using Ground Observations over Indonesia Region (2020). https://doi.org/10.1063/5.0000854
3. Kimball, H.: Measurement of solar radiation intensity. Mon. Weather Rev. **55**, 55 (1927). https://doi.org/10.1175/1520-0493(1927)55%3c155:MOSRIA%3e2.0.CO;2
4. Abbott, C.G.: Application of solar radiation measurements, Annals of the Astrophysical Laboratory of the Smithsonian Institution, IV. Smithsonian Inst., Washington, D.C. Chap. 6 (1922). https://www.jstor.org/action/doBasicSearch?Query=Abbot
5. Solar Energy Atlas of Nepal Based on regional adaptation of Solargis model Solargis reference No. 170–17 (2021). http://efaidnbmnnnibpcajpcglclefindmkaj/ https://energydata.info/dataset/ead8d2fa-de59-48d3-8693-817b36b3a470/resource/43740f99-4075-405b-861f-038e47097802/download/solarenergyatlas_nepal_esmap-wbg.pdf
6. Scharmer, K., Greif, J.: The European Solar Radiation Atlas, Les Presses de l'École des Mines, Paris (2000). https://doi.org/10.1007/978-3-642-80237-9
7. Global Solar Atlas 2.0 Interactive web site, solar potential data and maps Reference No. Date: 170–14/2019 , 29 November 2019, world data bank. https://globalsolaratlas.info
8. Hoyer-Klick, C., et al.: Solar Atlas for the Mediterranean, Conference: Solar Paces (2011). https://www.researchgate.net/publication/225025694
9. FrenchPI, Mossad, Surface Solar Irradiance in Egypt for energy production) project (#5404) co-funded by the Science and Technology Development Fund (STDF) of Egypt and the French AIRD, and of the MISRA (MIsr Solar Radiation Atlas) project funded by the French Ministry of Foreign Affairs and the UAE government, 2017. https://www.academia.edu/34012002/Egyptian_Atlas_for_Surface_Solar_Irradiation_Egyptian_Atlas_for_Surface_Solar_Irradiation

10. Mosalam Shatout, M. A "solar energy input to Egypt as a special publication of NRIAG, Egypt". Egyptian solar radiation atlas was, 1981. https://www.researchgate.net/profile/Mos alam-Shaltout-2

11. Mosalam Shaltout, M.A., and Hassen, A.H., "Solar energy distribution over Egypt using cloudiness from meteosat photo" Solar energy, Vol. 45, no.6, pp.345–351, (1990). https://www.sciencedirect.com/science/article/pii/0038092X90901556

12. Mosalam Shatout, M. A., "Atlas of Solar Radiation for the Arab Republic of Egypt", New and Renewable Energy Authority, US Agency for International Development, Professor of Solar Radiation at the National Institute for Astronomical and Geophysical Research, 1991 AD. https://www.worldcat.org/title/egyptian-solar-radiation-atlas/oclc/25011751

13. Khalil, S.A.: Prediction of the diffuse solar energy on horizontal at different selected locations, Energy Power Eng. **14**, 635–651 (2022). https://doi.org/10.4236/epe.2022.1410034

14. Khalil, S.A.: A comprehensive study of solar energy components by using various models on horizontal and inclined surfaces for different climate zones. Energy Power Eng. **14**, 558–593 (2022). https://doi.org/10.4236/epe.2022.1410031

15. Khalil, S.A., Rahoma, U.A.: Verification of solar energy measurements by(ERA-5) and its impact on electricity costs in North Africa. Int. J. Astronomy Astrophys. **12**, 301–327 (2022). https://www.scirp.org/journal/ijaaISSN Online: 2161–4725, ISSN Print: 2161–4717

16. Khalil, S.A.: Performance evaluation and statistical analysis of solar energy modeling: a review and case study. J. Nigerian Soc. Phys. Sci. **4**, 911 (2022). https://doi.org/10.46481/jnsps.2022.911

17. Iqbal, M.: An Introduction to Solar Radiation. Academic Pricss (1983). https://www.scienc edirect.com/book/9780123737502/an-introduction-to-solar-radiation

18. Karttunen, H., Kröger, P., Oja, H., Poutanen, M., Donner, K.J. (eds.): Fundamental Astron-omy. Springer (2006). http://efaidnbmnnnibpcajpcglclefindmkaj/ https://observatory.znu.ac.ir/files/uploaded/editor_files/observatory/files/Fundamental%2BAstronomy%2B5th%2BEdition.pdf

19. Musad Salama Musad Mandour - Solar radiation in Egypt - Arab Republic of Egypt - Man-soura University - Faculty of Arts - Department of Geography - 2002 AD. https://www.noor-book.com/-pdf

20. Abd-Wahhab, M.M.: Climate and its impact on standard evapotranspiration in Egypt - Beni Suef University (2015). https://www.maktabtk.com/blog/post/.html

21. Mosalam Shaltout, M.A., Hassan, A.H., Hosary, T.N.: The estimation of moisture over egypt from meteosat satellite observations. Adv. Spac Res. **18**(7), 741–746 (1996). https://www.sci encedirect.com/science/article/pii/0273117795002898

22. Mosalam Shaltout, M.A., Hassan, A.H., Fathy, A.M.: Studying the ultraviolet and visible solar radiation over Cairo and Aswan and their correlation with climatological parameters "; Proc. Indian Acod. Sci. (Chem.Sci.) **110**(3), 361–371 (1998). https://doi.org/10.1007/BF0 2870014

23. Rahoma, W.A., Rahoma U.A., Hassan, A.H.: Application of neuro-fuzzy techniques for solar radiation. J. Comput. Sci. **7**(10), 1605–1611 (2011). https://www.researchgate.net/pub lication/261672193_Application_of_Neuro-Fuzzy_Techniques_for_Solar_Radiation

24. Rahoma, U.A., Hassan, A.H.: Fourier transforms investigation of global solar radiation at true noon: in the desert climatology. Am. J. Appl. Sci. **4**(11), 902–907 (2007). https://thesci pub.com/abstract/ajassp.2007.902.907

25. Rahoma, U.A.: Utilization of solar radiation in high energy intensive of the world by PV system. Am. J. Environ. Sci. **4**(2), 121–128 (2008). https://thescipub.com/abstract/ajessp.2008.121.128

26. Rahoma, U.A.: Clearness index estimation for spectral composition of direct and global solar radiation. Appl. Energy **68**(4), 337–346 (2001). https://www.sciencedirect.com/science/art icle/pii/S0306261900000519

27. Khalil, S.A.: The effect of atmospheric transmittance of UVB solar irradiance to broadband solar radiation at different climate sites. Int. J. Appl. Phys. **9**(3), 1–18 (2022). https://www.internationaljournalssrg.org/IJAP/paper-details?Id=177

28. Khalil, S.A., Ali Rahoma, U.: Verification of solar energy measurements by (ERA-5) and its impact on electricity costs in North Africa. Int. J. Astron. Astrophys. **12**, 301–327 (2022). https://www.scirp.org/journal/paperinformation.aspx?paperid=120952

29. Khalil, S.A.: Evaluation of models for prediction of monthly mean hourly sky-diffuse solar radiation on tilted surface, Cairo, Egypt. Int. J. Pure Appl. Phys. (1), 97–107 (2010). ISSN 0973-1776. https://www.researchgate.net/publication/285879487_Evaluation_of_models_for_prediction_of_monthly_mean_hourly_sky-diffuse_solar_radiation_on_tilted_surface_Cairo_Egypt

30. Khalil, S.A., Shaffie, A.M.: Prediction of clear-sky biologically effective erythematic radiation (EER) from global solar radiation (250–2800 nm) at Cairo, Egypt. Adv. Space Res. **51**, 1727–1733 (2013). https://www.sciencedirect.com/science/article/pii/S0273117712007405

31. Khalil, S.A., Shaffie, A.M.: Performance of statistical comparison models of solar energy on horizontal and inclined surface. Int. J. Energy Power (IJEP) **2**(1) (2013). https://www.researchgate.net/publication/309174129_Performance_of_statistical_comparison_models_of_solar_energy_on_horizontal_and_inclined_surface

32. Khalil, S.A., Shaffie, A.M.: A comparative study of total, direct and diffuse solar irradiance by using different models on horizontal and inclined surfaces for Cairo, Egypt. Renew. Sustain. Energy Rev. **27**, 853–863(2013). https://www.sciencedirect.com/science/article/pii/S1364032113004152

33. Khalil, S.A., Shaffie, A.M.: Evaluation of transposition models of solar irradiance over Egypt. Renew. Sustain. Energy, Rev. **66**, 105–119(2016). https://www.sciencedirect.com/science/article/pii/S1364032116302921

34. Khalil, S.A., Khamees, A.S., Morsy, M., Hassan, A.H., Ali Rahoma, U., Sayad, T.: Evaluation of global solar radiation estimated from ECMWF-ERA5 and validation with measured data over Egypt, Turkish J. Comput. Math. Educ. **12**(6), 3996–401 (2021). https://turcomat.org/index.php/turkbilmat/article/view/8371

35. Khalil, S.A., Ali Rahoma, U., Hassan, A.H., Shaffie, A.M.: Prediction of doses biologically active ultraviolet solar radiation from measurements of global solar radiation. Int. J. Curr. Res. vol. **12**(05), 11493–11501 (2020). http://efaidnbmnnnibpcajpcglclefindmkaj/ https://www.arcjournals.org/pdfs/ijres/v6-i2/4.pdf

36. Khalil, S.A., Khamees, A.S., Morsy, M., Hassan, A.H., Ali Rahoma, U., Sayad, T.: Evaluation of global solar radiation estimated from ECMWF-ERA5 and validation with measured data over Egypt. Turkish J. Comput. Math. Educ. **12**(6), 3996–4012 (2021). https://turcomat.org/index.php/turkbilmat/article/view/8371

37. https://www.soda-pro.com/help/helioclim/helioclim-3-outputs

38. https://www.ecmwf.int/

39. https://www.ecmwf.int/en/forecasts/dataset/ecmwf-reanalysis-v5

Jurisprudential Reliance on Astronomical Calculations in Determining the Beginnings of the Hijri month

Mohammed Gharaybeh[✉]

School of Shariah-Department of Jurisprudence, The University of Jordan, Amman, Jordan
d.m.garaibah@gmail.com

Abstract. Determining the start of the Hijri month through astronomical calculations not only merges Islamic tradition with modern science but also governs religious events for global Muslims. Traditionally, the commencement of each month was heralded by the sighting of the crescent moon, but the advent of technological advancements has ignited debates regarding the utilization of scientific calculations for lunar month initiation. These astronomical calculations entail intricate computations of the moon's phases and angles, employing algorithms to pinpoint the start of each lunar month, a proposition purportedly aligned with Islamic principles. Despite this, controversy ensues, with traditionalists underscoring the historical and spiritual significance of moon sighting, which has long fostered unity among believers. This ongoing debate reflects broader discussions on the integration of science and tradition within the Islamic world, posing a formidable challenge in striking a balance between religious customs and scientific progress. Nevertheless, continuous dialogues aim to forge a harmonious coexistence between religious scholars and scientists, seeking to reconcile faith and reason in determining significant Islamic calendar events. Consequently, this paper emphasizes the precedence of visual sighting over astronomical calculations in confirming the start of lunar months, stressing the imperative of combining astronomical calculations with visual confirmation. Moreover, it advocates for the permissible use of optical aids such as telescopes for observing the crescent moon's birth, treating sightings through such devices as equivalent to naked-eye observations.

Keywords: Astrology · Astronomy · Lunar month · New moon calculations

The initiation of each Hijri month hinges upon a fusion of Islamic tradition and modern scientific inquiry. Crucial to Muslims globally, the Hijri calendar serves as a linchpin in orchestrating pivotal religious occasions. Historically, the advent of each lunar cycle was heralded by the sighting of the crescent moon, embodying a timeless tradition. However, in the wake of technological advancements, contentious debates have arisen regarding the feasibility of employing scientific calculations to demarcate lunar month commencements. Astronomical computations entail intricate analyses of lunar phases and angles, championed by scientists as congruent with Islamic principles. Nonetheless,

---

M. Gharaybeh—Chief Justice Department.

H. M. K. Al Naimiy et al. (Eds.): AUASS-CONF 2023, SPPHY 420, pp. 160–177, 2025.
https://doi.org/10.1007/978-981-96-3276-3_13

traditionalists staunchly advocate for the enduring historical and spiritual significance of moon sighting, fostering unity among adherents. This dialogue encapsulates broader deliberations on the integration of science and tradition within the Islamic paradigm, navigating the intricate terrain of religious customs vis-à-vis scientific progress. Ongoing dialogues aspire to reconcile faith and rationality in adjudicating significant Islamic calendar events, fostering a harmonious coexistence between religious scholars and scientists.

Throughout the millennia, Muslims have predominantly relied on visual moon sighting, especially during Prophet Muhammad's era. When direct observation isn't feasible, they resort to a standardized thirty-day month. However, a contentious debate has arisen regarding the use of astronomical calculations by astrologers to determine lunar months. This has led to doctrinal discrepancies among adherents. Despite advancements in astronomy, discussions persist regarding the validity of such calculations, affecting fasting practices, Hajj months, and legal matters concerning divorced women [1].

In contemporary society, Muslims embrace knowledge and societal progress, integrating modern scientific methodologies into various aspects of life. Paradoxically, those who reject astronomical computations for lunar months rely on them for essential religious practices [2]. This illustrates the complexity of navigating tradition and scientific progress within Islamic principles. Sheikh Mohsen Al-Asfour highlighted advancements in moon sighting technology, advocating for their adoption to prevent disagreements among Muslims. He emphasized the need to embrace progress and lamented the reluctance to engage with modern methods, urging deeper exploration of their benefits.

Henceforth, the principal aim of this research is to establish a criterion harmonizing evidence drawn from both moon sighting and astronomical calculation. Employing a descriptive methodology throughout, I meticulously gathered, analyzed, and scrutinized data to attain the desired outcomes. The findings underscore the significance of integrating astronomical calculations aligned with observation, facilitated by technologies such as telescopes, ground photography, or satellites.

1   A Historical Glimpse

Across the annals of antiquity, there emerges a narrative of profound astronomical and mathematical achievements. Ancient civilizations, such as the Sumerians nearly six millennia ago, meticulously crafted tables grounded in lunar observations, laying the groundwork for subsequent astronomical endeavors. The Chaldeans, celebrated for their astrological insights, meticulously recorded lunar appearances, while pre-Islamic Arabs displayed remarkable navigational prowess in the celestial realm. Armed with an intuitive understanding of celestial mechanics, they discerned lunar phases and extrapolated patterns in stellar and meteorological phenomena. This expertise, vital for agrarian societies, culminated in the development of a lunar calendar comprising 28 distinct segments, reflecting the moon's cyclical journey through time.

Despite their profound knowledge of lunar phenomena and celestial observations, ancient Arabs were not formally designated as astronomers or mathematicians; rather, they were often referred to as naturalists. Their methodologies were characterized by simplistic and intuitive calculations. The initiation of each month was determined by the visual confirmation of the moon's disappearance, marking the onset of a new lunar cycle. Notably, this practice predates the advent of Islam and was not explicitly codified within its teachings. Nevertheless, the lunar calendar became integral to temporal reckoning and societal affairs. Henceforth, Allah unequivocally states in the Quran, They ask you (O Muhammad peace be upon Him) about the new moons. Say: These are signs to mark the fixed period of time for mankind and pilgrimage [4]. At another juncture of the Quran, it specifies the precise count of months, as follows: "Indeed, the number of months with Allah is twelve [lunar] months in the register of Allah [from] the day He created the heavens and the earth; of these, four are sacred" [5], underscored the divine sanctity and ordained nature of the twelve lunar months, with four designated as sacred. Thus, this celestial framework served as the cornerstone of temporal organization and religious observance within Islamic civilization.

The astronomical knowledge inherited from the pre-Islamic epoch lacked organization, devoid of a systematic scientific framework or a specific theoretical underpinning. Instead, it constituted a patchwork of disparate information, observations, and a fusion of astronomy and astrology. Renowned Italian orientalist Nallino, specializing in astronomical heritage, underscores the rudimentary state of astronomy during this era. He asserts, "In general, during the first century of the Hijra and the early second century, Muslims remained distant from the field of astronomy and other mathematical and natural sciences" [6]. Notably, ancient commentators and scholars often relied on unreliable narratives concerning celestial phenomena. These accounts drew from the prevalent beliefs among the general populace, be they from the People of the Book or the Magians, whenever they sought to elucidate celestial occurrences.

The transmission of astronomical knowledge from neighboring civilizations, such as Persia, India, and Greece, burgeoned the Islamic world's astronomical acumen in the third century of the Hijrah. Texts such as Ptolemy's Almagest and Euclid's Elements were diligently translated, enriching Islamic scholarship with a trove of astronomical wisdom. Yet, despite these strides, the discipline remained entwined with vestiges of myth and astrology, permeating foundational texts and tempering its scientific rigor [7].

Nallino traces the roots of Arabic astronomical manuscripts to the "Miftah al-Najoom," attributed to Hermes the Wise, possibly dating back to around 125 Hijri. By the third century of the Hijra, astronomy witnessed significant progress, resulting in a plethora of astronomical compendiums and instruments. Muslim astronomers diligently enhanced and updated these compendiums, including notable tabulations like Sabi's, Alkhani's, and Ibn al-Shatir's, along with the Ghubaq table and the meticulously validated "Mumtahan." Luminaries such as Sufi, Batani, Biruni, Tusi, and Ibn al-Shatir introduced innovative theories and concepts, laying the groundwork for modern astronomy, eventually leading to Copernicus' heliocentric model. From the Abbasid era onwards, astronomy evolved into a mathematical discipline grounded in empirical evidence, with scholars like Ibn al-Haytham making significant contributions. Ibn al-Haytham's investigations shed light on star emissions and lunar illumination, while also pioneering the development of early magnifying lenses for text examination. His work coincided with the establishment of observatories across the Islamic world, marking a transformative period in astronomical inquiry and understanding.

During the Umayyad era, the establishment of an observatory in Damascus marked a turning point in Islamic astronomy, echoed by subsequent institutions under the Abbasids in Baghdad and the Fatimids atop Mount Mukattam in Egypt. Across regions like Syria, Isfahan, Samarkand, and Al-Andalus, observatories flourished, showcasing meticulous instruments like the solar quadrant and the Arabic astrolabe. The astrolabe, particularly, became integral to Arab observatories, laying the groundwork for modern theodolites widely used in geological mapping [8].

In astronomy, particularly in the 20th century, a profound advancement has enabled highly precise calculations regarding celestial movements. This evolution has led to meticulous tracking of planetary orbits and positions, notably the moon's trajectories across epochs. Consequently, it has decisively clarified the start of lunar months, revealing the complex relationship between the moon's orbit and the Earth's revolution around the sun. Utilizing this detailed understanding, astronomy accurately predicts the moon's emergence from the sun's vicinity, often specifying the exact time in hours, minutes, and seconds well in advance.

Furthermore, astronomy has adeptly computed pivotal astronomical events such as the crescent's sunset, the sun's sunset, the duration of the crescent's visibility post-sunset (if it remains above the horizon), and assorted parameters including its altitude, thickness, and angular deviation from the western horizon. This comprehensive comprehension extends across a myriad of scenarios and substantiating evidences. Consequently, it has facilitated the prognostication of the crescent's visibility or non-visibility with remarkable precision. In the contemporary milieu, access to such precise information is readily available, with international scientific and astronomical organizations disseminating meticulously compiled astronomical tables. These tables, hailing from esteemed institutions such as those in Britain, the United States, Germany, and beyond, furnish anticipatory values for these significant dates and times spanning forthcoming decades [9].

Amidst this paradigm shift in astronomy, proponents of the astronomical calculations viewpoint assert: "The incontrovertible reality, universally experienced, is the emergence of novel circumstances coinciding with the advent of the modern epoch, fostering a highly conducive environment for substantiating the onset of lunar months through a definitive methodology of crescent sighting, hitherto uncertain. Put differently, the contemporary era has engendered a potent mechanism for verifying the crescent's appearance, alongside the Prophet's emphasized visual sighting method, peace be upon him. Initially neither contradicted by the Prophet nor in opposition to the visual sighting method, it rather serves as a supplementary measure. With the harmonization of both approaches, we can confidently furnish the global community with a crescent determination that is more precise and dependable" [10].

2 Calculations in Astronomy and Their Accuracy

Before examining the efficacy of astronomical calculations, it's crucial to address the disparities among scholars regarding crescent moon sighting. In Ramadan 1431 H / 2010 CE, divergent views emerged, differing by up to three days. Abdulaziz Al-Shammari, an astronomer with the Arab Union for Astronomy and Space Sciences, highlighted these discrepancies. In Libya, Ramadan was expected to begin on Tuesday, 2010/9/10, with astronomers insisting on the conjunction preceding dawn. Conversely, in Egypt and other areas, the commencement was pegged to Wednesday, 2010/9/11, based on the conjunction preceding sunset and the moon setting after it, regardless of visibility. Saudi Arabia planned for crescent sighting on Tuesday evening, but in regions like Riyadh, Sudair, and Shaqra, where visibility was challenging, Wednesday evening would mark the start of Ramadan. Al-Shammari suggested that Morocco, Oman, and other nations would start Ramadan on Thursday, 9/12/2010, supported by astronomical calculations [11]. This highlights the need for calendar unification, as discrepancies deepen division. The variation in sighting across the Islamic world raises questions about the reliability of astronomical calculations in resolving fasting initiation disparities and determining the onset of lunar Hijri months.

In a comparative study between actual moon sightings ("Ruyat al-Hilal") and astronomical calculations over the years, Dr. Ayman Said Kurdi, a professor of astronomy at King Saud University, concluded the following between 1400 H and 1422 H:

- The calculation matched the sighting 14 times when the crescent was observed.
- The calculation matched the sighting 24 times when the crescent was not observed.
- The calculation did not match the sighting 18 times, where moon sighting and non-sighting were reported astronomically.
- The calculation did not match the sighting twice, as the crescent was astronomically confirmed but not reported.

The researcher concluded from his comparative study that the correlation between astronomical calculations and visual observation is 67% [12].

Advocates of astronomical calculations contend for their method's reliability over visual sightings, citing studies like the Sheffler experiment, which reported a 15% false sighting rate over four years, questioning visual accuracy. Atmospheric conditions, including weather clarity, purity, and pollution levels, are vital factors influencing crescent sightings. Both natural elements (winds, storms, clouds, dust, water vapor) and industrial pollutants pose significant challenges. Modern industrialization has notably degraded air quality, contrasting with earlier eras, complicating precise lunar observations [13].

While some astronomers assert the precision of astronomical calculations, Dr. Ali Abanda, former Director General of Jordan's Meteorological Department, affirms their unparalleled accuracy. Historically, discrepancies were minimal, often measured in fractions of a second. Present advancements have further enhanced precision to one part in a thousandth of a second. This precision extends to forecasting celestial phenomena like the moon's movement, planetary positions, eclipses, and timings, facilitated by electronic computers, sometimes years in advance [14].

Dr. Imad Muhajid emphasizes the crucial role of accurate astronomical calculations in space missions. They are essential for the success of humanity's exploration of space, including landing on the moon. He also highlights the contributions of Arab Muslim scientists to these endeavors, highlighting their dependence on precise calculations [15].

Skeptics doubt astronomical findings due to discrepancies in crescent sightings and month delineation among astronomers. They question astronomy's relevance and its speculative nature, citing instances like the unanimous refraining from crescent sighting in 1992 despite actual sightings, and similar discrepancies in 1994. Recent disparities during a lunar eclipse in Iran further fuel skepticism [16].

But before forming an opinion on the accuracy of astronomical calculations, it is worth noting the perspective of Sheikh Abdullah bin Suleiman bin Al-Munea. He suggests clarifying some terms before delving into the subject:

1. Conjunction (Al-Iqtiran): It is the meeting of the sun and the moon in one longitudinal line, such that the sunlight does not reflect on the moon or any part of it.
2. Birth (Al-Wiladah): It is the separation of the moon from the sun, with the sun appearing in the west and the moon behind it in the east. At birth, sunlight appears on a part of the moon.
3. Possibility of Visibility (Imkan Al-Ru'ya): It is the occurrence of the ability to see the crescent moon after its birth visually.

Astronomers globally agree on the exact moment of the crescent's birth, often pinpointing it to the second, with negligible discrepancies akin to mathematical calculations. However, contention arises among astronomers concerning its visibility post-birth. While some argue for pre-sunset emergence with specific horizontal angles and altitudes required for observation, others hold divergent views. This disparity leads to discord among astronomers who otherwise synchronize on the timing of birth. Consequently, this discord perplexes Sharia scholars and jurists across epochs, who struggle to distinguish between birth and visibility, presuming any discrepancy signifies inconsistency in birth determination [17].

While differentiating between the birth and visibility of the crescent moon is crucial for reducing disagreements, it does not entirely resolve the issue. The primary focus should be on visibility rather than birth.

In order to validate the commencement of a new month, all astronomical calendars, including the Umm al-Qura calendar, stipulate two prerequisites to authenticate the sighting of the crescent moon:

1. The occurrence of a conjunction between the sun and the moon, signifying the moon's positioning between the Earth and the sun. In this configuration, the sun illuminates one half of the moon while the other half, facing Earth, remains shrouded in darkness, unseen.
2. The moon's descent below the horizon post-sunset, even if it occurs only by a minute [18].
3. The crescent should not linger on the western horizon for more than 30 min after sunset.

Consequently, Astronomers argue that since 1859, the new crescent moon has never appeared west of the sun within 22 min after sunset. Factors like atmospheric conditions, observer health, and visual acuity are crucial in crescent sightings. Confusion may arise, especially among those lacking astronomical knowledge, regarding the difference between synodic and lunar months, typically two days apart. Misconceptions, like the premature declaration of a new month's arrival as seen in Libya, stem from ignorance of these astronomical facts. Sheikh Mohsen Al-Asfour emphasizes the importance of dispelling ignorance and rigidity to resolve crescent disputes. He advocates for adherence to established jurisprudential decisions, fasting upon sighting, and breaking the fast upon confirmation, stressing the significance of visual observation in dispelling doubts, particularly through sunlight reflecting on the moon's edge. To attain scientific consensus, he outlines four imperative steps:

1. Precise determination of the thirtieth night in the lunar month through scientific calculations and moon's apparent movement observation.
2. Vigilance over lunar phases, including the first quarter on the seventh night, full moon on the fifteenth night, and second quarter on the twenty-first night.
3. Refinement of scientific calculations concerning conjunction and lunar eclipse, ensuring compliance with Sharia descriptions and regulations to avert disagreement.
4. Harnessing modern astronomical tools for crescent moon documentation, image capture, and public accessibility to foster confidence and assurance [12].

3 Objections to Using Astronomical Calculations and Their Basis

The consensus among the four imams and numerous jurists advocates against relying on astronomical calculations, instead favoring naked-eye sightings or completing the month as thirty or twenty-nine days, contingent upon differences between the Hanbali school and the majority. Jurists across the four schools generally reject astronomers' opinions in determining lunar months, asserting fasting's association with a fixed factor unaffected by change, whether through crescent sighting or completing thirty days. This skepticism toward astronomers stems from their divergent views, as articulated by three imams. While the Shafi'i school permits reliance on astronomers' opinions for themselves and their adherents, the prevailing view deems it non-obligatory for the general populace [19]. Central to this argument are hadiths like "Fast upon sighting it, and break your fast upon sighting it" [20] and "Do not fast until you see the crescent and do not break your fast until you see it. If it is cloudy for you, then estimate for it" [21]. Furthermore, the Quranic verse, which states, "So whoever sights [the new moon of] the month…" [22] supports the emphasis on direct sighting. Additionally, the hadith of Ibn Umar highlights the Prophet's acknowledgment of the community's non-literate nature in lunar month determination, affirming a reliance on sightings rather than calculations or astronomy [23].

Imam al-Sarakhsi, Ibn Abidin, Ibn al-Hajib, Shams al-A'imma al-Halwani, and Sheikh Alish from the Maliki school reject reliance on astronomical calculations. The Islamic Fiqh Council, under the Organization of Islamic Cooperation, declared in 1401 that astronomical calculations are not valid for determining lunar month beginnings [24]. Professor Ahmed Abdel-Moneim Al-Bahi of Al-Azhar asserts that witness testimony holds greater weight than astronomical calculations. Advocates of this stance argue that skepticism towards astronomical calculations in lunar month determination doesn't arise from doubts about their accuracy but from Sharia's linkage, as expressed by Prophet Muhammad (Peace be upon him), between crescent sightings and lunar month commencements. In a confirmed hadith narrated by Ibn Umar, the Prophet Muhammad, peace be upon him, advised: "Fast when you see the crescent, and break your fast when you see it. If it's cloudy, then estimate" [25]. In an authenticated narration, the Prophet, peace be upon him, decreed: "Should cloud obscuration prevail, adhere to a complete thirty-day count" [26]. A subsequent narration by a Muslim stipulates: "In case of cloud cover, resort to a fixed estimation of thirty days," elucidating the absolute estimation prescribed in the initial narration. Another tradition relayed by Bukhari, Muslim, and An-Nasai, attributed to Abu Huraira, May Allah be pleased with him, directs: "Upon sighting the crescent, commence fasting, and upon its reappearance, terminate fasting. In the event of cloud cover, observe fasting for a duration of thirty days" [27]. Therefore, all Prophetic teachings tie fasting to the new crescent's sighting. "Estimate" means observing Sha'ban or Ramadan for thirty days. Default isn't 29 days unless seen. Worship rulings hinge solely on textual evidence, excluding reasoning or analogy.

In "Al-Aql wal-Fiqh," Professor Sheikh Mustafa Ahmed Al-Zarqaa explores objections against using astronomical calculations to start the Hijri month. He argues that Sharia doesn't require knowledge of calculations or writing for fasting and worship. Obligations rely on clear signs accessible to everyone, emphasizing Sharia's timeless application.

Al-Zarqaa delved into reasons for rejecting reliance on astronomical calculations, stressing its historical roots over modern relevance. Ibn Hajar, echoing Ibn Bizzah, deemed calculations invalid, citing Sharia's prohibition of astrology for its speculative nature. Discord arises from varied outcomes, as Ibn Bizzah noted. Al-Zarqani, citing An-Nawawi, explained the avoidance of calculations due to conjecture. Ibn Battal linked hadiths to disregarding astrology. Ibn Taymiyyah argued against calculation reliance, citing the crescent's unpredictability and Sharia's stance on hidden matters. He reiterated this stance throughout his discourse [28]. In "Al-Jami' li Ahkam al-Siyam," Professor Abu Iyas Mahmoud Awwadah asserts the impermissibility of relying on astronomical calculations for the Hijri month. Most scholars concur, citing difficulty and exclusivity of such knowledge. Ibn Sireen permits calculation for those with this expertise, citing "Faqaddiru lahu" in hadiths. However, Awwadah interprets this differently. The hadith advises relying on sighting the crescent, as narrated by Ibn Umar, may Allah be pleased with them, emphasizing observation over estimation [29]. The hadith applies universally to the entire Islamic nation, as fasting and sighting the crescent involve everyone. "Faqaddiru lahu" isn't exclusive but addresses all. Assigning it solely to those with knowledge is an error. Hadiths elucidate it differently from Ibn Sireen's interpretation. Another narration by Abu Huraira instructs fasting upon sighting the crescent and completing Sha'ban for thirty days if obscured [30]. The phrase "Aqdimu lahu" in the hadith means "Complete the count of Sha'ban as thirty days" and "Fast for thirty days." This interpretation refutes Ibn Sireen's error, as it doesn't rely on astronomy or calculations. Thus, it's for the entire Muslim community, as evident. Bukhari's hadith, "Do not fast until you see the crescent," strengthens this argument further [31]. Therefore, He didn't limit fasting to sighting alone; the prohibition is fasting without sighting, which includes fasting based on calculations, unmistakably bypassing sighting [32]. Professor Awaidha advocated utilizing astronomy and its calculations for moon sighting. He emphasized the permissibility of using tools like magnifying lenses and telescopes to aid observation, enhancing visibility without any objection [33].

Dr. Bakr bin Abdullah Abu Zaid debunked the reliance on astronomical calculations in his research, refuting arguments based on the hadith "We are an unlettered nation; we do not write or calculate." He clarified that the Prophet's statement simplifies matters, emphasizing reliance on sighting or completion rather than writing or calculation.

The phrase "complete it for him" applies universally to the entire Islamic nation, as fasting, sighting, and breaking fast are for everyone. Dr. Abu Zaid highlighted the error in attributing this phrase solely to those with knowledge and cited alternative hadith interpretations, like the one by Abu Huraira, supporting his argument. He concluded that reliance on sighting or completion obviates the need for writing or calculation, supported by scholars such as Ibn Taymiyyah, Ibn Arabi, and Ibn Hajar. The research stressed adherence to the Prophet's guidance regarding the crescent and emphasized the wisdom in relying on visible signs over calculations [33].

Subsequently, Dr. Abu Zaid presented further evidence, highlighting the speculative nature of astronomical calculations. He noted the lack of concrete evidence supporting this method, aside from claims by some astronomers. However, their assertions contradict material evidence, supported by the Fatwa Committee of Al-Azhar. Islamic law prioritizes trustworthy testimony, yet contemporary evidence shows computational errors, as seen in past announcements regarding crescent sightings. For instance, in 1406 AH, astronomers deemed it impossible to sight the Shawwal crescent on the 30th of Ramadan, but Sharia-compliant witnesses confirmed sightings. This inconsistency indicates the speculative and unreliable nature of modern astronomical results. Additionally, Abu Zaid pointed out that modern observatories, like any machinery, are prone to technical faults, affecting accuracy. Al-Azhar's Fatwa Committee, led by Sheikh Mahmoud Shaltout in 1979, echoed concerns about the speculative nature of modern astronomical calculations [34].

Opponents of using astronomical calculations argue it resembles the practices of soothsayers and astrologers, connecting earthly events to celestial movements. This perspective is attributed to Sheikh al-Islam Ibn Taymiyyah, interpreted by scholar al-Zarqani. Ibn Taymiyyah denounced astrological beliefs, stating they are intellectually false and prohibited in Islam. He emphasized the influence of spirits and angels over celestial bodies and warned against acquiring knowledge from stars, equating it to sorcery, as mentioned by the Prophet (peace be upon him) [35]. Sheikh al-Islam strongly opposed relying on calculations for lunar months, expressing firm disapproval. He emphasized that those who resort to writing or calculating deviate from the path of believers in this matter. [36] Scholars like Sheikh Muhammad Nasir al-Din al-Albani, Sheikh Abdul Aziz bin Baz, and Dr. Wahba al-Zuhayli advocate against reliance on astronomical calculations [37]. Dr. Wahba al-Zuhayli contends against relying on migrants' accounts or astronomical calculations. He argues this contravenes the Prophet's Sharia since only the actual sighting of the crescent obligates us religiously [38].

Sheikh Saleh Al-Luhaidan, President of the Supreme Judicial Council in Saudi Arabia, advocates for Sharia-based visual sighting to confirm sacred months' start. He cited the Prophet's statement, "We are an illiterate nation," to emphasize reliance on Allah's prescribed worship without calculations or writings. Trustworthiness, integrity, and fear of Allah are pivotal, not consensus or individual piety. They prioritize visual evidence, requiring trust, honesty, and perceptiveness [39]. The Ibadi school's stance remains unclear, though Sheikh Atfich suggested visual sighting using tools like telescopes is not valid [40]. Twelver Shia Imamiyya scholars, like Sunni counterparts, have varying opinions on this issue. Ancient scholars such as Sheikh Tusi and Allama Hilli, alongside contemporary figures like Ayatollah Muhammad Baqir al-Sadr, reject reliance on astronomical calculations. Al-Sadr emphasized the need for more than just modern means in confirming the Islamic lunar month, requiring visibility and individual jksatisfaction. Notable figures like Ayatollah Khomeini, Ayatollah Khoi, Ayatollah Klibaykani, Sheikh Zain al-Din, [41] and Ayatollah Ali al-Sistani share similar views [42].

4 Proponents of Astronomical Calculations

The second viewpoint considers the permissibility of relying on astronomical calculations. Dr. Mohammed Abdul Latif Farfour, in a study submitted to the Islamic Jurisprudence Council, outlined three perspectives of later scholars. Firstly, Al-Zahdi's opinion allows reliance on astrologers' words, as reported by Judge Abdul Jabbar and the compiler of sciences. Secondly, Ibn Maqatil's opinion suggests consulting astrologers and relying on their consensus. Thirdly, Imam al-Sarkhasi's explanation asserts that fasting and breaking fast hinge on moon sighting, not astrologers' opinions [43].

The endorsement of astronomical calculations is attributed to Matarf ibn Abdullah, Ibn Qutaybah, Al-Subki, and Baghdad scholars. Abu Al-Abbas ibn Sarij, a prominent Shafi'i scholar, prioritizes this view over Al-Muzani. Al-Qushayri, Al-Abadi, and Abu Khattab among Maliki scholars also support it [44].

Imam Taqi al-Din Al-Subki, in his fatwas, argues against accepting witness testimony if calculations refute moon sighting, deeming it speculative. He asserts that accepting witnessed impossibilities contradicts Sharia, which doesn't legislate impossibilities [45]. Contemporary scholars, including Sheikh Yusuf al-Qaradawi, Sheikh Mustafa al-Zarqa, Sheikh Ahmed Mohamed Shakir, and Dr. Mohamed Mukhtar al-Salami, advocate this opinion. Sheikh Mohamed Rashid Rida and Sheikh Tantawi Juhari also endorse it, interpreting the Prophet's saying "Estimate for him" as referring to calculation rather than traditional methods [46].

During Sheikh Al-Maraghi's tenure at Egypt's Supreme Sharia Court, he aligned with Al-Subki's stance, rejecting witness testimony when astronomical impossibility arises. Despite initial opposition, notably from Sheikh Ahmed Shakir and his father, Ahmed Shakir eventually withdrew his objections. Al-Maraghi emphasized commencing the month through calculations unless ignorance persists [47].

The debate continues over whether sightings or calculations determine the Hijri month's start. Explored at a 1984 London symposium, it concluded that even a brief crescent sighting post-sunset, determined through calculations, initiates the month [48].

Sheikh Mustafa Ahmed Al-Zarqa ardently advocates reliance on astronomical calculations, drawing from Islamic tradition. He argues that understanding rulings and their application is paramount when textual clarity exists. This includes the Prophet's emphasis on visual observation for initiating the month [49].

The discussion encapsulates Islamic legal perspectives on lunar month commencement, blending traditional moon sightings with accepting astronomical calculations. Scholars, ancient and modern, significantly shape this discourse, molding interpretations and methodologies.

Prophetic hadiths were studied, highlighting the illiterate community's reliance on visual moon sightings to determine month lengths. However, contemporary scholars like Sheikh Mustafa Ahmed Al-Zarqa advocate for astronomical calculations, offering a contrasting viewpoint.

Despite regional disparities, the emphasis on unity within the Islamic community regarding month commencements, festivals, and Hajj persists. This discourse involved an extensive analysis of hadiths and scholarly opinions, particularly focusing on the practicality of moon sightings in various regions.

Furthermore, the significance of unity in religious practices and ongoing dialogues within the Islamic community were explored. This exploration, spanning various viewpoints, historical contexts, and religious texts, yielded a comprehensive understanding of the complexity surrounding Islamic lunar months.

In terms of legal discussions, Al-Qalyubi, citing Al-Awza'i, asserted that definitive calculations override crescent sighting testimony, rendering fasting impermissible. This highlights the importance of adhering to established guidelines.

Moreover, it was clarified that completing thirty days due to obscured moon sightings doesn't reveal the reality of the previous or next month. The reliance on visual observation, rooted in historical illiteracy, doesn't constrain adopting precise astronomical calculations, offering a more reliable alternative.

In conclusion, Sheikh Al-Zarqawi's perspective, outlined in his research, encompasses four key points. Firstly, analyzing authentic hadiths pertaining to fasting reveals the Prophet's directive to rely on visual moon sightings due to the community's illiteracy and lack of precise calculation methods. This underscores the necessity of adopting modern astronomical calculations, providing certainty and relieving the confusion surrounding crescent confirmation.

Secondly, early scholars, denouncing reliance on calculations, attributed it to conjecture and intuition, grounded in the rudimentary state of astronomy during their time. They rejected speculative methods, emphasizing direct observation's reliability over estimations [50].

Thirdly, early scholars encountered a significant challenge: the historical association between astronomy and forbidden practices such as fortune-telling and magic. This connection led to skepticism about relying solely on astronomical calculations to determine lunar months. Firstly, this approach is speculative and based on guesswork, contradicting the principle of direct visual observation despite its potential uncertainties. [51] It's inconceivable that the Prophet, peace be upon him, would forbid understanding Allah's cosmic order, in line with the Quran's call to ponder the universe [52].

Therefore, Ibn Taymiyyah criticized those mixing legitimate astronomical calculations with forbidden practices, likening them to fortune-tellers. This parallels the Prophet's warning about astrology as a form of magic. It's important to note that this criticism doesn't apply to legitimate astronomical calculations, which weren't fully developed then.

Lastly, in the present era, modern astronomy has divorced itself from such associations, attaining maturity and trustworthiness. Contemporary astronomical calculations far surpass the intuitive methods of the past, offering precision and accuracy free from the taint of forbidden practices.

The evolution of astronomy is evident through modern observatories and sophisticated instruments detecting planetary movements with unparalleled precision. This advancement, measured against past speculative methods, underscores its accuracy and certainty [53]. Sheikh Muhammad Al-Ghazali lamented the rejection of astronomical calculations, criticizing those who benefit from technological advancements while disregarding the mathematical principles behind them [12].

Conversely, Sheikh Abdullah bin Manea contends that Sharia does not oppose utilizing astronomical calculations. He distinguishes between traditional speculative methods and modern precise tools, asserting the compatibility of astronomy with Islamic teachings. He emphasizes the importance of adhering to valid moon sightings while integrating scientific evidence [54].

Moreover, he emphasizes the importance of adhering to moon sighting and accepting valid testimonies, provided conditions are met. If astronomical calculations confirm the crescent's setting before sunset, testimonies claiming sighting after sunset are deemed invalid, as it contradicts scientific evidence. These scholars stress the compatibility of modern astronomy with Islamic teachings, discouraging the rejection of advancements based on outdated beliefs. [55].

5 Supporters of Astronomical Calculations in Ja'fari Shia School

Contemporary scholars within the Ja'fari Shia School, including Sheikh Bahauddin Muhammad ibn al-Hussein al-Amili (Sheikh al-Bahai) and Ayatollah Muhammad Hussein Fadlallah, advocate for employing astronomical calculations to determine lunar months. [43] Sheikh al-Bahai asserted the religious validity of this method, emphasizing the significance of visibility attainment, even if not universally observed. Others, such as Sheikh Muhammad Jawad Mughniyah, Ayatollah Muhammad Muhammad al-Sadr, and Sayyid Musa al-Sadr, support this stance, stressing the necessity of scientific approaches to avoid religious confusion and legal disruptions. [56]A clear fatwa endorsing astronomical calculations was issued by Lebanese Shia authority Sheikh Afif al-Nabulsi, setting the fasting start for 1431 AH (2010 CE) based on precise calculations [57].

5.1 Third View: Using Astronomical Calculations in Clear Sky Conditions

Dr. Mohammed Farfour suggests prioritizing visual or telescopic observation from Earth's surface over sky or mountain views, citing the Prophet's guidance. Clear sky sightings ensure certainty without resorting to calculations or neighboring observations. However, in uncertain cases, involving astronomers and meteorologists for precise calculations is permissible. Religion and knowledge align, especially nearing the month's end, where the possibility of the new month's beginning arises. Further concealment or sightings are considered within the realm of religious sighting when visual observation is impractical [58].

5.2 Necessary Clarification by Sheikh Mustafa Al-Zarqaa:

Sheikh Mustafa Al-Zarqaa presented a research paper to the Islamic Jurisprudence Assembly on using astronomical calculations for lunar months. He affirmed its permissibility religiously, clarifying it doesn't negate traditional sighting or completing months at thirty days when visibility is tough. Al-Zarqaa stressed that accepting calculations doesn't nullify Sharia's reliance on visual observation, which remains fundamental. He argued Sharia can't tie rulings to uncertain knowledge over time.

Accepting calculations doesn't forbid religiously; it's an alternative due to astronomy's precision. Though not as easy as visual sighting, it's permissible, with visual observation as the basis if knowledge is lost. Sharia allows alternatives while maintaining the original method. Thus, accepting calculations doesn't change fasting rulings based on direct visual sighting, in line with prophetic tradition [59].

Employing astronomical calculations to determine the start of the lunar month yields various legal perspectives, including:

1. Some permit the unrestricted use of astronomical calculations.
2. Others outright prohibit their use.

In scenarios where witnesses testify to crescent moon sightings, while calculations suggest otherwise, a question arises. According to "Al-'Adhb al-Zalal" by Al-Hattab, prevailing scholars often disregard calculations in favor of eyewitness testimony. Al-Sabki and other Shafi'i scholars argue against accepting testimony over calculations due to their definitive nature versus presumptive testimony. This reflects a Shafi'i difference of opinion.

The opinion permitting calculations in denial scenarios suggests they are valid when sighting is deemed impossible. They are not considered when affirming crescent sightings. Ibn 'Abidin and the Secretary-General of the Islamic Fiqh Council support this stance, indicating calculations override visual observations when conflicting. This approach prioritizes denial testimony over affirmation based on calculations, thus shaping the legal framework for utilizing astronomical calculations [60].

Despite reliance on visual sighting and witness testimony, Sheikh Abdullah bin Suleiman bin Munea argues for combining textual evidence with astronomical results to establish crescent sightings. He emphasizes that if astronomers claim crescent birth before sunset, but it's not observed, relying on astronomical affirmation isn't allowed. Similarly, restricting crescent visibility by a specific angle post-birth isn't permissible. Testimony of crescent sighting after sunset when born before sunset is valid due to categorical ruling out of sighting possibility.

In negation scenarios, astronomical calculations are considered, not in affirmation ones. Ibn 'Abidin and the Secretary-General of the Islamic Fiqh Council support this, emphasizing calculations' acceptance in such cases.

1. Witnesses' testimony before calculated crescent appearance isn't valid, whether before conjunction or coinciding with it, determined by accepted astronomical calculations, noted by scholars like Ibn Taymiyyah.
2. Testimony of crescent sighting after sunset when seen before sunrise isn't valid.
3. Crescent sighting remains primary for month beginning, supported by calculations for visibility conditions, but confirmation requires valid testimony. If calculations suggest possibility, but no hindrances exist, and crescent isn't observed, completing the month within thirty days is obligatory.

Sheikh Ahmad Shakir, known for his Hadith expertise, questioned whether Shariah permits confirming Arab months with astronomical calculations [62]. In his thesis,

Sheikh Ahmad Shakir reflected on a past debate sparked by Sheikh Mustafa Al-Maraghi's opinion, supported by Taqi al-Din al-Subki, about rejecting witness testimony if astronomical calculations negate moon sighting. This stance triggered intense debates, even within Sheikh Shakir's family, initially opposing the opinion. However, Sheikh Shakir now acknowledges its correctness. He emphasizes the need to confirm sightings through calculations in all cases, except when knowledge is inaccessible. Sheikh Shakir highlights this isn't an innovation; scholars understand rulings can vary based on circumstances [63].

Abdullah bin Muni' delineated four scenarios where adherence to prophetic tradition intersects with reliance on astronomical calculations:

1. If the crescent moon is sighted after sunset with the sun setting before it, both legal and astronomical principles corroborate the month's commencement.
2. When the sun sets after the moon without anyone sighting the crescent, both legal and astronomical viewpoints concur on "non-sighting," marking the night as the current month's final one.
3. Conflict arises if the sun sets before the moon, yet no crescent sighting is reported. Here, astronomical evidence supports the month's beginning, but legal rulings oppose it. In this instance, adherence to legal perspective is imperative.
4. If the moon sets before the sun, and someone claims to have sighted the crescent post-sunset, while astronomical data negates the month's beginning, emphasis is on astronomical reality.

This consensus aligns with views of various Muslim scholars, ancient and contemporary, including Ibn Taymiyyah, Ibn al-Qayyim, Al-Qarafi, Ibn Rushd, Al-Subki, Al-Maraghi, Ahmad Shakir, and Sheikh Mustafa Al-Zarqa [54].

6 Conclusion

- The default method for confirming the start of lunar months is through visual sighting.
- Relying solely on astronomical calculations is insufficient; it is necessary to accompany these calculations with visual confirmation.
- The use of magnifying devices such as telescopes is acceptable for observing the birth of the crescent moon.
- The sighting of the crescent moon through these devices is equivalent to a naked-eye sighting.
- It is essential to distinguish between the birth of the crescent moon (separation of the moon from the sun with the sun in the west and the moon in the east) and the possibility of visual observation, which occurs after a certain period.
- The acceptance of the testimony of witnesses is based on the actual sighting of the crescent moon, not its birth.
- Astronomical calculations are heeded in instances of negation rather than affirmation, a consensus reached by esteemed Muslim scholars during the Mecca conference in the year 2012 AD.
- Witnesses' testimony is rejected in situations where astronomical conditions make moon sighting impossible.

- The priority is given to visual sighting to establish the beginning of the lunar month.
- Astronomical calculations should be seen as a supplementary method and must be combined with visual confirmation.
- The utilization of optical aids such as telescopes and binoculars is permissible for observing the crescent moon's birth.
- Observing the crescent moon through these devices is treated on par with naked-eye sightings.
- A clear distinction should be made between the birth of the crescent moon (sun in the west, moon in the east) and the subsequent visibility, which occurs after a specific time.
- Witness testimony is contingent on actual sightings, not the birth of the crescent moon.
- Astronomical calculations are considered when visual sighting is deemed unfeasible (negation), not when it is possible (affirmation).
- Testimony is rejected when astronomical conditions make moon sighting physically impossible.
- The integration of technology, such as telescopes, is permissible to enhance the accuracy of moon sighting.
- The overarching principle is that the sighting of the crescent moon takes precedence over astronomical calculations.
- The consensus is on utilizing astronomical calculations in impossibility (negation) cases while relying on visual sightings for affirmation.
- The religious significance of the crescent moon sighting is emphasized, and astronomical calculations are regarded as supplementary tools.

References:

1. The magazine "Majma' al-Fiqh al-Islami" (Islamic Jurisprudence Assembly), Second Session of the Conference. Islamic Fiqh Assembly, Issue Number Two, Part Two, 1407 AH / 1986 CE. (Research on the Beginnings of the Arabic and Islamic Months by Sheikh Haroon Khalil Jile, p. 911)
2. Al-Ansari, A.H.: The Wandering Crescent, Al-Ayyam Newspaper / Issue 7795, Friday, August 13, 2010, corresponding to 3 Ramadan 1431 AH
3. Bahrain Middle Magazine (Online)
4. Surah Al-Baqarah: 189
5. Surah At-Tawbah (9:36)
6. Al-Mu'min, A.A.: Establishing the Crescent between the Era of the Prophet (PBUH) and the Modern Era, citing Carlo Nallino, Astronomy in the Quran Middle, Rome 1911 (Offsite Qasim Rajab) / p. 138
7. ibid
8. Azzab, K.: Debates on the Crescent Issue by Sight or Calculation, Saturday, August 14, 2010 (Online article)
9. Al-Mu'min Prince: Establishing the Crescent between the Era of the Prophet (PBUH) and the Modern Era 12
10. ibid
11. "Dunya Al-Watan" website, citing the "Face the Press" program broadcast by the "Al-Arabiya" channel before Ramadan 1431 / September - August 2010

12. Al-Ansari, A.H.: The Wandering Crescent, Al-Ayyam Newspaper / Issue 7795, Friday, August 13, 2010, corresponding to 3 Ramadan 1431 AH

13. Al-Mu'min Prince: Establishing the Crescent between the Era of the Prophet (PBUH) and the Modern Era

14. Abnda, A.: How do you legitimately establish the Crescent by sight? Al-Dustour Newspaper 1987/8/28

15. Ramadan Crescent between Legal Sight and Astronomical Calculations, Astronomer Emad Mujahid, Hadi Al-Islam Magazine, Volume 52 Ramadan 1429 AH, September 2008, p. 85

16. .

17. Lecture by Ayatollah Sheikh Muhammad Sannad, continuing at the beginning of Shawwal 1426 AH, titled "The Failure of Astronomical Calculations in Establishing the Crescent."

18. Online Sharia Scholars Website Astronomical Determination of the Beginning of the Lunar Months Rajab, Sha'ban, Ramadan, Shawwal, and Dhu al-Hijjah 1436 AH, Abdullah bin Suleiman Al-Munie

19. "Al-Wasat" Bahraini Magazine

20. Al-Jaziri, A.R.: Book of Jurisprudence on the Four Schools, Volume One, p. 551

21. Part of a noble Hadith narrated by Bukhari 1909, Muslim, Nasai, Ahmad, and Ibn Hibban

22. Narrated by Bukhari 1906, Muslim, Darimi, Abu Dawood, Nasai, and Ibn Majah

23. Al-Baqarah: 185

24. Narrated by Al-Khwarizmi (1913) and Muslim (1080)

25. Fatwa by Sheikh Dr. Salman Ouda on the Internet, reviewing the opinions of all factions

26. Decisions of the Islamic Fiqh Council of the Muslim World League from its first session in 1398 AH to the eighth session in 1405 AH, p. 66, p. 79

27. Nidal Qassoum, Mohamed Al-Otaibi, and Karim Mazyane: Establishing Lunar Months and the Problem of Islamic Timing, Dar Al-Tali'a for Printing and Publishing, Beirut, 2nd Edition / 1997, p. 105

28. Narrated by Muslim in the Book of Fasting, number 100

29. Al-Zarqaa, A.-M.: Reason and Jurisprudence in Understanding the Prophetic Hadith, chapter "Establishing the Crescent by Astronomical Calculation in this Era," p. 83

30. The Hadith narrated by Bukhari 1906, Muslim, Darqutni, Abu Dawood, Nasai, and Ibn Majah

31. The narrators are Bukhari 1909, Muslim, Nasai, Ahmad, and Ibn Hibban. According to Muslim 2514 with the wording, "If you see the crescent, fast, and if you see it, break your fast. If it is cloudy for you, then fast thirty days."

32. The Hadith narrated by Bukhari 1906, Muslim, Darimi, Abu Dawood, Nasai, and Ibn Majah

33. Abu Iyas Mahmoud bin Abdul Latif bin Mahmoud (Petition): The Comprehensive Compilation of Fasting Rules, p. 38

34. Dr. Bakr bin Abdullah Abu Zeid, a research titled "The Ruling on Establishing the First Lunar Month and Unifying the Sighting Presented to the Islamic Fiqh Council in its Second Session held in Jeddah in 1405 AH."

35. 30 Ruling on Establishing the First Lunar Month and Unifying the Sighting, a research by Sheikh Dr. Bakr bin Abdullah Abu Zeid

36. Narrated by Abu Dawood in Medicine (03905), Ibn Majah in Adab (3726), and authenticated by Al-Nawawi in "Riyadh al-Saliheen" number (1669) in the Understanding of Prophetic Hadith, chapter "Establishing the Crescent by Astronomical Calculation in this Era," p. 81, and Al-Dhahabi in "Al-Kaba'ir."

37. Mustafa Ahmed Al-Zarqaa, Reason and Jurisprudence:

38. See the opinion of Ibn Baz in Al-Dustour Jordanian Newspaper 1987/8/21 p. 5

39. Wahba Al-Zuhaili: Islamic Jurisprudence and Its Evidence, Vol. 2, p. 528

40. Middle East Newspaper: Tuesday, 17 Sha'ban 1429 AH, 2008/8/19 "Legitimacy of Astronomical Calculations in Saudi Arabia: Ongoing Debate... Still Standing."

41. Nidal Qassoum, Mohamed Al-Otaibi, and Karim Mazyane: Establishing Lunar Months and
42. the Problem of Islamic Timing, Dar Al-Tali'a for Printing and Publishing, Beirut, 2nd Edition / 1997, p. 19, citing the book "Pure Gold."
43. To view the opinions of some Shiite scholars, you can refer to the "Forums of Lights: Truth, the Daughter of Dialogue" website Words of Contemporary Jurists
44. Al-Mustani Website
45. Whoever approaches a soothsayer and believes in what he says has indeed disbelieved in what was revealed to Muhammad (PBUH), narrated by Abu Dawood and mentioned by the people of the four Sunan, and authenticated by Al-Hakim from the Prophet (PBUH) with the wording: Whoever approaches a soothsayer or a magician and believes in what he says has indeed disbelieved in what was revealed to Muhammad (PBUH)
46. Journal of the Islamic Fiqh Council Second Session of the Islamic Fiqh Council Conference, Issue Two, Part Two, 1407 AH 1986 CE (Research by Dr. Muhammad Abdul Latif Farfour, titled: A Reading Message in the Language of Reading on Clarifying Calculation and Reading, p. 897)
47. Muhammad Abdul Latif Farfour's article and that of Sheikh Abdullah Al-Munie, cited by Al-Khateeb Al-Sharbini in Mughni Al-Muhtaj, Vol. 2, p. 154
48. Khalid Azzab / Debates on the Crescent Issue by Sight or Calculation, Saturday, August 14, 2010 (Online article)
49. Al-Zarqaa, M.A.: "Reason and Jurisprudence in Understanding the Prophetic Hadith," chapter on "Establishing the Crescent by Astronomical Calculation in this Era" from p. 71
50. Al-Zarqaa, M.A.: "Reason and Jurisprudence in Understanding the Prophetic Hadith," chapter on "Establishing the Crescent by Astronomical Calculation in this Era" p. 71
51. Sahih Ibn Majah, compiled by Muhammad Nasir al-Din al-Albani, No. 3017, and its degree is authentic
52. Surah Yunus: 101
53. Al-Zarqaa, M.A.: "Reason and Jurisprudence in Understanding the Prophetic Hadith," chapter on "Establishing the Crescent by Astronomical Calculation in this Era" p. 84
54. Middle East Newspaper: Tuesday, 17 Sha'ban 1439 AH, 2008/8/19 "Legitimacy of Astronomical Calculations in Saudi Arabia: Ongoing Debate... Still Standing."
55. The previous reference
56. Lecture by Sheikh Musa Sadr on the Internet titled "Relying on Scientific Means is the Preferred Way to Establish the Crescent."
57. Cultural Iraq Network
58. Journal of the Islamic Fiqh Council Second Session of the Islamic Fiqh Council Conference, Issue Two, Part Two, 1407 AH 1986 CE (Research by Dr. Muhammad Abdul Latif Farfour titled: A Reading Message in the Language of Reading on Clarifying Calculation and Reading p. 897)
59. Journal of the Islamic Fiqh Council Second Session of the Islamic Fiqh Council Conference, Issue Two, Part Two, 1407 AH 1986 CE (Comment by the Secretary-General of the Council, Dr. Muhammad Al-Habib bin Khogah, p. 1006)
60. The Pure Sweetness in the Matters of Crescent Sighting, authored by Sheikh Muhammad bin Abdul Wahab bin Abdul Razaq Al-Marrakshi, edited by Professor Abdullah bin Ibrahim Al-Ansari, Publications of the Religious Affairs Administration in Qatar, 1397 AH 1977 CE, p. 467
61. Journal of the Islamic Fiqh Council Second Session of the Islamic Fiqh Council Conference, Issue Two, Part Two, 1407 AH 1986 CE (Research by Dr. Muhammad Abdul Latif Farfour titled: A Reading Message in the Language of Reading on Clarifying Calculation and Reading, p. 897)

62. Supplement to the Recommendations of the Symposium on Hilaal and Covenants in the Book by Dr. Yusuf Al-Qaradawi: Facilitating Jurisprudence in the Light of the Quran and Sunnah (Fasting Jurisprudence) p. 22
63. From an online article by Sheikh Abdullah bin Suleiman bin Munie

Sun Vertical Depressions and Their Effects on the Morning Twilight Phases in Egypt

B. A. Marzouk[1], Nasser M. Ahmed[1], K. A. Edris[2], R. A. Mawad[2], M. M. Beheary[2], A. Bakry[2], A. H. Ibrahim[2], A. R. Mouner[2], Ahmed M. Abdelbar[2], A. Abulwfa[1], Samy A. Khalil[1], M. G. Rashed[1], Yasser A. Abdel-Hadi[1(✉)], A. H. Hassan[1], A. E. Ghitas[1], M. A. El-Sadek[1], U. A. Rahoma[1], M. A. Semeida[1], G. A. Goma[1], M. Sedek[1], R. H. Hamid[1], S. S. Khodairy[1], S. Hamzawy[1], E. Abdel-Wahed[1], M. M. Hussien[1], M. A. Shahat[1], A. A. Elminawy[1], Hussien M. Farid[3], W. A. Rahoma[3], M. Yossef[4], and K. Ali Eden[4]

[1] National Research Institute of Astronomy and Geophysics (NRIAG), Cairo, Egypt
yasser_hadi@yahoo.com
[2] Faculty of Science, Astronomy and Meteorological Department, Al Azhar University, Cairo, Egypt
[3] Faculty of Science, Astronomy and Meteorological Department, Cairo University, Cairo, Egypt
[4] Faculty of Science, Physics Department (Space Science), Helwan University, Cairo, Egypt

Abstract. This research is a new addition to the previous work carried out in Egypt to determine the altitude of the sun (sun vertical depression) corresponding to the beginning of the twilight. The observations of this issue began since more than 60 years in our institute and were attended by many astronomers and astrophysicists. Naked-eye (N.E) twilight observations of morning twilight were carried out in good viewing conditions by four instruments: two digital cameras, CCD cameras and SQM during the time interval (Aug. 2015–Dec. 2019) in many regions in Egypt by several research groups. The regions in Egypt were: Kottamia observatory, Kharga, Aswan, Hurghada, Marsa-Alam and Fayum. More than 30 scientific observers and amateurs participated in these observations. The purpose of these measurements was to determine the sun vertical depression of the sun below the horizon, D_o (D_o = -altitude of the sun) at which the normal eye can discriminate the true dawn (strong horizontal white thread). The results of these measurements indicated that the true dawn by the naked eye was observed at sun vertical depression $D_o = 14.56°$ (mean + 1SD)), while it was ranged from $D_o = 14°$ to $15°$ by measuring the light intensity using multi instruments depending on three different criteria for eye threshold of M, M1 and M2 (in the moonless sky conditions). The full hierarchical shape of the false dawn does not occur regularly in every morning. Generally, the interval time of the false dawn was found to be within $15° \leq D_o \leq 18°$. In the most morning twilight days, the percentage of color portion was Blue > Green > Red before $D_o \approx 13.5°$ and it changes its behavior after this degree.

Keywords: Sun vertical depression (D_o) · Atmospheric optics · True dawn · False dawn · CCD · Canon and Nikon cameras · Image processing · SQM · Naked eye observation

H. M. K. Al Naimiy et al. (Eds.): AUASS-CONF 2023, SPPHY 420, pp. 178–209, 2025.
https://doi.org/10.1007/978-981-96-3276-3_14

1 Introduction

According to Hassan et al. [1], it was concluded that there is a big agreement between the photoelectric measurements and the naked eye observations for the morning of twilight (true dawn) in four regions in Egypt. Also, many researches of the true dawn and false dawn were published. It is well known that observing the faint light (zodiacal light or false dawn or pseudo dawn) and strong white thread (true dawn) by the naked eye depends on the contrast threshold of the human eye during the evening or morning night sky [2–6].

The eye control of rays entering by a process is known as adaption. The iris (the colored part of eye) changes its area to control the size of the eye pupil. This can adapt the eye to a range of brightness of about 1:1 million. But the range of movement of eye pupil diameter is only four-fold, and so the control of human pupil eye size is not the only compensatory mechanism for brightness. The human eye retina itself is responsible for the adaption. The eye responses the intensity of the light logarithmically. This means that the eyes response is directly small for any given change in real intensity. The minimum energy rate for light incident into the eye must be at least 10^{-6} Watts. The number of photons arriving the eye each second is 250 photons. Many of these original photons do not arrive the human retina and are "wasted". Some 2% of the light incident on the cornea is lost by reflection. For light of wavelength 500 nm, 50% of the remaining photons are lost by absorption and scattering within the various media on the way to the human retina. The quantum efficiency is the ratio of the number of photons which reach into a rhodopsin molecules and cause electro genesis to the numbers of photons incident of the cornea [6, 7].

Studies of twilight (false dawn and true dawn) are very crucial to broad fields such as: the sighting of the new moon for any lunar month to be on the background of evening and morning twilight, calculation the total energy of the stars and zodiacal light must be taken into account because it is the dividing line between the end of the night and morning twilight or the end of evening twilight and the appearance of the stars [8–12].

In Egypt, during the winter of 1908 in Aswan, observations of the morning and evening twilight by a camera were carried out. The conclusion of the results was that the first thread light appeared from the east direction was at $D_o = 17.35°$ (false dawn) and the first appearance of color difference was at $D_o = 14.25°$ (true dawn), while the dusk in the evening twilight disappeared at $D_o = 14.9°$ [13].

Thomas, L and M. R. Boivman studied the penetration of solar radiations of wavelength 100- 600 nm to heights atmosphere below 100 km during the pre-sunrise period numerically. Their studies have shown the relative importance of absorption by molecular oxygen and ozone, and Rayleigh scattering in limiting the atmosphericpenetration of solar radiations during the pre-sunrise period [14].

Rozenberg reported that when the sun vertical depressions reaches of $D_o = 10–15°$ below the horizon, the intrinsic glow of the upper atmospheric layers begins to appear together with starlight and the illumination conditions gradually approach those of night sky. Also, he reported that the transition to complete night is usually when the sun vertical depression is between $D_o = 17 -19°$ [15].

In Yemen (Sana'a, $\varphi = 15.4°$ N, $\lambda = 44.2°$ E, Elev. 2200 m), Sultan studied the morning twilight in the autumn of 2003, when the excellent conditions of observing was during the period 23–28 November after the heavy rain, clear with dry sunny days and very clear nights sky. These data have been taken by the naked eyeobservations. The false dawn (the beginning of the astronomical twilight) started to glow faintly when the sun vertical depression was $D_o = 18.95°$, while the developing twilight starts to show colors divergence at $D_o = 13.2°$. The true dawn (the beginning of nautical twilight) started when $D_o = 12°$ [16].

At KSA (Riyadh, $\varphi = 25° 45' 41'' $ N, $\lambda = 47° 12' 10'' $ E, Elev. 540 m) the twilight of the true dawn in the deep desert was studied by the naked eye observations and camera (Nikon D70) detection. Both methods were carried out in parallel to each other. The observations were recorded during one year (twice in every month) by four groups consisting of two observers. These results indicated that the true dawn can be determined by naked eye at sun vertical depression $D_o = 14.6° \pm 0.3$, while it can be detected by the camera at $D_o = 14.5° \pm 0.62$ [17].

The UBVRI twilight sky brightness was estimated on more than 2000 FORS1 archival images, which included both flats andstandard star observations taken in twilight and covering a sun vertical depression range $4° \le D_o \ge 22°$. Twilight studies have been proved to be important tools for analyzing the atmospheric structure with interesting consequences on the characterization of astronomical sites. Active discussions of this topic have started again recently in connection with the evaluation of Dome C, Antarctica, as a potential astronomical site and several site-testing experiments, including twilight brightness measurements. One of the most interested results from this work is detecting of the beginning of the true dawn at $D_o = 14.8°$ below the horizon. It was found that there is a cardinal point of reversing the spectrum of the reflected radiation from the earth surface, while the Rayleigh scattered sun flux contributes in the blue-wards region of 5000Å until $D_o = 15°$, after which the Pseudo continuum of the night sky emission takes over [18, 19].

Many published several researches for measured the true and false dawn at different sites in Egypt (2008–2019), namely; Bahria, Kottamia, Matrouh, Aswan, Sinai, Assiut and Wadi Al Natron. It was concluded from of those studies that the sun vertical depressions for the beginning of true dawn ranges between $13.5° \le D_o \le 15°$, while the false dawn ranges between $15.5° \le D_o \le 18.5°$ according to the photoelectric measurements and naked eye observations, where the background were desert, sea and agriculture [20–28].

In Libya (Tubruq, $\varphi = 32° 05'$ N, $\lambda = 23° 59'$ E, Elev. 40 m) during the period (2007–2008) in the Mediterranean Sea conditions (429 day), the sun vertical depressions for the beginning of true dawn was $D_o = 13.48°$, while during period (2009–2013) as the desert background (624 day), it was $D_o = 14.7°$ [29, 30].

In Indonesia (Depok, $\varphi = 6° 27'$, $\lambda = 106° 48'$ E, Elev. 50–140 m) using the SQM instruments to study 26 cloudless morning twilight measurements, it was concluded that the true dawn announces itself at $D_o = 14° \pm 0.6$, where the observations were taken when the sensor was pointing toward the zenith [31].

In Malaysia, the morning twilight was studied by SQM in five regions to measure the brightness of the night sky (false and true dawns). The measurements were taken when the device was directed to the east direction toward at the horizontal angle of the sunrise and at five degrees above the horizon. The sun vertical depression (D_o) and the illuminance of night sky in magnitude (m, mag./arcsec^2) for the beginning of the true dawn was at $D_o = 14.19° \pm 0.52$ (for high confidence $D_o = 14.71°$) and of magnitude $m = 21.22^m \pm 0.25$, while the begin of the pseudo dawn was at $D_o = 18.62° \pm 0.82$ and of magnitude $m = 22.17^m \pm 0.104$. The hierarchical shape of the pseudo dawn did not occur regularly. The difference in the illuminance of light magnitude (m) between the true dawn and the full night was found to be 0.83^m as a relative value from the sky background [32].

In KSA (Hail, $\varphi = 27° \, 31' \, N$, $\lambda = 41° \, 42' \, E$, Elev. 1015 m) and different locations (in the deep desert), the beginning of the appearance of the true dawn was studied as a strong light from among the darkness that increases over time with the naked eye. It was found that it appears at $14.66° \le D_o \le 14.8°$ [33].

In Egypt (Fayum, $\varphi = 29° \, 17' \, N$, $\lambda = 30° \, 03' \, E$, Elev. 50 m) the morning twilight throughout good seeing condition in 2018-2019 was studied for four different applied criteria. The true dawn was found to be appearing between $D_o \approx 14°$ and $14.8°$ according to observations of the naked eye and SQM. The illuminance of night sky radiation in both directions of vertical and east directions became equal at the nautical twilight $Do = 12°$ [34].

Table 1 summarizes the previous published works of the true dawn using naked eye (*N.E*), photoelectric (*P. E.*) and SQM in the many different regions in the world through the interval 1909–2022 including the observations in the different countries, namely: Egypt, Libya, KSA, Indonesia and Malaysia, D_o (degree) and authors. The final results of the sun vertical depressions D_o of the beginning of the twilight) are ranging from $D_o = 13.5°$ (in the coastal and agricultural background) to $D_o = 14.7°$ (in the desert background), since it depends mainly on the transparency of the atmospheric conditions. Note that the absolute values without the standard deviation mean that these values are added to the standard deviation (SD) as the highest degree of confidence (mean + 1SD).

The aim of this present work is determining the true and false dawn in different regions at desert, agricultural and coastal backgrounds in Egypt by multi methods (four devices) and naked eye depending on the observations of group trips.

Table 1. The important previous research of sun vertical depression (D_o) for the true dawn using, SQM, photoelectric (P.E), camera instruments and naked eye observation (N.E).

Place	Publication Year	Tools	Do (degree)	Reference
Egypt, Aswan	1909	Camera	14.25	[13]
Yemen, Sana'a	2004	Camera, N.E	12°	[16]
KSA, Riyadh	2005	Camera, N.E	14.6 ± 0.3	[17]
KSA, Hail	2018	N.E	14.66 -14.8	[33]
Libya, Tubruq (2 researches)	2007–2008 and 2009–2013	N.E	13.5–14.7	[29, 30]
Egypt (9 researches)	2008–2019	N.E and P.E	13.5–14.7	[20–28]

(*continued*)

Table 1. (*continued*)

Place	PublicationYear	Tools	Do (degree)	Reference
Indonesia, Depok	2020	SQM	14 ± 0.6	[31]
Malaysia (5 regions)	2021	SQM	14.71	[32]
Egypt, Fayum	2022	N.E and SQM	14- 14.8	[34]

2   Site Observation and Methodology

Table 2 represents the location of observation (Lat. (*N*), Long. (*E*), Elev., *NL*, numbers of observation group trips (G.T) and the tools of observation for morning twilight days in Egypt (Kottamia, Aswan, Kharga, Hurghada, Fayum and Marsa-Alam). The tools of observation were: two cameras (Canon and Nikon), CCD camera, Sky Quality Meter (SQM) and naked eye (*N.E*). These observations were in the desert and coastal background during the time interval from 2015 to 2019. Each trip consisted of an average member from 4 to 9 members of different ages from 15 to 65 years and was of different diversity of people (professional and amateur astronomers).

Table 2. The locations of observation groups (G.T) at different places and the tools of observation.

Location	Lat. N	Long. E	Elev. (m)	N. L	G.T	Tool
Kottamia	29°:55.9′	31°: 49.5′	411	Desert	3	Canon, SQM, N.E
Aswan	23°:48.2′	32°: 29.5′	210	Desert	2	Nikon, Canon, N.E
Kharga	25°:18′	30°: 10′	40	Desert	1	Canon, N.E
Hurghada	27°:06′	33°:51′	30	Sea	1	N.E
Fayum	29°: 17′	30°: 03′	50	Desert	2	Nikon, Canon, SQM, CCD, N.E
Marsa-Alam	25°: 53′	34° 25′	20	Sea	1	N.E

All observations in this research were taken horizontally towards the east direction. The observations of morning twilight were taken when the moon was below the horizon, except for two observation days (26th Dec. 2015 in Aswan, when the moon was full (phase of the moon is 0.993) and 25th Nov. 2016 in Fayum when the moon was after the last quarter (phase of the moon is 0.164)). The Python ver. 2.7 program (numpy and scipy package, BGR as: B (blue color: 400–480 nm), G (green color: 480-600 nm), R (red color: 600–700 nm)) was used to analyze images of the camera into BGR colors. Measurements of the dawn have been recorded by the following devices in parallel:

1- Canon EOS 20D, 8.2M COMS Sensor, 5 fps Hi-Speed Shooting, Hi-Precision 9-point AF, DIGIC II Image processor.
2- Nikon D5200F-stop: f/2.5 Exposure time: 20s, ISO speed: ISO-100, Focal Length: 18 mm, Max aperture: 1.6

3- Apogee U8300, CCD Kodak KAF-8300 or KAF-8300CE, Array Size (pixels) 3326 × 2504, Pixel Size 5.4 × 5.4 microns, Imaging Area 18 × 13.5 mm (243 mm2)
4- Sky Quality Meter (SQM-LU-DL).

Additionally, the tools of calculations used in our analysis were:

1- Python ver. 2.7 (program to analysis image process of camera into B, G and R colors): using numpy and scipy package, B, G and R as: B (blue color: 400–480 nm), G (green color: 480–600 nm), R (red color: 600–700 nm).
2- M: The eye's light threshold is 2% of the celestial background [15] and $\Delta m = 0.75^{m}$ [25], $\Delta m = 0.83^{m}$ [32], where we calculate 2% at the time of $D_o \geq 20°$ because it represents the true night away from effected of the zodiacal light or the false dawn. The 2% percentage represents a distinct scientific proposal, because in reality the celestial background cannot be fixed or similar on all days, but it is variable according the weather conditions from one night to another and from one season to another.
3- M1: The threshold of light perceived by the eye that distinguishes the true dawn from the heavenly background, based on [32], where the corresponding degree for the threshold of eyes is 0.83^{m} (as the relative magnitude values). So, the subtraction of this value from the full night sky gives the magnitude value of the true dawn, which gives the true dawn degree (D_o) corresponding to this light magnitude.
4- M2: The threshold of eye for the true dawn as the light intensity equivalent to the height of zodiacal light (false dawn) in the curve and the corresponding degree of the vertical sun depression (D_o).

To calculate the percentage necessary for the eye to see from the difference in light magnitude between complete night and the beginning of the appearance of light that the eye senses [36, 37];

$$\Delta m = m_B - m_A = 2.5 \log_{10}(I_A/I_B), I_A/I_B = (2.512)^{(m_B - m_A)} \tag{1}$$

$$2.512^{0.75} = I_A/I_B = 1.995(\approx 2\%), 2.512^{0.83} = I_A/I_B = 2.148(\approx 2\%) \tag{2}$$

where: Δm represents the minimum difference in the amount of illumination between the amount of illumination of the sky background and the amount of illumination required for the eye to sense it.

1- To convert the local mean time of the morning twilight stages $(B.T)$ to sun vertical depressions D_o, we use the equations of [25, 27], checked with the Moon calculator, Moon 6 [38].
2- Optical properties calculations are carried out as follows:
   a- The color intensity (arbiter unit) is C.I (a.u): blue (B), green (G) and red (R).
   b- The total intensity is T.I (a.u) $= B + G + R$.
   c- The color portion is C.P (%) for the three colors B, G and R calculated as:

$$C.P (B) = (B/T.I) \times 100, \ C.P (G) = (G/T.I) \times 100 \text{ and } C.P (R) = (R/T.I) \times 100$$

3 The Results

Based on the proposed hypotheses and observations taken by different tools (five tools, four devices and N.E) for the morning twilight and in different regions, we reached the following results:

Figures 1 and 2 show the variation of the color intensity C.I (a.u) for the Blue – Green – Red (BGR) colors, the color portion C.P (%) for the same bands and the total radiation intensity T.I (a.u) reduced from a Canon camera image according the sun vertical depression (D_o) on 20th Aug. 2015 in Kottamia. It is clear that the gradient in radiation of C.I (a.u) and C.P (%) are B > G > R for the interval time $D_o = 6°–12°$, while for the interval time $D_o = 12°–18°$, it is R > B and G. The general profile of the C.P of B (%) is corresponding to T.I (a.u) and reversed with R values. The intensity of all colors begins with a perceptible increase at $D_o = 14°$ (M: The eye's light threshold is 2% over the night sky background) and the true dawn by N.E group observations was found to be $D_o = 13.85°$.

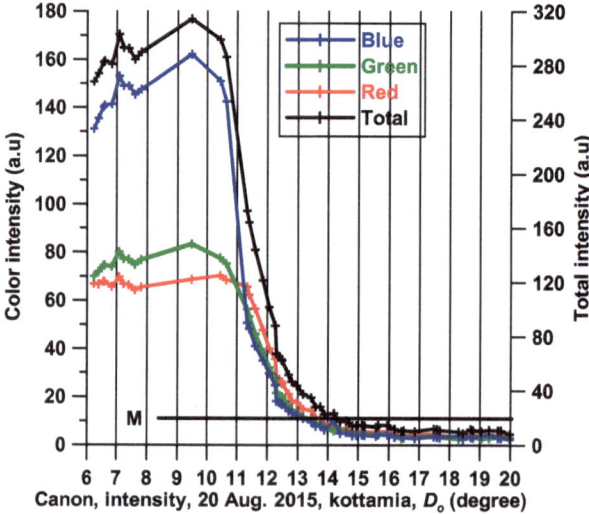

Fig. 1. The variation of the color intensity C.I (a.u) for the BGR colors and the total radiation intensity (T.I) with the sun vertical depression (D_o) at 20th Aug. 2015 in Kottamia.

Figure 3 shows the variation of the total intensity T.I (a.u) and the color portion C.P (%) reduced from a Canon camera image for the BGR colors against the sun vertical depression (D_o) on 19th Sep. 2015 in Kottamia. The gradient of C.P before $D_o \approx 13°$ (C.P $\approx 33\%$) is: B > G > R, while at the interval time $D_o = 13°–18°$ it is reversed. The sun vertical depression $D_o \approx 18°$ was found to be the inverse point direction of three colors, the point $D_o = 13.5°$ was found to be the beginning of color divergence, the increase rate of T.I (a.u) starts clearly at $D_o = 15°$ (M: The eye's light threshold is 2% over the night sky background), while the true dawn was detected by $N.E$ at $D_o = 14.5°$, at which the total light intensity starts to exceed the threshold of the eye.

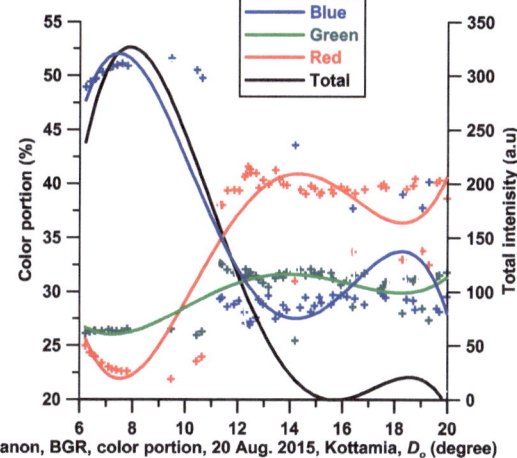

Fig. 2. The variation of color portion C.P (%) for the BGR colors and the total radiation intensity (T.I) with the sun vertical depression D_o (degree) at 20th Aug. 2015 in Kottamia.

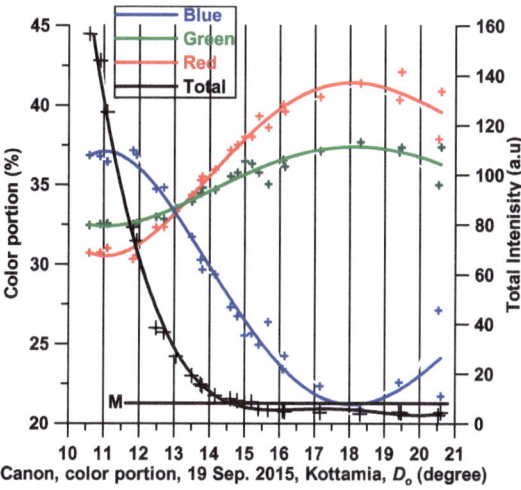

Fig. 3. The variation of T.I (a.u) and C.P (%) for the BGR colors with D_o (degree) at 19th Sep. 2015 in Kottamia.

Figures 4 and 5 show the variation of BGR colors for T.I (a.u), C.I (a.u) and C.P (%) reduced from the Canon camera image against the sun vertical depression (D_o) at 11th Feb. 2016 in Kottamia. The interval time from $D_o = 15.5°$ to $18.5°$ represents the zodiacal light rang (false dawn). The T.I (a.u) after $D_o = 18°$ represents a constant intensity of the full night. It is clear that the depression angle of the true dawn (at M values) is $D_o = 14.8°$. The values of C.I (a.u) and C.P (%) are: G > R > B in the interval of the false dawn and full night ($D_o \approx 15°–20°$), while for the true dawn region at $D_o < 15°$ it is B > G > R. In all pass bands, the blue color behavior is similar to the total intensity trend. The point of $D_o = 13.\overset{.}{5}°$ (C.P $\approx 35\%$) represents the beginning of the color divergence for G and B, while the eye begins to discriminate the white thread (*N.E*) at $D_o = 14.55°$.

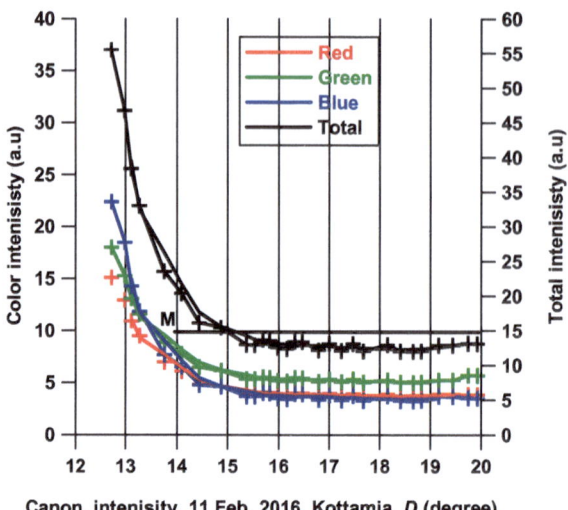

Fig. 4. The variation of BGR colors for T.I (a.u) and C.I (a.u) with D_o (degree) at 11th Feb. 2016 in Kottamia.

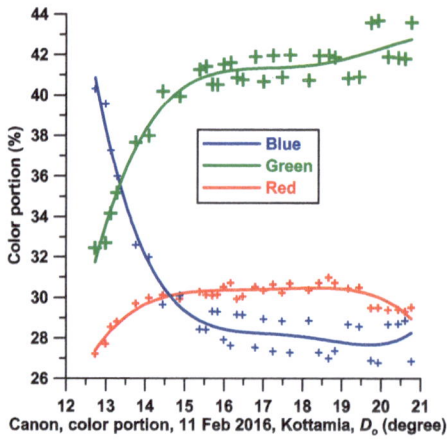

Fig. 5. The variation of C.P (%) with D_o (degree) at 11th Feb. 2016 in Kottamia.

Figure 6 shows the relation between the sun vertical depression D_o and both the illuminance radiation (mag./arcsec^2) and air temperature (°C) at Kottamia from the full night to sunrise for the morning twilight of 28th June 2019. The full night sky is represented as the straight line of a magnitude value at 19.5^m and occurs from sun vertical depression $D_o \approx 20°$ going to upper values. So, the true dawn has to be of magnitude 18.67^{m} (subtracting 0.83^{m} from the full night) which corresponds to $D_o =$ $13.5°$. It is noticed that the minimum measured temperature was ≈ 21 °C at the sun vertical depression $D_o \approx 6°$ and there is no significant variation in the temperature from the beginning of the true dawn until the sunrise. For Kottamia (Figs. 1, 2, 3 and 4) on the different days from the different curves of the radiation intensity (a.u) and the color portion (%), it was clear that the $D_o = 14°$ indicates the beginning of the true dawn according the threshold of eye (M1).

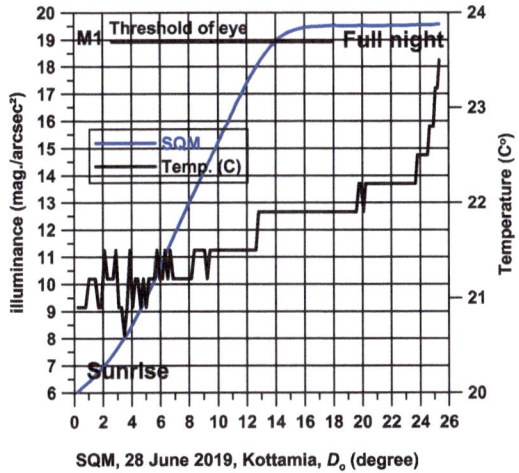

SQM, 28 June 2019, Kottamia, D_o (degree)

Fig. 6. The relation between the sun vertical depressions D_o (degree) and both the illuminance (mag./arcsec^2) and air temperature (C°) at Kottamia for the morning twilight on 28th June 2019.

Figure 7 shows the radiation variation of C.P (%) and T.I (a.u) reduced from the Canon camera image for the BGR colors against the sun vertical depression (D_o) on 20th Nov. 2015 in Kharga. Generally, the gradient values of C.P are G > R > B before $D_o = 13.5°$, the red color has an opposite behavior to the blue color over the observation periodwhile there are two turning points between them at $D_o \approx 12°$ and $D_o \approx 19°$. At D_o $= 13.25°$, there is an intersection point between the green and blue color portions which represents the beginning of color divergence. The increasing rate of T.I starts clearly at $D_o = 17°$ which represents the start of the zodiacal light (false dawn), while it starts at $D_o = 14.5°$ for the true dawn. The point at $D_o = 12°$ (nautical twilight) represents point of reverse direction for the B and G colors.

Figure 8 shows the radiation variation of C.P (%) and T.I (a.u) as the reduction from the Canon camera image for the BGR colors against the sun vertical depression (D_o) at 22th Nov. 2015 in Kahrga. It shows that the gradient values of C.P (%) are B > G > R when the sun vertical depression is $D_o \le 13.5°$ (C.P $\approx 33\%$), while in the interval time between the sun vertical depression values 13.5°and 18° it is G > R > B. At $D_o = 16.5°$, the curves of the blue and green get reversed, while the point $D_o = 13.5°$ represents the beginning of color divergence. The increasing rate of T.I start clearly at $D_o = 15°$, while the depression value of $D_o = 18.5°$represents the maximum intensity of zodiacal light (false dawn).

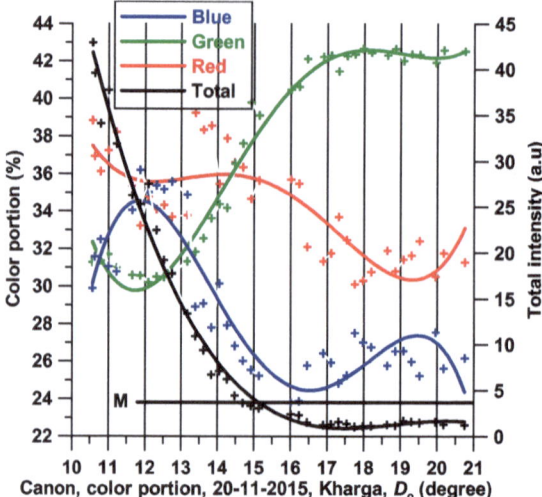

Fig. 7. The variation of C.P (%) and T.I (a.u) for the BGR colors against the sun vertical depression D_o (degree) at 20th Nov. 2015 in Kahrga.

Figure 9 shows the variation of C.I against D_o of B, G and R colors for the morning twilight of mean values of four days (19th - 22th Nov. 2015) at Kharga. For all colors, the increasing rate of C.I starts at $D_o \approx 17°$ (false dawn), while the contrast of the colors due to the threshold of the eye (M) starts at $D_o \approx 15°$.

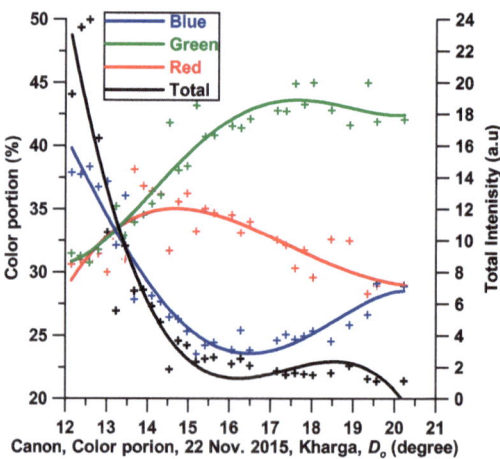

Fig. 8. The radiation variation of C.P (%) and T.I (a.u) values for the B, G and R colors against the sun vertical depression D_o (degree) at 22th Nov. 2015 in Kahrga

Figure 10 shows the variation of C.I (a.u) against D_o (degree) of B, G and R colors reduced from the Nikon camera image for the morning twilight on 23th Dec. 2015 in Aswan. The time corresponding the sun vertical depression beginning from $D_o \approx 15.5$

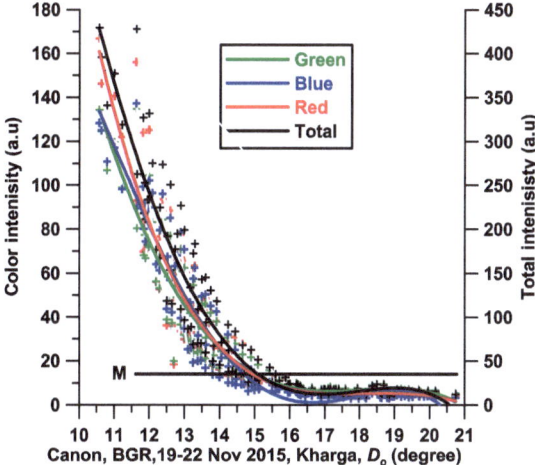

Fig. 9. The variation of C.I (a.u) against D_o(degree) of BGR colors for the morning twilight of mean values of the four days (19th-22th Nov. 2015) at Kharga.

going to the deep night, the order of C.I is R > G > B, the range $D_o \approx 14.5°$–$15.5°$ represents a period of inversion in the illuminance, while in the interval time corresponding to $D_o < 14.5°$ the color order is B > R > G.

Figure 11 shows the variation of C.P (%) against D_o (degree) of B, G and R colors for the morning twilight on 23th Dec. 2015 in Aswan (moonless condition). The point $D_o = 14°$ (C.P $\approx 35\%$) represents the intersection point of the two colors red and blue. The dominant color at $D_o > 14°$ is red, while the green color (C.P $\approx 35\%$) gets changed slightly over the time period $D_o = 12°$–$23°$.

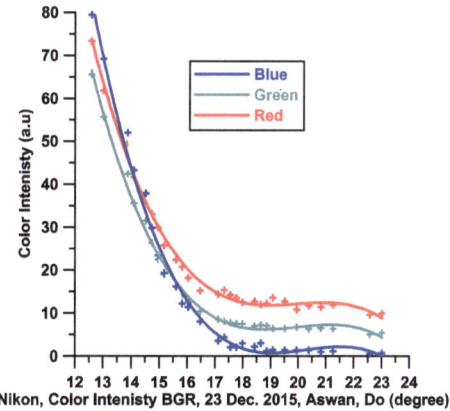

Fig. 10. The radiation variation of C.I (a.u) values against D_o (degree) of B, G and R colors for the morning twilight of 23th Dec. 2015 at Aswan.

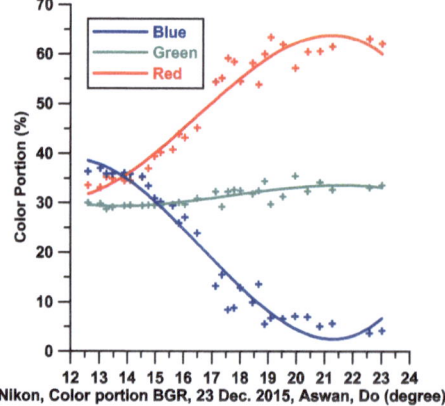

Fig. 11. The radiation variation of C.P (%) values for the B, G and R colors against the sun vertical depression D_o (degree) of 23th Dec. 2015 at Aswan.

Figure 12 shows the variation of C.I (a.u) against D_o (degree) of B, G and R colors from Nikon image for the morning twilight at 26th Dec. 2015 in Aswan. It is obvious that the minimum colors intensity are in the interval time corresponding to $D_o \approx 15°$–$16°$ which represents the period of darkness (about one degree) between the end of the false dawn and the beginning of the true dawn, and the C.I (a.u) shows that the R > G > B in all the twilight interval (N.E, $D_o = 13.84°$). The maximum intensity of the false dawn is at $D_o \approx 19.5° \pm 1$.

Figure 13 shows the variation of C.P (%) against D_o (degree) of B, G and R colors from Nikon camera image for the morning twilight on 26^{th}Dec. 2015 in Aswan. The point $D_o = 14°$ represents the inflection point of two colors R and B. The dominant color after $D_o = 14°$ is red. The green color still gets changed slightly over the time period corresponding to $D_o = 12°$-$23°$. The means of the color portions distribution during this interval are R (37%), G (33%) and B (30%).

Table 3 represents the state of the sun and the moon during the morning twilight observation on 26th Dec.2015 in Aswan. The sun vertical depression D_o is calculated from the full night at $D_o \approx 20°$ to civil twilight $D_o \approx 6°$ (sun vertical depression = -altitude of the sun), SAz (sun azimuth), Ma (moon altitude) and MAz (moon azimuth). The moon is on the second day after the full moon (illuminated percentage is ≈ 0.993) and at $D_o = 14°$, the relative altitude between the sun and moon $\Delta a = Ma\text{-}D_o = 33°$, the altitude of the moon ranges from $45°$ to $17°$, while ΔAz gives the gradient approximately constant around $171°$ ($\Delta Az = MAz\text{-} SAz$) since the moon was on the opposite side of the sunrise and the illuminance of the moon is around -12.48 mag./arcsec^2 (at full moon is -12.78^m). This means that the measured intensity represents the intensity of the twilight plus the intensity of the light reflected from the moon. Comparing between Fig. 11 (moonless condition) and Fig. 13 (moon above horizon), we have to notice that the new moon happened on 25th Dec. 2015 at 11h: 12m: 30s (UT) and its conditions are given in Table 3. To compare between the two days 23th Dec. and 26th Dec., 2015 in terms of C.I (a.u) and C.P (%), we find that the distribution of the color intensities are R > G > B for $D_o > 14$. This means that the appearance or absence of the false dawn does not change the color distribution, but changes the shape of color portion (C.P) over the time period and the color divergence cannot be seen in the case of moon above the horizon and the point $D_o \approx 14$, which represents a pivotal point (true dawn).

Table 3. Thestate of sun and moon during the observation of morning twilight on 26th Dec. 2015 at Aswan, D_o (sun vertical depression = -altitude of the sun), SA_z (sun azimuth), Ma (moon altitude), MA_z (moon azimuth), $\Delta a = Ma\text{-}D_o$ and $\Delta A_z = MA_z\text{-}SA_z$.

D_o degree	SA_z degree	Ma degree	MA_z degree	Phase of the moon	Δa degree	ΔA_z degree	Illuminance of the moon mag./arcsec^2
20.04	107.04	25.233	279.233	0.993	45.25	171.82	-12.48^{m}
18.08	108.19	23.267	280.03	0.993	41.55	171.85	-12.48^{m}
15.05	109.29	20.22	281.15	0.993	35.27	171.86	-12.48^{m}
14	109.68	19.05	281.5	0.993	32.99	171.82	-12.47^{m}
12.04	110.44	17.188	282.28	0.992	28.46	171.84	-12.47^{m}
6.09	112.89	11.185	284.6	0.992	17.27	171.71	-12.47^{m}

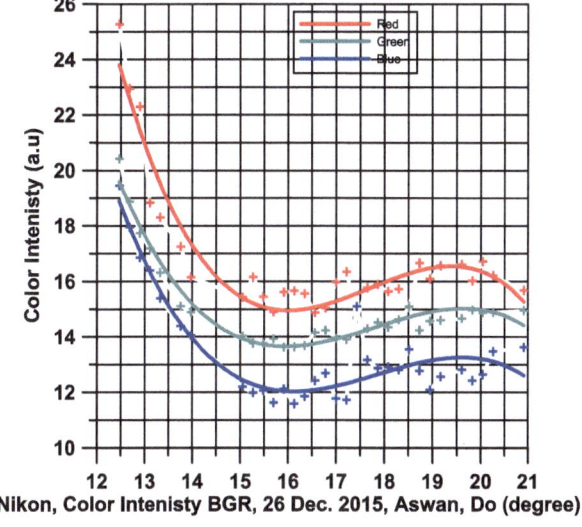

Fig. 12. The variation of C.I (a.u) with D_o (degree) of BGR colors for the morning twilight on 26th Dec. 2015 in Aswan.

Figure 14 shows the variation of total intensity T.I (a.u) against the sun vertical depression D_o on 23th Dec. 2015 (moonless condition) and the increase of total intensity of light which is semi-constant in the regions $D_o \approx 15°–21°$ and the shape of the false dawn is invisible.

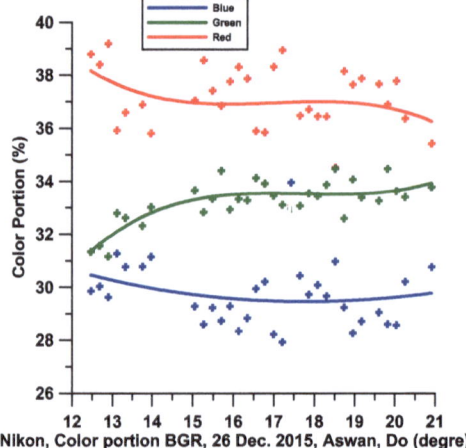

Fig. 13. The variation of C.P (%) against D_o (degree) of BGR colors for the morning twilight on 26th Dec. 2015 in Aswan.

Figure 15 shows the variation of the total intensity T.I (a.u) against the sun vertical depression D_o on 26th Dec. 2015 (moon above horizon), where the variation of false dawn (hierarchical form) ranges 4° in the region $D_o \approx 15.5°–19.5°$. Therefore, adding an equivalent amount of light above the threshold of the eye (the horizontal line in the figure) gives the condition which the human eye can discriminate the true dawn. From this figure, the sun vertical depression corresponding to this amount of the light intensity is $D_o = 14.4°$. This means that the false dawn with its well-known hierarchical form does not appear regularly on all days. On 26th Dec., the interval time about of $D_o = 1°$ in the range $D_o \approx 15.5°–16.5°$ represents the dark interval between the end of the false dawn and the beginning of the true dawn.

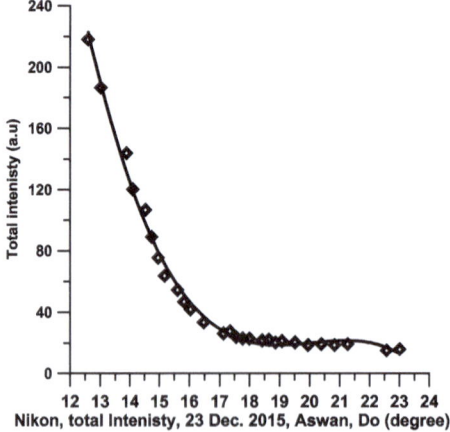

Fig. 14. The variation of T.I (a.u) against D_o (degree) of B, G and R colors for the morning twilight on 23th Dec. 2015 at Aswan.

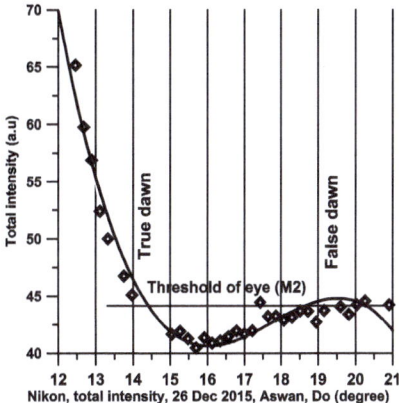

Fig. 15. The variation of T.I (a.u) with D_O (degree) of B, G and R colors for the morning twilight of 26th Dec. 2015 at Aswan.

Figure 16 shows the variation of C.I (a.u) and T.I (a.u) against D_o (degree) of B, G and R colors for the mean values on four days 12th-15th Jan. 2016 at Aswan, which show for all colors the maximum intensity of zodiacal light (false dawn is around $D_o = 19.5° \pm 1.5$) and the increasing rate of C.I (a.u) and T.I (a.u) start at $D_o \approx 16°$, while the contrast of colors begins at $D_o \approx 15°$ (true dawn). The distribution of the color intensity of the three-colors are B > G > R. The general behavior in terms of temporal change of illuminance has the same direction throughout the entire twilight.

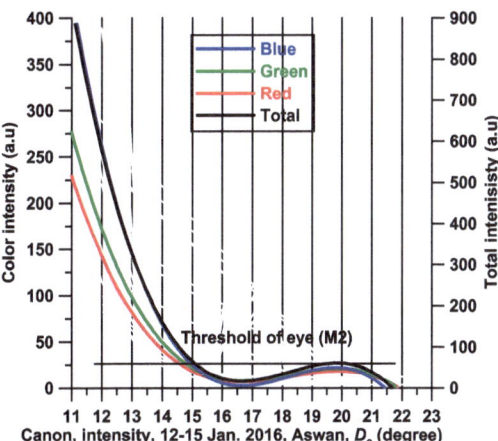

Fig. 16. The variation of the mean values of BGR colors for C. I (a.u) and T.I (a.u) against D_O (degree) on the four days 12th-15th Jan. 2016 at Aswan.

Figure 17 shows the variation of C.P (%) and T.I (a.u) reduced from the Canon camera image for the B, G and R colors against the sun vertical depression (D_o) on 12th Jan. 2016 in Aswan. The C.P (%) distribution are G > R > B before $D_o = 14°$, while for the interval time $D_o > 14°$ it has an opposite trend. There are two points of inversion at $D_o \approx 21°$ and $D_o \approx 18°$. The increasing rate of T.I start clearly at $D_o = 15°$, while at $D_o = 13.75°$ (C.P 34%) the beginning of colors discrimination (the intersection of three colors) begins. Hence, the true dawn D_o starts to get seen in the sun vertical depression range between 13.75° and 15°.

Figure 18 shows the variation values of both the C.P (%) and T.I (a.u) against D_o for B, G and R colors on 13th Jan. 2016 in Aswan. The Percentage of C.P are B > G > R. Both B and G reflect their behavior at $D_o = 16°$ and the maximum points of inflection for B, G and R colors are at $D_o = 13°$ and $D_o = 18°$, where the increase of T.I starts at $D_o = 17.5°$.

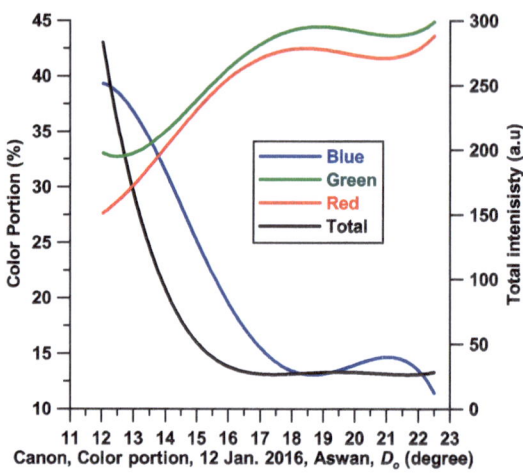

Fig. 17. The variation of C.P (%) and T.I (a.u) for B, G and R colors against the sun vertical depression D_o (degree) on 12th Jan. 2016 in Aswan.

Figure 19 shows the variation values of T.I (a.u) reduced from the Canon camera image against the sun vertical depression (D_o) on 14th Jan. 2016 in Aswan. The interval time from $D_o = 15°$ to 18° represents the zodiacal light (false dawn). The interval time $D_o > 18°$ represents a straight line and the constant intensity which means that the light region is the full night (scotopic vision). It is clear that, the true dawn is at the intersection point on the line of the threshold of eye (M2) with the intensity curve which is equivalent to the amount of light intensity of the false dawn after disappearance at a time corresponding to $D_o = 13.3°$as the marked increase of the illuminance which means definitely the true dawn. It is also noted that after the disappearance of the hierarchical shape, which represents the false dawn, the level of light intensity (M) does not return to the state of complete darkness, but exceeds the amount of light, due to the continuous state of dispersion arising from the collapse of the light of the false dawn and the preservation of the atmosphere near the surface of the earth with a percentage of these dispersed lights before the appearance of the true dawn.

Figure 20 shows the variation values of C.I (a.u) for BGR colors reduced from the Canon camera image with the sun vertical depression (D_o) at 14th Jan. 2016 in Aswan. The time interval corresponding to $D_o = 15°$ to 18° represents the zodiacal light (false

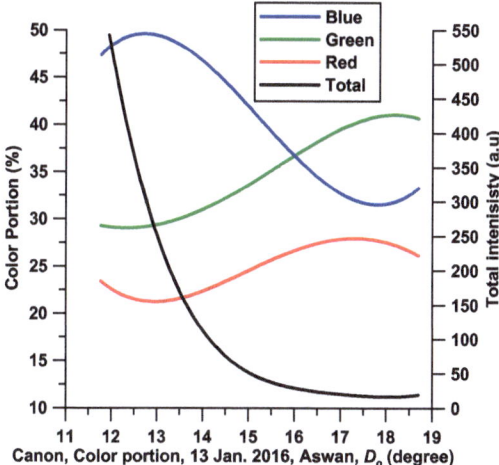

Fig. 18. The variation values of both the C.P (%) and T.I (a.u) against the sun vertical depression D_o(degree) for B, G and R colors on 13th Jan. 2016 in Aswan.

dawn) for the three colors. The time interval corresponding to $D_o > 18°$ represents a straight line of a constant intensity, which means that this light region is the full night. The distribution of C.I (a.u) is B > G > R, especially in the light regions for false and true dawn. It is clear that the blue color is dominant, especially in the region of less than 18°, which means that the air is clear of pollutions (natural and artificial), since the existence of dust increases the frequent collisions and converts short wavelengths into long wavelengths, which causes the turn from blue to red.

Figure 21 shows the variation values of B, G and R colors for C.P (%) reduced from the Canon camera images against the sun vertical depression D_o on 14th Jan. 2016 in Aswan. The distribution of color percentage values C.P (%) are B > G > R for the false and true dawn. The B and G behaviors get reversed at $D_o = 17.75°$, while the R color has the same direction against different values before and after $D_o = 18°$ and the maximum value at $D_o = 17.5°$. This means that point $D_o = 18°$ (astronomical twilight) is considered as a pivotal point, at which the false dawn gets separated from the full night.

Table 4 represents the state of the sun and moon properties during the observation of the morning twilight on 25th Nov. 2016 at Fayum (4 days after the last quarter, where the last quarter was on 21th Nov. 2016 at 08h: 33m UT). The moon is on the second day after its full state. The Δa (relative altitude between the sun and moon $= Ma-D_o$) has a range of one degree from 46° to 45°, while ΔAz ranges between 8° and 13° ($\Delta Az = MAz-SAz$), which means that the moon azimuth was in the nearest apparent place from the sunrise azimuth and the illuminance of the moon is around -8 mag./arcsec^2. Hence, the effect of the moon on the twilight light in this case is weak.

Figure 22 shows the variation of color intensity C.I (a.u) for B, G and R colors against the sun vertical depression D_o (degree) on 25th Nov. 2016 in Fayum in a large scale (C.I $\approx$ 0–14 (a.u) and $D_o \approx$ 18.5°–9°) and the true dawn appears at $D_o = 12°$. Figure 23 shows the variation of the color intensities C.I (a.u) for B, G and R colors

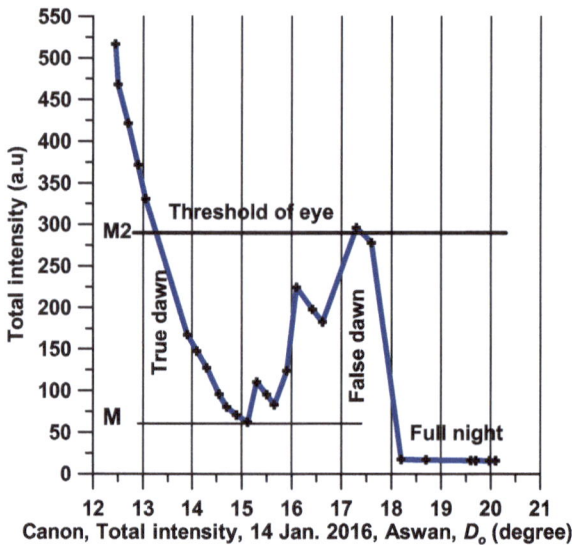

Fig. 19. The variation values of T.I (a.u) against the sun vertical depression D_o (degree) on 14th Jan. 2016 in Aswan.

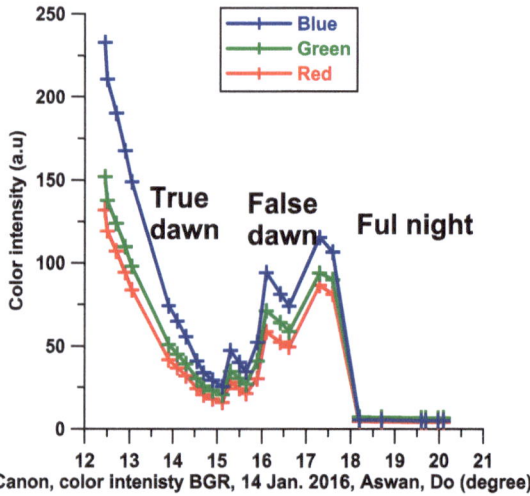

Fig. 20. The variation values of B, G and R colors for C.I (a.u) against the sun vertical depression D_o (degree) on 14th Jan. 2016 in Aswan.

against the sun vertical depression D_o (degree) on 25th Nov. 2016 in Fayum in a small scale (C.I $\approx$ 0.2–1.9 (a.u) and $D_o \approx$ 18.5°–11°). From Figs. 22 and 23, the interval time according to $D_o \approx$ 18.5°–13.5° is a straight line, which is the full night. The false dawn didn't appear in this morning. The color intensity C.I (a.u) distribution was G > R > B.

Figure 24 shows the variation values of B, G and R colors between C.P (%) and T.I (a.u) against D_o on 25th Nov. 2016 in Fayum. The starting point of the color difference is

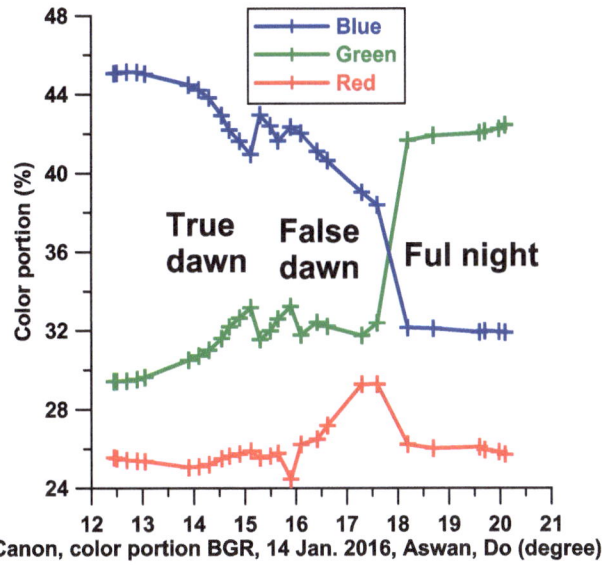

Fig. 21. The variation of B, G and R colors for C.P (%) against the sun vertical depression D_o (degree) on 14th Jan. 2016 in Aswan.

Table 4. The state of the sun and moon during the observation of the morning twilight on 25 Nov. 2016 at Fayum, D_o (sun vertical depression = -altitude of the sun), SAz (Sun azimuth), Ma (moon altitude), MAz (moon azimuth), $\Delta a = Ma\text{-}D_o$ and $\Delta Az = MAz\text{-}SAz$.

D_o degree	SAz degree	Ma degree	MAz degree	Phase of the moon	Δa degree	ΔAz degree	Illuminance of the moon mag./arcsec^2
20.03	103.24	26.28	111.23	0.164	46.31	7.995	−8.07
18.13	104.18	28.04	112.7	0.164	46.17	8.523	−8.07
14.97	105.76	30.94	115.27	0.164	45.91	9.51	−8.06
12.04	107.27	33.59	117.83	0.162	45.63	10.56	−8.04
6.06	110.5	38.86	123.72	0.161	44.93	13.22	−8.03

at $D_o = 9°$, the sun vertical depression at $D_o = 15°$ marks the beginning of the stability in the curve of C.P (%), while at $D_o = 12°$ the marks the increase beginning of the illuminance. The point of $D_o = 9°$ represents the beginning of color divergence for the three colors (C.P ≈ 33%).

Figure 25 shows the variation values of T.I of the light from CCD on 25thNov. 2016 in Fayum. The part of the curve before $D_o = 17°$ represents the full night, while at $D_o = 14.8°$ represents the marked increase of the illuminance, which means the beginning of the true dawn.

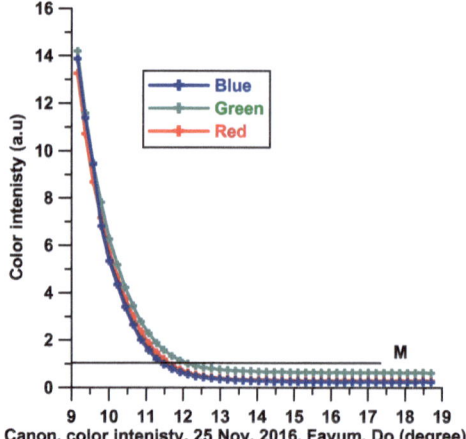

Fig. 22. The variation of color intensity C.I (a.u) for B, G and R colors against the sun vertical depression, D_o (degree) on 25th Nov. 2016 at Fayum in a large scale (C.I ≈ 0–14 (a.u) and D_o ≈ 18.5°–9°).

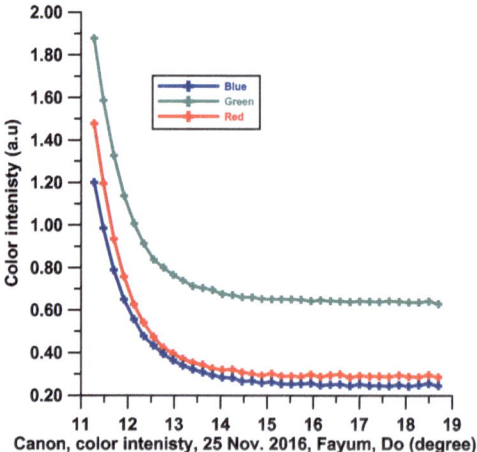

Fig. 23. The variation of color intensity C.I (a.u) for B, G and R colors against the sun vertical depression, D_o(degree) on 25th Nov. 2016 at Fayum in a small scale (C.I ≈ 0.2–1.9 (a.u) and D_o ≈ 18.5°–11°).

From Figs. 22 and 24 for the image process reductions and Fig. 25 as resultant from the data intensity of CCD camera, it is clear that the moon at this phase (0.164) has no effect on the light of the true dawn, since there is no false dawn on this morning. This means that the false dawn does not appear on all days.

Figure 26 shows the light variation of T.I (a.u) from CCD (a.u) on 8th Dec. 2016 in Fayum. It is clear that the full night is in the time period greater than that is corresponding to D_o ≈ 17°, while at that is corresponding D_o ≈ 14° there is a marked increase of the illuminance.

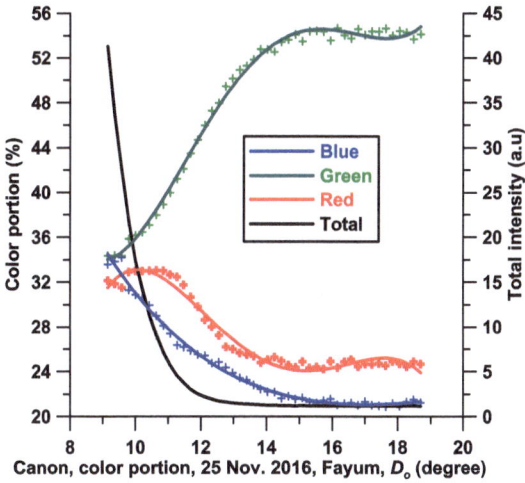

Fig. 24. The variation of B, G and R colors between C.P (%) of B, G and R and T.I (a.u) against D_o (degree) on 25th Nov. 2016 in Fayum.

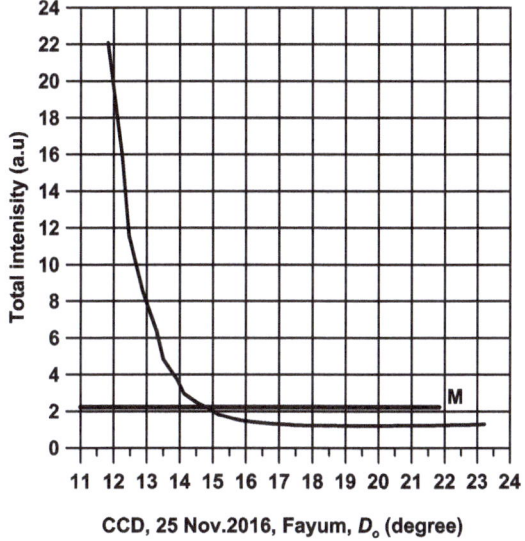

Fig. 25. The variation of T.I against D_o (degree) from CCD on 25th Nov. 2016 in Fayum,

Figure 27 shows the light variation of B, G and R colors for the C.P (%) and T.I (a.u) against D_o (degree) on 8th Dec. 2016 in Fayum reduced from Canon camera image. The starting point of the color discrimination is at sun vertical depression $D_o = 13°$ (C.P ≈ 33%). The point of $D_o = 16°$ marks the beginning of the stability of the curve of C.P (full night). The point $D_o = 12°$ the marked increase of the illuminance begins, while the stability of three colors B, G and R is recorded at $D_o ≈ 18°$. The distribution of the individual C.P (%) is G > R > B, while for the T.I (a.u) there is a noticeable increase of the illuminance starting from $D_o = 14°$ (true dawn).

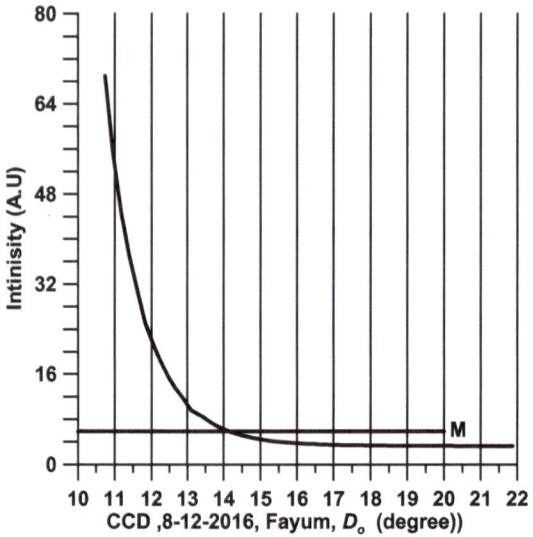

Fig. 26. The twilight variation of T.I (a.u) from CCD (a.u) against the sun vertical depression D_o (degree) on 8th Dec. 2016 in Fayum

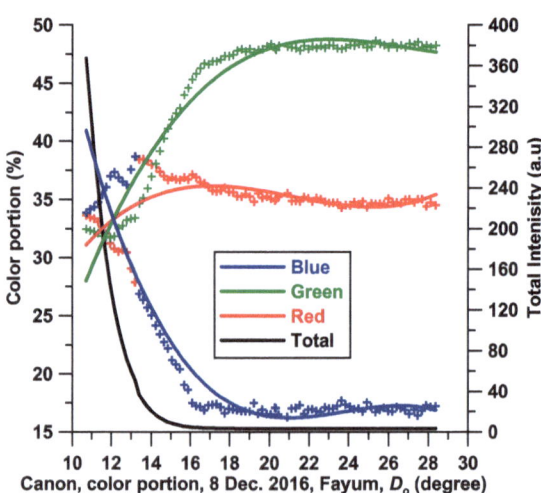

Fig. 27. The light variation of B, G and R colors for the C.P (%) and T.I (a.u) against the sun vertical depression D_o (degree) on 8th Dec. 2016 in Fayum.

4 Discussion

We have some interpretations of the results got from our work.

Interpretation 1:

To compare between two days on which the moon was above the horizon:

The first day was on 12th Dec. 2015 in Aswan (Figs. 12, 13, 14 and 15 and Table 3), the moon phase was 0.993, for the gradient of C.I (a, u), C.P (%) are R > G > B and no point of color discrimination. The moon light has no effect on the shape of the false dawn, which means that the false dawn is not affected by the presence of the moon. But only in the presence of the moon, the true dawn can be distinguished clearly especially by the naked eye.

The second day was 25th Nov. 2016 in Fayum (Figs. 22, 23, 24 and 25 and Table 4), the moon phase was 0.164, the distribution of C.I (a, u), C.P (%) was G > R > B, the point of color divergent is at $D_o \approx 9°$ and the saturation color portion it was at $D_o \approx 14°$.

In the presence of a similarity between Aswan and Fayum regions, where the desert background is dominant, and the difference is in the percentage of moon Ilumination, the difference in the order of colors is between red and green, and in both cases the blue was the last. In the case of the full moon, there was no point of separation for the color portion, while in the case of the moon; it was 0.164 divergence point at $D_o \approx 9°$. The presence of the moon above the horizon for some days created a delay of about a one degree in the appearance of the true dawn, especially in camera observations. This is due to the fact that the human eye is more sensitive in spectral response than all devices (the human eye can sense about million shades of color [6]). All of the observers distinguished the light of the true dawn from the light of the moon by their own eyes. The degree of the true dawn in the presence of the moon was exactly equal to the degree in the absence of the moon, since there is no man-made device that equals the sensitivity of the naked eye. So, the effect of moonlight on the stages of twilight, especially in the case of true dawn, requires further studies.

Interpretation 2:

To compare between the morning twilight of the two days on which a strong and clear hierarchical shape of the false dawn occurred at Aswan in 26th Dec. 2015 (the moon was present during twilight phase) in Figs. 12 and 15, the true dawn beginning was at sun vertical depression $D_o \approx 14.5°$, while the false dawn ranged between the sun vertical depression values $D_o \approx 17.5°–20.5°$ ($\Delta D_o = 3°$).

On 14th Jan. 2016 (Figs. 20 and 21) as moonless condition, the true dawn occurred at sun vertical depression $D_o \approx 14.5°$, while the false dawn ranged between sun vertical depressions $D_o \approx 15°–18°$ ($\Delta D_o = 3°$), yet the hierarchical shape was clear in both cases. The directions of blue and green lights were inflected at the point $D_o = 18°$, while the percentage of red light remains in the same direction until the end of the observation. So, in all cases, the true dawn occurred at sun vertical depression $D_o \approx 14.5°$ because it is an earthly (terrestrial) phenomenon caused by the incidence of the solar rays on the upper edge of the earth's atmosphere and is not affected by the zodiacal light whose source is supposed to be the asteroid belt between Mars and Jupiter.

Interpretation 3:

Rozenberg (p. 52) gave an explanation of the starting point of discrimination of the colors (especially Figs. 3, 5, 7, 8, 11 and 17 in our work), which represents the normal morning twilight at sun vertical depression $D_o \approx 13.5°$ as: In practice the intensities of two orthogonally polarized components were measured, one of which (I_{II}) corresponded to the vibrations of the electric vector of the light wave in the plane of the solar meridian. Figure 39 in Rozenberg, p. 53 [15], shows the relation between the intensities of the two components of polarizations (parallel polarization, I_{II} and vertical polarization, $L_{\perp}$ (averaged over several days of observation)) and the sun vertical depression D_o. The weaker component I_{II} reaches its limiting (night) value significantly earlier than the stronger component $L_{\perp}$, for which the electric vector vibrates normally to the scattering plane. The tail of the $I_{II}(D_o)$ curve begins at $D_o \approx 13°.5$, while the tail of the $L_{\perp}(D_o)$ curve does not set in until $D_o \approx 15°$. This gives some insight into the meaning of the drop in polarization between $D_o = 13°-14°$ and $17°-18°$, corresponding to the effective scattering levels from 160–200 km to 300–350 km (see Rozenberg, Figs. 34, 35, 37) [15].

Interpretation 4.

The appearance of the zodiacal light strongly (false dawn) in Aswan indicates that the best place to observe the false dawn is in the tropical region, as it is confirmed by previous observations in this region [39–41].

Figure 28 shows the daily variation of the true dawn for all observations by the naked eye (N.E) against D_o (degree), while Fig. 29 shows the distribution of the sort of D_o between $13.23° \le D_o \le 14.91°$. From both figures, we can notice that the most populated results of the sun vertical depression corresponding to the beginning of the true dawn lay between $14° \le D_o \le 14.5°$. Also, the mode of the sun vertical depression is found to be $14.29°$.

Figure 30 shows the statistical distribution of the true dawn observation by the naked eye (N.E) and the bar width of $D_o = 0.2°$ covered by Gaussian distribution (G.D), which shows the high frequency of D_o is ranged $14.2° \le D_o \le 14.4°$ and the maximum G.D is at $D_o = 14.1°$. The number of data points used is 43 and the mean values is $D_o = 14.14° \pm 0.424$.

Table 5 represents the statistical parameters of the naked eye (N.E) observations of D_o for the true dawn, which indicates its occurrence between $13.28° \le D_o \le 14.91°$ and the mean value is at $D_o = 14.14° \pm 0.429$ (for the high confidence is $D_o = 14.57°$ (mean + 1SD)). The true dawn recorded by the reduction of Nikon and Canon cameras occurs in the range $13.91° \le D_o \le 15.27°$ and the mean value is at $D_o = 14.513° \pm 0.356$. The difference between mean and median is small in all parameters. It is noticed that the range value between the minimum and maximum values of D_o for the naked eye is $1.63°$, while it is $1.36°$ for the Canon and Nikon cameras. This means that the high correlation between the data and the condition observation is relatively good. These results agree with the previous work in Table 1. The measurements of the white thread which is a reduction from the images of Canon and Nikon cameras gives $D_o = 14.87°$ (mean + 1SD) as a sun vertical depression corresponding to the time point for the beginning of the true dawn. This difference is about $D_o = 0.3°$ between the naked eye and the cameras monitoring. The beginning of the zodiacal light (ZL1, false dawn) was at sun vertical depression $D_o = 19.568° \pm 1.35$ and the end of zodiacal light (ZL2) was at sun vertical depression $D_o = 14.95° \pm 0.498$. Then, the zodiacal light (false dawn) has a beginning and an end according to a sun vertical depression which ranges in about 5°. This result agrees with both [17, 28].

Figure 31 represents the final results of aggregation of 10 observation groups in different places in Egypt (6 sites). The true dawn declares itself when the sun is below the horizon by $D_o = 14.57°$. There is a difference of $D_o \approx 1°$ between the high visibility in the desert background (Aswan, Kottamia, Fayum, and Kharga) and the low visibility in Hurghada and Marsa-Alam (coastal background). The compatibility between the naked eye observations and almost complete photometric observations (especially in the desert areas) opens the way for us to expand extensive observations of that kind of observations in the future.

These results agree with Miethe and Lehmann (1909) [13], where the first appearance of color difference was at $D_o = 14.25°$ (true dawn) and also with Patat et al. (2006) [18] who studied the *UBVRI* twilight brightness at dome C (the night sky brightness levels is reached at $D_o = 14.8°$ as the reflected point between the rays incident from the sky and the rays reflected from the earth). When we apply the threshold of eye as 0.83^m [21] between the full night and the true dawn, it will correspond to sun vertical depression $D_o = 14°$.

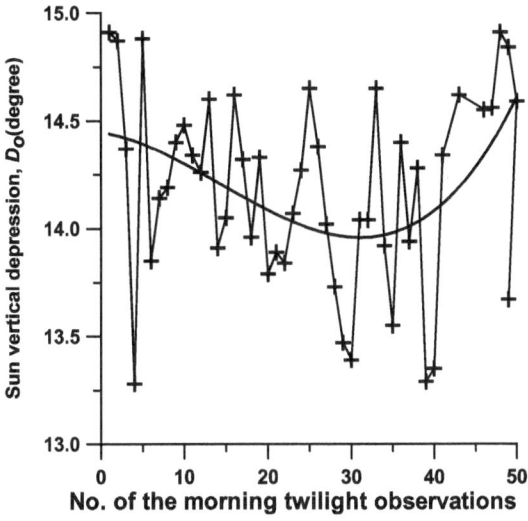

Fig. 28. Number of the morning twilight observations

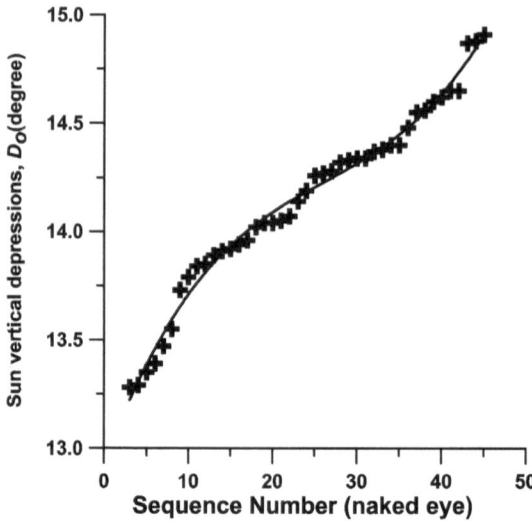

Fig. 29. The sort distribution of the naked eye observations for true dawn against D_o (degree).

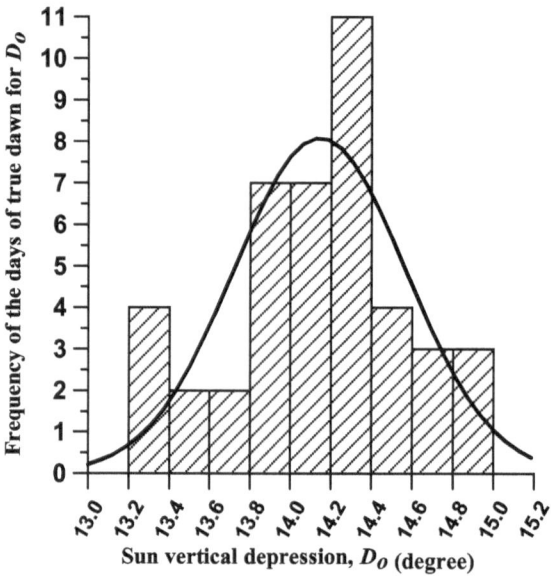

Fig. 30. The frequency of true dawn (N.E) values covered by G.D against the sun vertical depression D_o (degree).

Table 5. The statistical parameters of morning twilight stages values of sun vertical depression D_O (degree) with naked eye (N.E) and camera (canon) resulted for true dawn and limits of false dawn (begins at ZL1 and ends at ZL2).

Statistical Parameters	D_O, N.E	D_O, Camera	D_O, ZL1	D_O, ZL2
No. of values	43	13	5	5
Minimum	13.28	13.91	18.36	14.27
Maximum	14.91	15.27	20.0	15.67
Range	1.63	1.36	3.54	1.4
Mean	14.14	14.513	19.5	14.95
Median	14.19	14.48	19.23	14.91
S.D	0.429	0.356	1.354	0.498

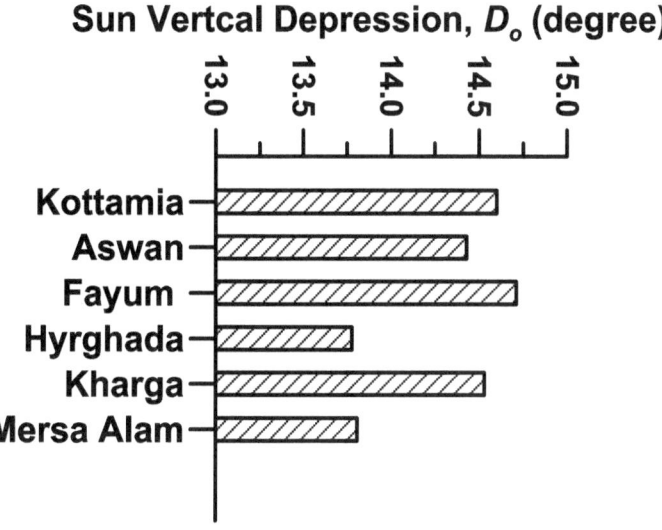

Fig. 31. The high confidence value of true dawn of D_O (degree) for the locations of observations of group of observations through four years in Egypt.

5   Future Work

In the future, we need more photometric observations about the effect of moonlight phases on the different phases of the twilight.

6 Conclusion

From this work, the following points can be concluded:

- According to the final results of observation by many groups in different places in Egypt (6 sites), the true dawn observation by the naked eye (for 43 selected observed clear days) is below the horizon by $D_o = 14.566°$ (mean + 1SD).
- From the three calculated methods according to the threshold of eye the M, M1 and M2 in the moonless conditions:
- M: The eye's light threshold is 2% as the relative values higher than the celestial background: The sun vertical depression corresponding to the beginning of the true dawn is between $D_o = 14°–15°$ [15, 21, 28, 29, 31, 35].
- M1: The illuminance energy of 0.83^m as the relative value between the full night ($D_o < 18°$) and the true dawn: The sun vertical depression corresponding to the beginning of the true dawn is $D_o = 14°$.
- M2: The threshold of eye for the true dawn as the light intensity equivalent to the brightness of the zodiacal light (false dawn) added to the curve part of the true dawn: The sun vertical depression corresponding to the beginning of the true dawn is $D_o = 14°$.
- The difference between the high visibility in the desert background (Aswan, Kottamia, Fayum, and Kharga) and the low visibility in Hurghada and Marsa- Alam (coastal background) for the sun vertical depression angle (D_o) is about 1°.
- The full hierarchical shape of the false dawn does not occur regularly. Generally, the interval time of the false dawn is $15° \leq D_o \leq 18°$.
- The difference between the observation manners of monitoring the stages of the dawn (Canon, Nikon and CCD Cameras) is not more than 0.5°, since the eye is considered as the primary determiner for sensing the beginning of the true dawn.
- Our research is quite consistent with our previous research and also with Patat et al. (2006) as well [35].
- The sun vertical depressions (D_o) for the beginning of the false dawn is $D_o = 19.56°$ ± 1.35, while the end of the false dawn will be at sun vertical depression $D_o = 14.95°$ ± 0.5.
- In the most twilight morning days, the percentage of color portions arrangement is B > G > R before $D_o \approx 13.5°$ and it changes its behavior after this degree.
- Generally, the point at sun vertical depression $D_o \approx 13.5°$ (color portion 33%) represents the beginning of color divergence for most normal morning twilight of B, G and R colors from the image process reduction of the images of Canon and Nikon camera.
- Generally, the various methods for determining the sun vertical depression (D_o) proved that the true dawn begins at the sun vertical depression range $D_o = 14°$ to 15° at moonless conditions.
- The light hierarchical shape that represents the false dawn ends in a straight line and it stays for a period of time about one degree of sun vertical depression without increasing and the darkness does not became completely before the false dawn, but there is a higher level of illumination than the complete night.

- Moonlight does not affect the shape of the false dawn, but only affects its intensity, which means that the false dawn is seen in the presence of the moon (with less intensity), while the true dawn is seen clearly and is not affected by moonlight especially by the naked eye.
- The presence of the moon above the horizon for some days created a delay of about a degree in the appearance of the true dawn, especially in camera observations.

Acknowledgment. The authors would like to thank Nawar S, Morcos A. B. and Hatem Odah from NRIAG for their participation in two of the largest scientific trips to observe the dawn in Fayum desert, one of them included more than 25 members of the researchers of the department of Astronomy at NRIAG, from Al-Azhar University and from Cairo University. Alsom the thanks have to go to Prof. Dr. Hosni Hamdan Hamama, a professor at Mansoura University for participating with us in observing dawn in Kottamia observatory, to engineers T. A. Aggag and K. M. Hassan from Egyptian Survey Authority (Egypt) for their participation in three field observation trips to the deep desert and to all the heads of Astronomy departments in Cairo University, Al-Azhar University and Helwan University for their contribution and scientific cooperation with the NRIAG in the observations by participating with distinguished scientific cadres to make the arduous scientific trips in the depths of the desert successful.

Great thanks have also to be given to all of the participants in field of observations from NRAIG technicians, especially from seismic center in Aswan and to the young students for their help in observations (especially the naked eye observations): Loay Ahmed Ghitas, Baraa Ahmed Ghitas, Marwan Taha Matar and Mohamed Amir Hussein.

References:

1. Hassan, A.H., Hassanin, N.Y., Abdel-Hadi, Y.A., Issa, I.A.: Naked eye observations for morning twilight at Different Sites in Egypt. NRIAG J. Astron. Geophys. **3**, 23–26 (2014)
2. Knoll, H.A., Tousey, R., Hulburt, E.O.: Visual thresholds of steady point sources of light in fields brightness from dark to daylight. J. Opt. Soc. Am. **36**(8), 480–482 (1946)
3. Pirenne, M.H.: Vision and the Eye. Chapman and Hall, London (1967)
4. Richard, H., Blackwell: Contrast threshold of the human eye. J. Opt. Soc. Am. **36**(11), 624–643 (1946)
5. Roach, F.E., Gordan, J.L.: The Light of the Night Sky. D. Reidel Pub.Com. U.S.A. (1973)
6. Crumey, A.: Human contrast threshold and astronomical visibility. Mon. Not. R. AstronSoc. **442**, 2600–2619 (2014)
7. McGillivray, D.: Physics and Astronomy. Macmillan Education LTD, London (1987)
8. Tousey, R., Koomen, M.J.: The visibility of stars and plants during twilight. J. Opt. Soc. Am. **41**, 3 (1952)
9. Samaha, A.E., Assad, A.S., Mikhail, J.S.: Visibility of the new moon. H. O. B. No. 48 (1969)
10. Assad, A.S., Mikhail, J.S., Nawar, S.: The first visibility of the new moon at Helwan and Daraw. H. O. B. No. 138 (1976)
11. Garstang, R.H.: Model for artificial night-sky illumination. Publ. Astron. Soc. Pac. **98**(601), 364 (1986)
12. Levasseur-Regourd, A.C.: Zodiacal light, certitudes and questions. Earth Planets Space **50**(6–7), 607–610 (1998). https://doi.org/10.1186/BF03352155
13. Miethe, A., Lehmann, E.: Dämmerungsbeobachtungen in Assuan in Winter 1908. Meteorologische Zeitschrift 97–114 (1909)

14. Thomas, L., Boivman, M.R.: Atmospheric penetration of ultra-violet and visible solar radiations during twilight periods. J. Atmospheric Terrestrial Phys. **31**, 1311–1322 (1969)
15. Rozenberg, G.V.: Twilight. Plenum Press, New York (1966)
16. Sultan, A.H.: Sun Apparent Motion and Salat Times. al-Irshaad **8** (2004)
17. Al Mostafa Z.A., et al.: Studying of Twilight Project. First part, Abdul-Aziz city of the science and technology, Institute of Research Astronomy and Geophysics, Saudi Arabia (2005)
18. Patat, F., Ugolnikov, O.S., Postylyakov: UBVRI twilight sky brightness at ESO-Paranal, Astron. Astrophys. **455**, 385–393 (2006)
19. Kenyon, S.L., Storey, J.W.V.: A review of optical sky brightness and extinction at Dome C, Antarctica. Publ. Astron. Soc. Pac. **118**, 489–502 (2006)
20. Hassan, A.H., Hassanin, N.Y., Abdel-Hadi, Y.A., Issa, I.A.: Time verification of twilight begin and end at Matrouh of Egypt. NRIAG J. Astron. Geophys. **2**, 45–53 (2013)
21. Hassan, A.H., Hassanin, N.Y., Abdel-Hadi, Y.A., Issa, I.A.: Brightness and Color Variation for Evening and Morning Twilight at Bahria of Egypt IV. NRIAG J. Astron. Geophys. **3**, 37–45 (2014)
22. Hassan, A.H., Issa, I.A., Mousa, M., Abdel-Hadi, Y.A.: Naked eye determination of the dawn for Sinai and Assiut of Egypt. NRIAG J. Astron. Geophys. **5**, 9–15 (2016)
23. Issa, I.A., Hassan, A.H.: Transparency of the night sky at Bahria/Egypt. I. NRIAG J. Astron. Astrophys. Special Issue Egypt 383–397 (2008)
24. Issa, I.A., Hassan, A.H.: Evening and morning twilights at Bahria/Egypt. II. NRIAG J. Astron. Astrophys. Special Issue Egypt 399–411 (2008)
25. Issa, I.A., Hassan, A.H.: Eye Criteria and times of end and begin of twilights. Bahria/Egypt. III. NRIAG J. Astron. Astrophys. Special Issue Egypt 413–423 (2008)
26. Issa, I.A., Hassanin, N.Y., Hassan, A.H., Abdel-Hadi, Y.A.: Verification of end of twilight and beginning prayer times at matrouh of Egypt VI. In: SEAC 2009, Alexandria, Egypt, pp. 125–145 (2009)
27. Issa, I.A., Hassanin, N.Y., Hassan, A.H., Abdel-Hadi, Y.A.: Atmospheric transparency, twilight brightness and color indices at Kottamia of Egypt. NRIAG J. Astron. Astrophys. Spec. Issue Egypt 379–398 (2011)
28. Semeida, M.A., Hassan, A.H.: Pseudo dawn and true dawn observations by Naked Eye in Egypt. Beni-Suef University. J. Basic Appl. Sci. (BJBAS) **2018**(7), 286–290 (2018)
29. Hassan, A.H., Abdel-Hadi, Y.A.: Naked eye determinations of the dawn at Tubruq of Libya through four years observations. Middle-East J. Sci. Res. **23**(11), 2627–2632 (2015)
30. Hassan, A.H., Abdel-Hadi, Y.A.-F., Rahoma, U.A., Issa, I.A.: Naked eye estimations of morning prayer at Tubruq of Libya. Al-Hilal: J. Islamic Astron. **3**(2), 75–98 (2021)
31. Saksono, T., Fulazzaky, M.A.: Predicting the accurate period of true dawn using a third-degree polynomial model. NRIAG J. Astron. Geophys. Egypt **9**(1), 238–244 (2020)
32. Abdel-Hadi, Y.A., Hassan, A.H.: The effect of sun elevation on the twilight stages in Malaysia. Int. J. Astron. Astrophys. (IJAA) **12**, 7–29 (2022)
33. Khalifa, N.S., Hassan, A.H., Taha, A.I.: Twilight observation by the naked eye of the dawn sincere at Hail and other areas in Saudi Arabia. NRIAG J. Astron. Geophys. Egypt **7**, 22–26 (2018)
34. Rashed, M.G., et al.: Determination of the true dawn by several different ways at Fayum in Egypt. Int. J. Mech. Eng. Technol. (IJMET) **13**(10), 8–24 (2022)
35. Kelsey, L.J., Hoff, D.B., Neff Astronomy, J.S.: Activities and Experiments, Kendall/Hunt (1974)
36. King-Smith, P.E., Kulikowski, J.J.: Pattern and flicker detection analyzed by sub threshold summation. J. Physiol. **249**, 519–548 (1975)
37. Smith, P.D.: Practical Astronomy with Your Calculator. Cambridge University Press, New York (1989)

38. http://www.Monzur@starlight.demon.co.uk
39. Nawar, S.: Wide band spectral photometry of the zodiacal light. Faculty of science, Cairo University (1978)
40. Nawar, S.: Sky twilight brightness and colour during high solar activity. Moon Planets **29**, 107–116 (1983)
41. Nawar, S., Morcos, A.B., El Agmy, R.M., Gad, G.M.A., Elgohary, S.: Sky twilight brightness at zenith expressed in magnitudes. NRIAG J. Astron. Geophys. **9**, 63–70 (2020)

Thermal Analysis of an Educational Nano Satellite

Mohamed Maher Elsayed[1]([✉]) [ID], Amir Ashraf[2] [ID], Ahmed Farag[2] [ID],
and G. M. Abdo[1] [ID]

[1] Mechanical Engineering Department, Faculty of Engineering and Technology, Badr
University in Cairo (BUC), Cairo, Egypt
mohamedmaher2001@outlook.com, Gamal.abdo@buc.edu.eg
[2] Faculty of Engineering, Suez University, Suez, Egypt

Abstract. CubeSats, being small satellites used for educational and scientific
purposes, face significant thermal management challenges due to their compact
size. This research paper focuses on conducting a thermal analysis of a 1U educational nanosatellite (CubeSat) utilizing Thermal Desktop and SINDA/FLUINT
software. The study involves the development of a detailed 3D thermal model of
the CubeSat using Thermal Desktop software, followed by thermal simulations
using SINDA/FLUINT software. By evaluating the CubeSat's thermal performance under various operating conditions, the analysis aims to identify potential
thermal issues that may arise. The results gained offer significant understanding of the CubeSat's thermal behavior and suggest ways to improve the design
for increased dependability and performance. This research contributes to the
understanding of CubeSat thermal management and facilitates the development
of improved design strategies for effective thermal control. By addressing thermal
challenges, CubeSats can achieve enhanced performance and ensure the success
of educational and scientific missions. The results of the thermal investigation
show that every CubeSat component is functioning within allowable temperature
ranges, meeting an essential requirement for any space mission to be successful.
By offering a framework for the creation of better design techniques for efficient
thermal control, this study advances the subject of CubeSat thermal management.

Keywords: CubeSats · Thermal control · Educational satellites · Thermal
behavior · Design optimization

1 Introduction

CubeSats are Miniaturized Satellites with a Wide Range of Applications that have revolutionized the way we access space based on a standardized form factor of 10 cm ×
10 cm × 10 cm cubes which is known as 1U [1]. CubeSats are typically deployed from
a larger launch vehicle. There are many different sub-sections involved in the development of a CubeSat [2]. One of the most known 1U CubeSat is Swiss cube which was
launched in 2009 [3]. For a number of years, it has been transmitting various kinds of
flight data. S. Corpino et al. (2015) [4] studied thermal analysis as it relates to a Nano

© The Author(s) 2025
H. M. K. Al Naimiy et al. (Eds.): AUASS-CONF 2023, SPPHY 420, pp. 210–220, 2025.
https://doi.org/10.1007/978-981-96-3276-3_15

satellite created at Politecnico di Torino. The study's main goal was to communicate the analysis's methodology and findings. The two primary mission parameters and the spacecraft architecture were discussed by the authors in order to set the boundary conditions and thermal environment for the study. Next, they gave a thorough explanation of the thermal model that was created to address the thermal balancing issue. The study included the presentation of the numerical simulation code as well. The obtained results were thoroughly discussed and analyzed. The temperature distributions were verified with commercial software, and the new tool showed outstanding modeling skills. In the 2006 study, Millan Diaz-Aguado and his colleagues [5], Carried out thermal analysis and cycling for tiny satellites. The worst-cases in hot and cold conditions were used in Chamber-N at the Johnson Space Center in Texas for the thermal cycling studies. However, the Finite Elements (FE) approach were used to conduct the thermal analysis, and the results were validated by comparing them to the test data. The satellite under investigation, FASTRAC, was planned to be placed in a Low Earth Orbit (LEO), with altitudes ranging between 300km and 500km. To assess the satellite's survivability, the researchers simulated the LEO orbits and analyzed their characteristics [6]. The purpose of this investigation was to ascertain the kind of environmental circumstances that FASTRAC could function in while on mission. In reference to a 1U educational Nano satellite, this research study highlights the field of thermal analysis. For scholarly purposes, the analysis comprises a fundamental thermal model [7] it divides the Cube Sat into two nodes: an external node representing the satellite's exterior surfaces and an internal node repre4senting the satellite's interior components [8]. The analysis included consideration of heat flow between the exterior and interior nodes as well as external heat sources from space. The temperature readings were confirmed using the temperature information that the CubeSat sensors collected during its flight. The ultimate objective of this research is to evaluate the thermal performance of the educational 1U CubeSat using Thermal Desktop and SINDA/FLUINT software [9]. The study seeks to detect potential thermal difficulties that may develop during different operating situations by means of a thorough thermal analysis [10]. In order to get important understanding of the CubeSat's thermal behavior, the research looks at temperature distributions and hotspot locations [11], enabling the identification of areas that require design modifications for improved performance and reliability. Our CubeSat's thermal analysis is crucial because it aids in the creation of thermal insulation that shields the interior systems from the harsh conditions of space travel. Due to the increased risk of harming the internal components, the mission's failure could result from the absence of a property-designed thermal layer [12]. Due to the budget constraint, it was decided to come up with a passive thermal control [13] that could satisfy the mission requirements. Using simulations and analytical techniques, a variety of materials will be examined in this section.

2 Governing Equations

Thermal regulation within the satellite [14] is interacted with the heat transfer concept conduction and radiation [15]

- Conduction between thermal node and its adjacent nodes
- Radiation among thermal nodes (include space)

- Qin: Heat Dissipation
- Qorbit: Orbital Fluxes

The thermal balance equation [16, 17] in transient case was solved by Thermal Desktop and SINDA/FLUINT software [18]. The energy balance terms in the equation are stated as

$$c_i \frac{dT_i}{dt} = A + B + C + D + E + F \tag{1}$$

$$A = A_{si} \times F_i \times Q_s \tag{2}$$

$$B = A_{si} \times F_i \times Q_a \tag{3}$$

$$C = \varepsilon_i \times F_i \times Q_t \tag{4}$$

$$D = Q_i \tag{5}$$

$$E = \sum_j k_{ij} \times F_i \times (T_j - T_i) \tag{6}$$

$$F = \sum_j \sigma_0 \varepsilon H_{ij} \times (T_j^4 - T_i^4) \tag{7}$$

where

$c_i \frac{dT_i}{dt}$	Accumulated energy
A	Absorbed solar flux
B	Albedo Flux
C	infra-Red
D	Heat Rejection
E	Heat Transfer by Conduction
F	Radiation Heat transfer

3 Thermal Model

The satellite components were chosen from ISISPACE [19] due to the readily available CAD models, datasheets, and literature examples. The ISISPACE 1U CubeSat structure, VHF/UHF transceiver, 1U solar panel, and Electronic Power System (EPS) were the industry standard Thermal Desktop software used for thermal modeling [20]. The thermophysical and optical Properties were entered. The final thermal model is shown in Fig. 1 and Fig. 2 shows the satellite cad model.

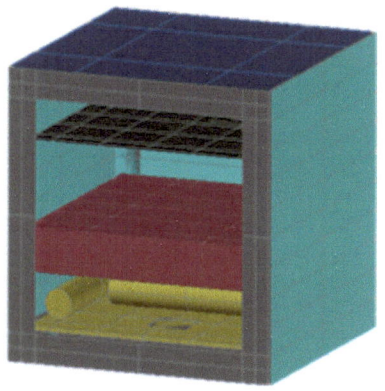

Fig. 1. Satellite Thermal Model

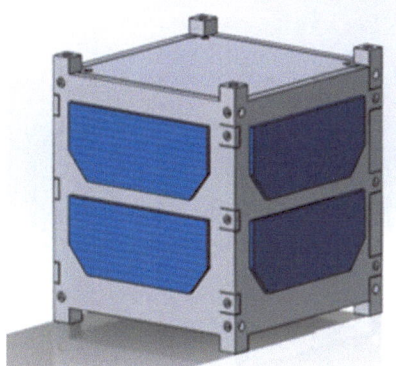

Fig. 2. Satellite cad model

3.1 Case Study

The two worst possible scenarios are investigated, representing extreme hot and cold conditions, by selecting two different beta angles: 75.1° and 0°. The choice of these angles allows us to explore the effects of the temperature extremes on the system under consideration. The 75.1-° beta angle represents the hot case, while the 0-° beta angle represents the cold case. By examining the system under these contrasting conditions, we aim to gain insights into the performance and behavior of the system in both extreme environments. This analysis will provide valuable information for understanding the system's robustness and adaptability, enabling us to make informed decisions and develop strategies to mitigate potential risks.

3.2 Cold Case Study

The cold case thermal model and cold case orbit are shown in Fig. 3, 4 respectively. The satellite subsystems' heat loads were added to the software as in Table 1.

Table 1. The satellite subsystems heat loads

Component	OBC	Transmitter	Magnetorquer	EPS	Antenna
Heat load (W)	0	0	0.18	0.01	0

3.3 Hot Case Study

The hot case thermal model and hot case orbit are shown in Fig. 5, 6 respectively. The satellite subsystems' heat loads were added to the software as in Table 2**.**

Nomenclatures

Symbol Quantity

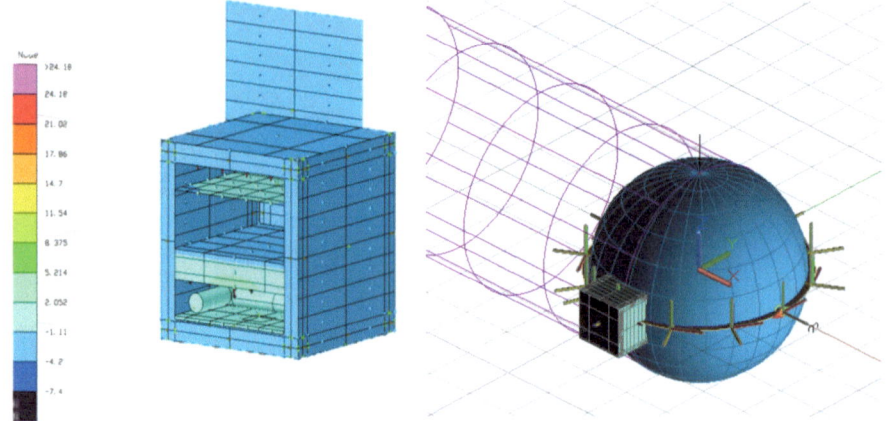

Fig. 3. Satellite Thermal Model cold case   **Fig. 4.** Cold Case Orbit

Table 2. The satellite subsystems heat loads

Component	OBC	Transmitter	Magnetorquer	EPS	Antenna
Heat load (W)	0.4	0.48	1.2	0.13	0.03

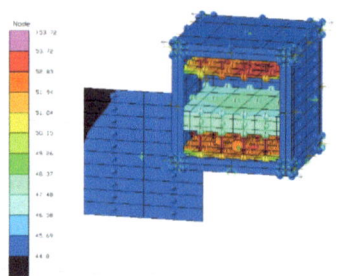

Fig. 5. Hot Case Thermal Model.

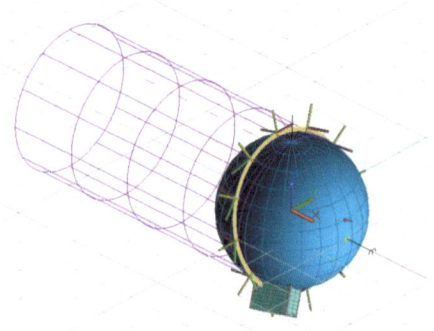

Fig. 6. Hot Case Orbit

F_i	Area (node i) m^2
ε_i	Emissivity factor $[-]$
K_{ij}	Heat transfer by conduction [W/K]
σ_0	Stefan-Boltzmann constant $[5.67 \times 10^{-8} \text{ W/m}^2/\text{K}^4]$
Q_{diss}	Heat rejection from equipment [W]
Q_I	Heat rejection (node i) $\left[\text{W/m}^2\right]$
Q_a	Solar albedo $\left[\text{W/m}^2\right]$
Q_T	Earth infrared (IR)
εH_{ij}	Heat transfer by radiation $\left[\text{m}^2\right]$

Q_s Incident Sun flux $\left[W/m^2\right]$

4 Results

The Results section, presents the findings obtained from the study on the two worst-case scenarios with different beta angles: 75.1° and 0°, representing extreme hot and cold conditions, respectively. This section aims to provide a comprehensive analysis of the system's performance, behavior, and any observed differences between the two scenarios and present the data collected.

4.1 Cold Case Results

The Temperature profiles presented here depict the thermal analysis for a cold case scenario in orbit with a beta angle of 0 degrees. The graphs provide a visual representation of the system's performance and behavior under extreme cold conditions shown in Figs. 7, 8, 9, 10, 11, 12 and 13.

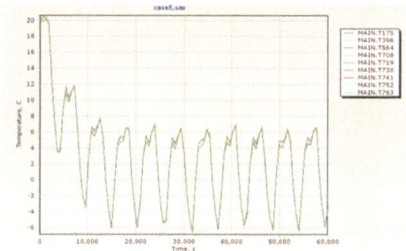

Fig. 7. Antenna Temperature Profile Note the cyclical heating and cooling experienced by the Antenna. It takes approximately four orbits for the Antenna's average temperature to stabilize and it Shows that minimum temperature at −7 c°.

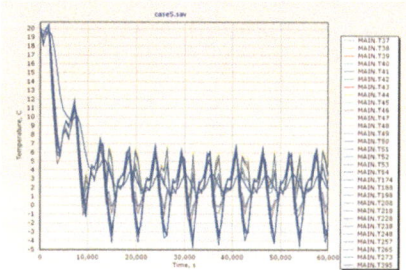

Fig. 8. EPS Temperature Profile Note the cyclical heating and cooling experienced by the EPS. It takes approximately four orbits for the EPS's average temperature to stabilize and it shows that minimum temperature at −5c°.

4.2 Hot Case Results

The Temperature profiles presented here depict the thermal analysis for a hot case scenario in orbit with a beta angle of 75.1°. The graphs provide a visual representation of the system's performance and behavior under extreme hot condition. Note that even though the satellite is always in sunlight, it still experiences cyclical heating and cooling. This is due to the nature of the orbit, specifically the difference in the amount reflected radiation and Earth IR received at different latitudes shown in Figs. 14, 15, 16, 17, 18, 19, 20.

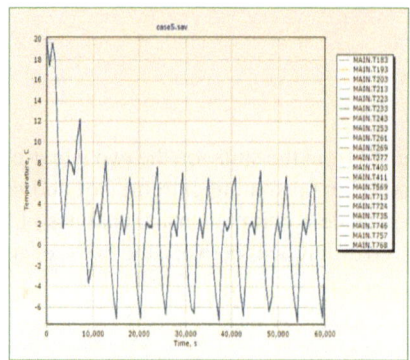

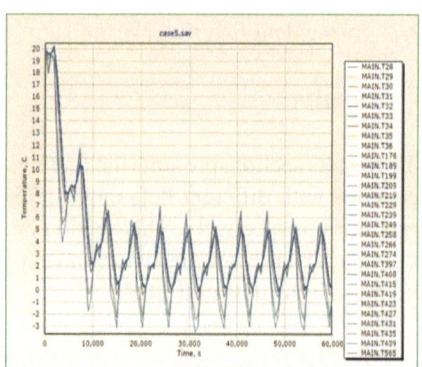

Fig. 9. SP Temperature Profile Note the cyclical heating and cooling experienced by the SP. It takes approximately three orbits for the SP's average temperature to stabilize and it shows that minimum temperature at −7 c°

Fig. 10. Magnetorquer Temperature Profile Note the cyclical heating and cooling experienced by the Magnetorquer. It takes approximately four orbits for the Magnetorquer's average temperature to stabilize and it shows that minimum temperature at −4 c°

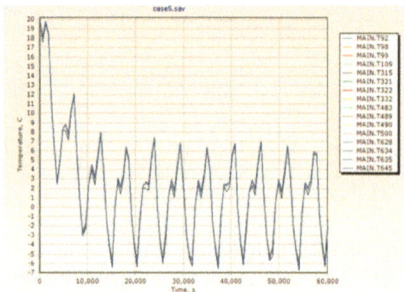

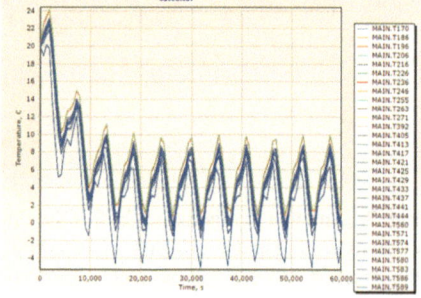

Fig. 11. STR Temperature Profile Note the cyclical heating and cooling experienced by the STR. It takes approximately four orbits for the STR's average temperature to stabilize and it shows that minimum temperature at −7 c°.

Fig. 12. OBC Temperature Profile Note the cyclical heating and cooling experienced by the OBC. It takes approximately four orbits for the OBC's average temperature to stabilize and it shows that minimum temperature at −5 c°.

4.3 Results Compression

In this section, a brief comparison is presented between the study results obtained in the hot and cold case scenarios and the standard values. The primary area of concern is the large thermal gradient of the magnetorquer and OBC. Since no maximum thermal gradient is reported for the magnetorquer and OBC, this is an area of possible investigation. Furthermore, the results are evaluated the results within a range where all components are functioning normally as shown in Table 3.

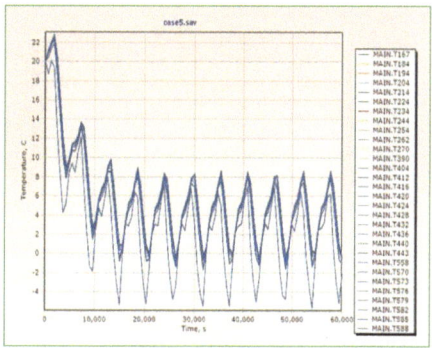

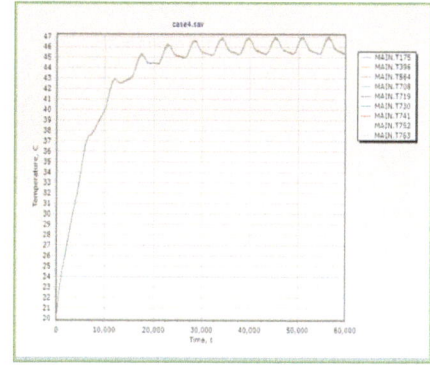

Fig. 13. Transmitter Temperature Profile Note the cyclical heating and cooling experienced by the Transmitter. It takes approximately five orbits for the Transmitter's average temperature to stabilize and it shows that minimum temperature at −6 c°.

Fig. 14. Antenna Temperature Profile Note the cyclical heating and cooling experienced by the Antenna. It takes approximately six orbits for the Antenna's average temperature to stabilize and it shows that maximum temperature at 47c°

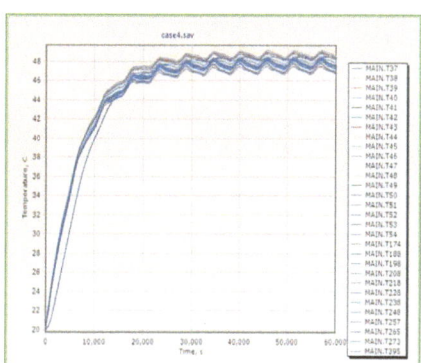

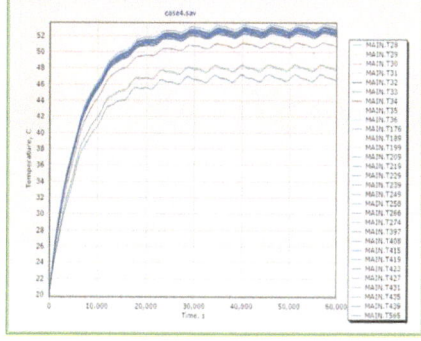

Fig. 15. EPS Temperature Profile Note the cyclical heating and cooling experienced by the EPS. It takes approximately five orbits for the EPS's average temperature to stabilize and it shows that maximum temperature at 49 c°.

Fig. 16. Magnetorquer Temperature Profile Note the cyclical heating and cooling experienced by the Magnetorquer. It takes approximately five orbits for the Magnetorquer's average temperature to stabilize and it shows that maximum temperature at 53 co.

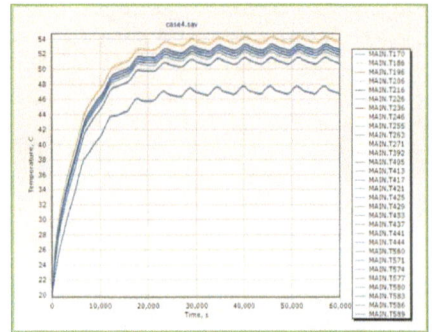

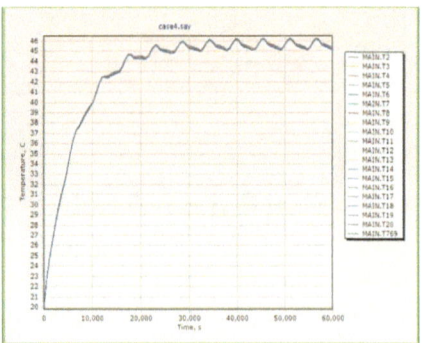

Fig. 17. OBC Temperature Profile Note the cyclical heating and cooling experienced by the OBC. It takes approximately six orbits for the OBC's average temperature to stabilize and it shows that maximum temperature at 54 c°

Fig. 18. SP Temperature Profile Note the cyclical heating and cooling experienced by the SP. It takes approximately six orbits for the SP's average temperature to stabilize and it shows that maximum temperature at 46 c°.

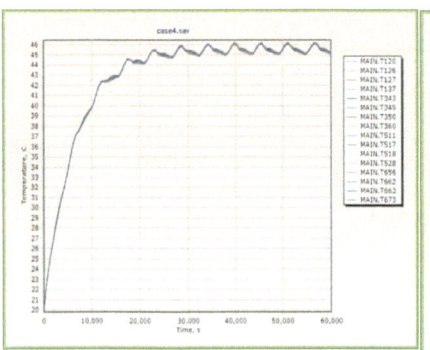

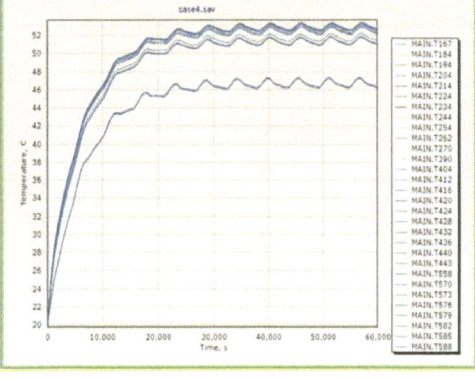

Fig. 19. STR Temperature Profile Note the cyclical heating and cooling experienced by the STR. It takes approximately six orbits for the STR's average temperature to stabilize and it shows that maximum temperature at 46 c°.

Fig. 20. Transmitter Temperature Profile Note the cyclical heating and cooling experienced by the Transmitter. It takes approximately five orbits for the Transmitter's average temperature to stabilize and it shows that maximum temperature at 53 c°.

Table 3. Comparison between Standard values and Results

	Component	STR	OBC	Transmitter	Magnetorquer	EPS	Antenna	SP
Standard	Min (C)	−40	−25	−20	−40	−20	−20	−40
	Max (C)	80	65	60	70	70	60	125
Result	Min (C)	−7	−5	−6	−4	−5	−7	−7
	Max (C)	46	54	53	53	49	47	46

5 Conclusion

This research paper presents a comprehensive thermal analysis of a 1U educational Nanosatellite using Thermal Desktop and SINDA/FLUINT software. The study involved the development of a simplified thermal model of the CubeSat, which was subsequently subjected to thermal simulations under various operational conditions to assess its thermal performance and identify potential thermal issues. The outcomes of the thermal analysis indicate that all components of the CubeSat are operating within acceptable temperature limits, satisfying a critical requirement for the success of any space mission. This research contributes to the field of CubeSat thermal management by providing a foundation for the development of improved design strategies for effective thermal control. By addressing thermal challenges, CubeSats can achieve enhanced performance and ensure mission success, furthering our ability to explore and utilize space for educational and scientific purposes. The components are within the range, so no changes is the thermal control system. It also helps in the easiness of the satellite thermal control. No action need during lifetime, because there's no hotspots or extreme temperature.

Acknowledgement. First and foremost, praises and thanks to ALLAH for his showers of blessings throughout my research work to complete this research successfully. Additionally, I would like to express my profound gratitude and appreciation to Dr. Mahmoud Abu-Bakr and Eng. Aya Salah for all of their support and assistance.

References

1. Zosimovych, N.: 1U CubeSat platform design. Int. J. Aerospace Sci. **8**, 1–7 (2020)
2. Elhefnawy, A., Elmaihy, A., Elweteedy, A.: A university small satellite thermal control modeling and analysis in the post-mission phase. FME Trans. **49**, 1014–1024 (2022)
3. Noca, M., et al.: Lessons learned from the first Swiss Pico-Satellite: SwissCube. In: 23rd Annual AIAA/USU Conference on Small Satellites, Utah, USA (2009)
4. Corpino, S., Caldera, M., Nichele, F., Masoero, M., Viola, N.: Thermal design and analysis of a nanosatellite in low earth orbit. Acta Astronaut. **115**, 247–261 (2015)
5. Diaz-Aguado, M.F., Greenbaum, J., Fowler, W.T., Glenn Lightsey, E.: Small satellite thermal design, test, and analysis. In: Defense and Security Symposium, Orlando (Kissimmee), Florida, United States (2006)
6. Boushon, K.E.: Thermal analysis and control of small satellites in low Earth orbit, ProQuest LLC (2018)

7. Karam, R.D.: Satellite Thermal Control for System Engineering. AIAA (1998)
8. Kim, T.Y., Hyun, B.-S., Lee, J.-J., Rhee, J.: Numerical study of the spacecraft thermal control hardware combining solid–liquid phase change material and a heat pipe. Aerospace Sci. Technol. **4**(1), 10–16 (2013)
9. Foster, I.: Air Force Research Laboratory, "Small Satellite Thermal Modeling Guide," Defense Technical Information Center (2022)
10. Gilmore, D.G.: Spacecraft Thermal Control Handbook Volume I: Fundamental Technologies. American Institute of Aeronautics and Astronautics, Segundo, CA: VA (2002)
11. Elgendy, Y.A.M.: Preliminary design for satellite thermal control system. In: Proceeding of the 12-th ASAT Conference, Cairo, 29–31 May 2007 (2007)
12. da Silva, D.F., Muraoka, I., Garcia, E.C.: Thermal control design conception of the Amazonia-1 Satellite. J. Aerospace Technol. Manag. **6** (2014)
13. Yang, L., Song, G., Zhang, L., Li, Q., Kong, L.: Quasi-all-passive thermal control system design. Sensors **827**, 1–18 (2019)
14. Ferziger, J.H., John Wiley, et al.: Numerical Methods for Engineering Applications, New York (1981)
15. Farag, A., Elfarran, M.: Thermal design and analysis of a low earth orbit micro-satellite. J. Adv. Eng. Trends **41**, 1–12 (2022)
16. Holman, J. P.: Heat Transfer, 4th ed., McGraw-Hill Inc., New York (1976)
17. Kreith, F.: Principles of Heat Transfer, 3rd edn. Intext Educational Publisher, New York (1976)
18. Cullimore, B.A.: Computer code SINDA '85/FLUINT System Improved Numerical Differencing Analyzer and Fluid Integrator, Version 2.3 (Martin Marietta) (2010)
19. ISISPACE (2006). https://www.isispace.nl
20. Sundu, H., Döner, N.: Detailed thermal design and control of an observation satellite in Low Earth Orbit. Eur. Mech. Sci. **4**(4), 171–178 (2022)

Twenty-Five years of the Foundation of the Arab Union for Astronomy and Space Sciences (AUASS)

Awni Kasawneh[1,2,3](✉), Hamid Al-Naimy[2,3], Mashhour A. Al-Wardat[1,2,3], and Dalal Al-Lala[3,4]

[1] University of Sharjah, P.O Box: 27272, Sharjah, United Arab Emirates
akasawneh@sharjah.ac.ae
[2] Al-Al-Bayt University, P.O Box:130040, Mafraq, Jordan
[3] Arab Union for Astronomy and Space Sciences, P.O. Box: 782, Amman 11941, Jordan
[4] Jarash Private University, P.O Box: 311, Jarash, Jordan

Abstract. This paper provides a review on the activities and achievements of the Arab Union for Astronomy & Space Sciences (AUASS) and the Arab astronomy in general, on the occasion of the 25th anniversary of the official foundation of AUASS (Silver Jubilee). The paper describes and discusses AUASS membership, AUASS electronic newsletters (AUASS News), AUASS webpage, AUASS Annual Meetings, and the international relations of AUASS. It also showcases the contribution of AUASS, whether by participation or organization, into international meetings, regional and international summer schools, science camps, and all events in relation to astronomical education, astronomy outreach, amateur astronomy, and astronomical heritage.

Keywords: Arab Union for Astronomy & Space Sciences · Sharjah Academy for Astronomy · Space Sciences and Technology · Institute of Astronomy and Space Sciences · office of astronomical education · office of astronomy for development · national outreach coordinators · Asia-Pacific Space Cooperation Organization · astronomical application in Islamic sharia

1 Introduction

The Arab Union for Astronomy & Space Sciences (AUASS) was Founded on the 30th of August 1998, with its headquarters in the Jordanian Capital, Amman. AUASS was established as an outcome of the second Arab Conference organized by the Jordanian Astronomical Society and the Institute of Astronomy and Space Sciences (IAASS) in Al-Bayt University, in Amman Jordan in 1998. More than 100 astronomers and scientist from 14 Arab countries, France, Italy and the USA participated in this conference. A resolution of the conference was to establish the Union and its headquarters to be in Amman, Jordan. The union's objective is to foster the advancement of Astronomy, Astrophysics, and Space Science (AASS) in Arab nations via meetings, conferences, publications, and collaborative research endeavors, facilitated in partnership with international AASS organizations.

H. M. K. Al Naimiy et al. (Eds.): AUASS-CONF 2023, SPPHY 420, pp. 221–230, 2025.
https://doi.org/10.1007/978-981-96-3276-3_16

AUASS was officially established pursuant to the Honorable Council of Ministers Resolution No. 76/4/1/4631 dated 2/19/1998, with its permanent headquarters in the capital, Amman. It is now a member of the League of Arab States within the specialized federations. AUASS includes most of Arab scientists, researchers, and amateurs in the field of Astronomy, Space Science, and atmospheric sciences and technology. The Union is considered one of the specific unions that has been affiliated with the Arab Economic Unity Council of the League of Arab States since 2001. It is also a member of the International Astronomical Union. It is currently headed by the Iraqi-origin astronomer Professor Dr. Hamid Majoul Al-Naimiy, the chancellor of the University of Sharjah. The position of Secretary-General of the Union is held by the Jordanian origin Dr. Eng. Awni Al-Khasawneh.

The honorary president of the Supreme Council of the Union is his highness Sheikh Sultan bin Muhammad Al Qasimi, Ruler of Sharjah-UAE (Fig. 1).

Fig. 1. Arab union for astronomy and space sciences logo

2   Membership

AUASS is a space astronomical organization based in the city of Amman-Jordan. It was officially established in 1998 and includes most Arab scientists, researchers, and enthusiasts in the fields of space science technology, amateur astronomer, as well as astronomical researchers in addition to more than 73 Arabic astronomical Societies. It is one of the specialized unions affiliated with the Economic Unity Council of the Arab League, under the General Secretariat of the League of Arab States since 2001. Additionally, it is a member of the International Astronomical Union. AUASS currently has more than 750 participants from all Arab state member represented by higher council from 18 Arab countries https://auass.com/members-of-the-supreme-council/ UAE, Jordan, Bahrain, Lebanon, Syria, Oman, Libya, Tunisia, Saudi Arabia, Kuwait, Morocco, Iraq, Palestine, Egypt, Mauritania, Algeria, Sudan, Yamen.

AUASS has 2 deputies for the presidents: Shawki Dallal and Mohammad Assiry and it is president by Hamid Mjoul Al-Naimi, an Iraqi astronomer. The Secretary-General

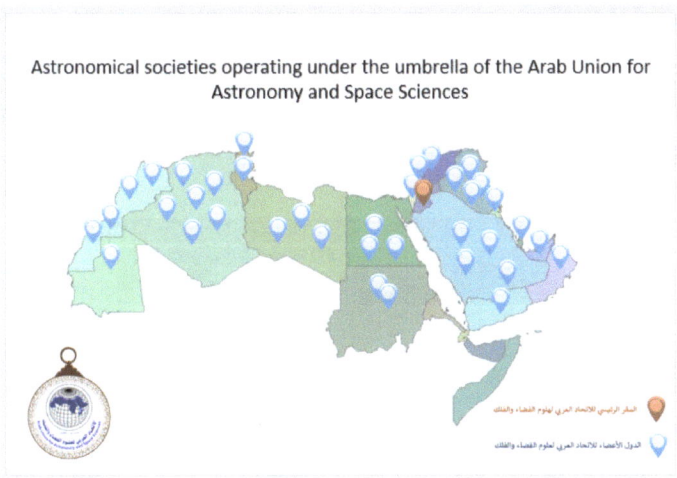

Fig. 2. Astronomical Societies operating under the umbrella of the Arab Union for Astronomy and Space Sciences

is Awni Al-Khasawneh from Jordan. The Honorary President of the Union is Sheikh Sultan bin Muhammad Al Qasimi, the Ruler of Sharjah (Fig. 2).

3   AUASS Journal Magazine and Electronic Newsletter

AUASS is publishing quarterly Journal under the name (Alkwn) (Awni Khasawneh et al., 2015) typically 4 times in a year. The Editors have been Khalil Qonsol (2006–2010), unfortunately from (2011–2015) the Journal was not published, Awni Khasawneh (2016–2023). In all AUASS-News, more than 1000 articles have been published in different public newspaper and others. AUASS and JAS published a monthly newsletter "Al-Thuraya-magazine" (AUASS & JAS, 2011) to share news, announcements, articles on astronomy, international and local astronomical meetings and activities, summer and winter schools, astronomical education in Arab states, archeoastronomy and astronomy in culture (Fig. 3).

Fig. 3. AUASS Journal, magazine, and electronic newsletter

4   AUASS Webpage

The webpage (https://auass.com/) was created in Jun 2005 for providing all activities and necessary information concerning AUASS. It includes the objectives and all activities of AUASS, list of AUASS members, annual meetings, and other events (Fig. 4).

Fig. 4. AUASS webpage

The webpage was renewed and updated many times, enriched with most of AUASS activities and information about all astronomical societies and concerned astronomy and space agencies in Arab states.

The webpage includes the history, activity, and achievements of many astronomical institutions in the Arab world such as: the National Research Institute of Astronomy and Geophysics (NRIAG)-Egypt, Oukaïmden Observatory- Morocco, and the institute of astronomy and space sciences (Maragha Observatory)- Jordan, and the Sharjah Academy for Astronomy, Space Sciences and Technology (SAASST).

There is information about AUASS, foundation, aims, current projects, publications, international collaboration, and a database of more than 700 Arab astronomers throughout the world, office of astronomical education (OAE), astronomical societies,

Fig. 5. AUASS activities

archeoastronomy locations, and the office of astronomy for development (OAD). The webpage is considered to be a reliable and up-to-date reference for all Arab astronomers in terms of the new and updated news and information it provides. AUASS also updates it website calendar to keep up to date with all astronomical events (Fig. 5).

5 Participation in International Organizations

Jordan represented by AUASS is one of 96 member-states of the international astronomical Union (IAU).

Hamid Al-Naimi, and Awni Khasawneh, respectively as they are president and, Secretary General of AUASS, were selected as national outreach coordinator (NOC) representing their countries in the International Astronomical Union. Additionally. Many other AUASS members from different Arab – states are representatives of their countries in IAU as national outreach coordinators (NOC). Dr. Awni Khasawneh is also a member of the IAU memberships committee.

AUASS is hosting The Arab World Regional Office of Astronomy for Development, IAU (AW-ROAD) and its associated Arabic Language Expertise Centre (AW-LOAD) which was officially inaugurated on December 2015 in Amman, Jordan.

It aims to improve the cooperation in the field of space science and technology to take advantage of the benefit of human sciences effectively in the sustainable development of natural resources through education, training and research.

AUASS in cooperation with IAU (AW-ROAD), signed the first regional agreement with South West and Central Asia, Regional Office of Astronomy for Development (IAU- SWCA- ROAD).

The last MEARIM held in 2021, was hosted by AUASS in cooperation with IAU through online platforms due to the COVID-19 pandemic. It was attended by participants from more than 40 countries, with more than 100 participants from all over the world.

AUASS was approved as one of the observers of Asia-Pacific Space Cooperation Organization (APSCO), in addition to that AUASS has an agreement with Chinese national space agency (CNSA).

6 AUASS Conferences, Workshop, Meetings and Other Activities

Since foundation, AUASS organizing regular conferences, workshop, annual meetings, most of the meetings usually used to be organized coincide with the general conference of AUASS and other events.

The 12th AUASS, Arab conference on Astronomy and space science & the7[th] Islamic Astronomical conference dated on, 1–3 May, 2018 (Amman/Jordan) was distinguished, many important resolutions were taken by the higher council.

Many themes were discussed during the conference.it includes, cosmology, astrophysics, search for life in the universe, astronomical observatories and instrumentation, solar activity and space weather, artificial satellites and application, celestial mechanics, spherical astronomy and geodesy, planetology and astrogeology, archeoastronomy and the Arab-Islamic history of astronomy, astronomical application in Islamic sharia, administrative axis on the occasion of the twentieth anniversary of AUASS.it brought It brought together astronomers, astronauts, archeologists, historians, philologists, artists and representatives of other fields.

Table 1. Conferences organized by AUASS.

Years	Name	Country
1997	The foundation Arab Astronomy conference	Amman / Jordan
1998	The 2nd AUASS conference	Amman / Jordan
2000	The 3rd AUASS conference	Amman / Jordan
2002	The 4th AUASS conference	Amman / Jordan
2004	The 5th AUASS conference	Amman / Jordan
2005	The 6th AUASS conference	Tripoli /Libya
2006	The 7th AUASS conference	Djerba /Tunis
2008	The 8th AUASS conference	Amman / Jordan
2009	The 9th AUASS conference	Khartoum /Sudan
2009	The First Arab Conference on Astrogeology	Amman/ Jordan
2011	The 5th Islamic-Astronomical Conference	Amman / Jordan
2012	The 10th AUASS conference	Sultanate of Oman
2013	The 1st Conference of Arab Surveying and Geographical Names Experts	Amman/ Jordan
2014	The 11th AUASS conference	Sharjah/ UAE
2014	The 6th Islamic-Astronomical Conference (March)	Amman / Jordan
2014	The 7th Conference of Arab Experts in Geographical Names	Amman / Jordan
2016	The 2nd Conference of Arab Surveying and Geographical Names Experts (November)	Amman / Jordan
2018	The 12th AUASS conference	Amman / Jordan
2019	United Nations / Jordan Workshop: Global Partnership in Space Exploration and Innovation Support provided by: AUASS, Inter Islamic Network on Space Sciences and Technology	Amman / Jordan
2020	The 13th AUASS conference coincide with /The fifth MEARIM V 2020″ Astronomy Education and Research for the Future Generations 10- 12 Nov 2020 conference	Amman / Jordan
2022	INTERNATIONAL SYMPOSIUM-2022 Space Exploration: Moon and Beyond	Mafraq / Jordan
2023	The 14th AUASS conference	Sharja /UAE

7   AUASS School Lectures

AUASS is playing very important rule in contributing to the astronomical education matters in Arab states.

AUASS endeavors to enhance the professionalism of astronomy education by aiding and backing astronomers and astronomy educators across the Arab world in acquiring the necessary skills for effectively communicating astronomy at both school and university levels, employing a professional and efficient approach.

AUASS in collaboration with the Ministry of Education is involved in supporting a special astronomy lecture program for Arab astronomical societies to facilitate and provide access to quality resources by preparing an excellent study material resource for teaching astronomical topics and supporting efforts to include more astronomy curricula in the national curriculum.

AUASS has also developed a ten-year strategic plan through which it was able to spread and enhance the culture of astronomy in schools and universities, contribute to developing school curricula, and translate major texts and resources in astronomy into the Arabic language.

AUASS signed many agreements with ministry of education and other educational institutions for partner establishment between astronomy organizations, research centers, and schools. Which include guest lectures, workshops, and field trips to observatories in different Arab member states.

AUASS has organized several training programs for teachers to expand their knowledge of astronomy and space science. To enable them to teach the subject easily, AUASS has also organized through astronomical societies several public lectures, stargazing events and astronomy festivals to engage students and the community.

8   Other Matters of Astronomical Education

Jordan and many Arab member states are actively participating in the IAU Astronomy Education programs. The AUASS through Arab member states is represented in the National Astronomy Education Coordinator Team (NAEC Team), for example Jordan is represented by Dr. Awni Al-Khasawneh, Dr. Mashhoor Al-Wardat, and Ms. Dalal Al-Lala.

AUASS has contributed to solving problems in astronomy education in Arab countries through various initiatives and collaborations.

Many programs and platforms were implemented in Arab states to improve astronomical education such as: teacher training programs, workshops and outreach programs, summer schools, promoting research opportunities, collaboration with educational institutions, public awareness campaigns, regional and international partnerships, development of online learning platforms, and the NASE program (Tables 1 and 2).

9   Archeoastronomy and Astronomy in Culture

AUASS played a significant role in promoting Archeoastronomy and integrating astronomy into the cultural fabric of Arab societies through various initiatives. The following steps was taken by AUASS.

Table 2. Curriculum Development.

Year	Summer School Name
2011	Special summer school for teachers
2013	First AUASS summer school for university students
2015	Second AUASS summer school for university students
2017	Third AUASS school for amateurs
2019	Fourth AUASS school for young astronomers
2022	Regional Summer School on Space Sciences and Technologies
2023	3rd regional astronomical summer school (3 RASS)- joint school between AUASS & BAO

Forming a committee consist of one representative from each Arab member states, the purposes of this committee to coordinate issues related to Archeoastronomy and to raise awareness on the importance of astronomical heritage in Arab member states and to strengthen the efforts and to protect and preserve such heritage since member state countries are rich in Archeoastronomical sites.

This committee offers member states the ability to evaluate and recognize the importance of astronomical heritage, through the enrichment of astronomical history in the member states and the promotion of astronomical culture. The committee was also able to identify and highlight astronomical sites and sites with astronomical history in the member states, which helped in keeping the memory of these sites alive. The committee has also provided reports on how to preserve these sites from deterioration and oblivion.

Furthermore, the committee was able to develop an educational program which highlighted the connections between astronomy, archaeology, and cultural heritage. These programs included workshops, lectures, and interactive sessions that explore ancient astronomical practices and their cultural significance.

The committee has already held several related meetings in Jordan, Egypt, and the UAE, in 2016, 2017, and 2023. The topics and discussions of the committee were also documented in many articles publishes in AUASS AL-Kwan magazine, and many lectures delivered under title of "Archeoastronomy and Astronomy in Culture".

AUASS is also part of the IAU Working Group "Astronomy and World Heritage," which has the objective of bridging Science and Culture to identify monuments and sites associated with astronomical observations across all geographical regions within AUASS member states. This initiative aligns with the aim of AUASS to preserve these sites, uphold their scientific significance, and raise public awareness and understanding of their importance.

10   Astronomy Outreach

AUASS is playing a very important role in hosting the IAU Astronomy for development, The AW-ROAD and its associated AW-LOAD which was officially inaugurated on December 2015 in Amman, Jordan.

The AW-ROAD is preparing and training Arab cadres specialized in the field of space and astronomy. Now the AW-ROAD became the center for training and scientific research for the Arab region. The AW-ROAD has achieved many accomplishments during this short period, in effective cooperation with member states, including many agreements with various countries and similar organization and institutions.

The IAU Office for Astronomy Outreach (OAO) has been established in Tokyo, Japan. The Jordanian National Coordinator of OAO is Dr. Awni Al-Khasawneh.

AUASS serves as the overarching entity for all Arab Astronomical societies, recognizing the significance of amateur astronomy development. Consequently, AUASS actively encourages Arab Astronomical societies to establish dedicated webpage sections. Furthermore, AUASS produces a seasonal astronomical journal featuring articles in Arabic, facilitating accessibility and outreach to the Arab public. It regularly issues press releases, organizes press conferences, interviews, and conducts seminars on scientific journalism. These press releases cover updates on astronomical events and activities hosted by AUASS and (AW-ROAD). AUASS members also deliver public lectures at various institutions such as universities, schools, and ministries, both in-person and virtually. Additionally, AUASS publishes the proceedings of its Annual Meetings, along with booklets, calendars, books, sky maps, and other promotional materials. AW-ROAD Consists of all Arab countries served by the office and considered as official members including Algeria, Bahrain, Egypt, Iraq, Jordan, Kuwait, Lebanon, Libya, Morocco, Oman, Palestine, Qatar, Saudi Arabia, Somalia, Sudan, Syria, Tunisia, UAE, and Yemen.

11   Future Plans to Promote Astronomy in Arab Region

AUASS has to put a great effort to develop Astronomy in the Arab region through professionalizing astronomy education by helping astronomers, astronomy education and to enable teachers, trainers, and amateurs gain the required skills to communicate about astronomy in primary and secondary schools with a professional and an effective approach. On the other hand, more serious attention should be paid to help implement the IAU OAO projects on a national level, share astronomy news and events within member states, and bring the IAU closer to local communities by Involving these communities in astronomy initiatives, ensuring that the benefits of astronomy education and research are widely distributed, through organizing public outreach events such as lectures, stargazing sessions, and workshops, to raise awareness and interest in astronomy, leverage traditional and digital media to disseminate information about astronomical events and discoveries.

AUASS has already taken steps to initiate an educational program on Archeoastronomy, History of Astronomy, and Astronomy in Culture. The purpose is to include the list of available astronomical items in Arab countries as Petra in Jordan, Abu Simbel, Playa Nabatieh, pyramids of Egypt.

Since AUASS is hosting the (AW-ROAD), enhanced collaboration is imperative to bolster and implement measures that advance the utilization of astronomy, encompassing its practitioners, expertise, and facilities, thereby leveraging requisite human and financial resources to unlock the field's scientific, technological, and cultural advantages for societal development.

AUASS has also already signed an MOU with APSCO and many astronomical organizations, space agencies. Therefore, training will be activated between these agencies and AUASS.

By considering these steps and tailoring them to the specific needs and contexts of the Arab region, future plans for promoting astronomy can be more effective and impactful.

References

1. Auass, M., Jas, M.: Al- Thuraya-Magazine **64**, 20 (2011)
2. Awni Khasawneh, A., Al-Naimy, H., AUASS, M.: "Alkwn" Magazine **2659**, 64 (2015)
3. Harutyunian, H.A., Mickaelian, A.M., Parsamian, E.S.: Astronomical heritage in the national culture. In: Proceedings of the Archaeoastronomical Meeting Dedicated to Anania Shirakatsi's 1400th Anniversary and XI Annual Meeting of the Armenian Astronomical Society, Held 25–26 Sep 2012 in Byurakan, Armenia (2014)
4. Arab Union for astronomy and Space Sciences Conferences, workshops, symposium, weekly lectures

Legal Challenges in Establishing Human Settlements in Space

Zeina Ahmad$^{(\boxtimes)}$ and Shadi A. Alshdaifat

Sharjah University, Sharjah, UAE
{U22105769,salshdaifat}@sharjah.ac.ae

Abstract. This paper delves into the legal complexities of establishing human settlements in space, a frontier rapidly transitioning from speculative to tangible due to advancements in aerospace technology. Despite the progression, significant legal challenges arise, particularly concerning the applicability and adequacy of current space law, primarily governed by treaties like the 1967 Outer Space Treaty (OST). These treaties, formulated when permanent human habitation in space seemed implausible, now reveal notable gaps in addressing property rights, governance, and commercial exploitation of space resources.

This paper first examines existing legal frameworks, highlighting their relevance and limitations in space settlements. Key focus areas include property rights on celestial bodies, jurisdiction and governance of extraterrestrial settlements, and liability and safety in these habitats. The paper then explores recent initiatives, like the Artemis Accords, which attempt to bridge some legal gaps but lack the standing of formal international treaties.

The research underscores the pressing need for evolving legal landscapes to accommodate the emerging requirements of space settlements. It emphasizes the importance of international cooperation in developing comprehensive legal frameworks that can navigate the intricacies of space settlements, addressing issues such as resource utilization, dispute resolution, and the integration of private sector activities. The paper concludes by underscoring the necessity of an adaptable legal framework, harmonizing with technological advancements and ethical considerations, to ensure responsible and sustainable human settlements beyond Earth.

Keywords: Outer Space Treaty · Celestial Property Rights · Space Settlement Governance · Liability in Space Habitats · Artemis Accords

1 Introduction

The once speculative idea of human settlements in space is rapidly becoming a reality, thanks to significant advancements in aerospace technology driven by government space agencies and private corporations. This progression towards colonizing celestial bodies like the Moon, Mars, and potentially others, heralds a new era of human exploration and advancement. However, it also introduces complex legal challenges that must be navigated with care.

© The Author(s) 2025
H. M. K. Al Naimiy et al. (Eds.): AUASS-CONF 2023, SPPHY 420, pp. 231–247, 2025.
https://doi.org/10.1007/978-981-96-3276-3_17

As initiatives such as NASA's Artemis program and various private Mars settlement ventures transition from visionary concepts to active projects with definitive timelines, we find ourselves at the threshold of a new epoch in space exploration. These technological feats, while groundbreaking, unveil a Pandora's box of legal issues, particularly regarding the application and adequacy of current space law for permanent human settlements [1].

The field of space law, evolving since the early days of space exploration, is governed by a framework established by treaties like the Outer Space Treaty (OST) of 1967. The OST, along with subsequent agreements such as the 1968 Rescue Agreement, the 1972 Liability Convention, the 1975 Registration Convention, and the 1979 Moon Agreement, sets fundamental principles for outer space activities. Nonetheless, these treaties, formulated at a time when permanent human habitation in space was not considered feasible, provide limited guidance for the intricate legalities of space settlements [2].

There are notable gaps in this legal framework, especially concerning property rights, governance, and the commercial exploitation of space resources. The emergence of private space enterprises and the growing reality of human settlements in space necessitate a reevaluation and enhancement of these.

As initiatives such as NASA's Artemis program and various private Mars settlement ventures [3] transition from visionary concepts to active projects with definitive timelines, we find ourselves at the threshold of a new epoch in space exploration. These technological feats, while groundbreaking, unveil a Pandora's box of legal issues particularly regarding the application and adequacy of current space law for permanent human settlements [4].

In summary, while existing space laws and treaties have established a basic framework for activities in space, there's a pressing need for this legal landscape to evolve. This evolution is necessary to address the unique challenges of establishing human settlements beyond Earth. The complex interplay between existing laws and the emerging requirements of space settlements presents a multifaceted legal arena that demands careful consideration and international cooperation. This paper will systematically explore these legal challenges in the context of establishing human settlements in space. Initially, it will provide a detailed analysis of the existing legal framework, including the Outer Space Treaty and subsequent treaties, highlighting their relevance and limitations for space settlements. Following this, the discussion will shift to specific legal issues such as property rights on celestial bodies, jurisdiction and governance of extraterrestrial settlements, and the management of liability and safety in these new habitats. Through this exploration, the paper aims to contribute to the ongoing discourse on space law and policy, providing insights and recommendations for a future where human settlements in space are not just a possibility but a reality.

2   Property Rights and Resource Utilization

The Outer Space Treaty (OST) of 1967, which forms the backbone of international space law, expressly prohibits national appropriation of outer space, including the Moon and other celestial bodies, as stated in Article II. While the treaty succeeds in preventing the extension of national sovereignty into space, its language is silent on the rights of individuals or private entities, particularly concerning the ownership and use of space resources. This silence is becoming more pronounced as private companies move forward with plans for the extraction of extraterrestrial resources and potential settlements on other planets [4].

Article I of the OST declares that the use of outer space shall be carried out for the benefit and the interests of all countries, and shall be the province of all mankind [5]. This article sets a tone of communal usage and benefit, which contrasts sharply with the concept of private property ownership, suggesting that the extraction and utilization of space resources should be approached as a collective human endeavor. However, what constitutes "the benefit and in the interests of all countries" remains open to interpretation, and whether this allows for private exploitation under a regulatory framework is a subject of intense debate [6].

Furthermore, Article VI places responsibility for national space activities, including those by non-governmental entities, on the respective states. This stipulation has implications for the oversight of private companies by their national governments, potentially leading to a form of indirect appropriation through private enterprise, which would contravene the spirit of the treaty.

As we advance toward the reality of commercializing space resources, the lack of clarity in the OST has led to differing national interpretations. For instance, the United States passed the Commercial Space Launch Competitiveness Act in 2015, which grants U.S. citizens the right to possess and sell extraterrestrial resources they obtain. This act seems to challenge the OST's framework, though proponents argue it does not claim sovereignty over celestial bodies, thus remaining within the treaty's boundaries [7].

The question of whether the OST allows for the private appropriation of space resources is deeply intertwined with ethical considerations. The principle of space as the "province of all mankind" is evocative of a common heritage of humanity, akin to the legal concepts governing international waters. The extension of this principle to celestial bodies suggests that space resources should benefit humanity as a whole, rather than individual private interests. However, without an international regulatory body or a more detailed framework to enforce this principle, the risk of a 'space rush', where resources are depleted or claimed by those with the technological capability to do so, remains a possibility.

Another concern is the potential for conflict over resources, which Article III of the OST attempts to mitigate by stating that states shall carry out their activities in the exploration and use of outer space by international law, including the United Nations Charter. This implies that any disputes over space resources should be resolved peacefully and through international cooperation [7]. However, without clear legal definitions and regulations regarding property rights and resource utilization in space, the potential for conflict remains [8].

Given these complexities, there is a growing consensus among scholars and policy-makers that the OST needs to be supplemented with additional treaties or agreements to specifically address the utilization of space resources. One proposed solution is the development of an international regulatory framework, which would establish clear guidelines for the exploitation of space resources, ensuring equitable access and sustainability. This framework could also provide a mechanism for sharing the benefits derived from space resources, in keeping with the OST's principles.

"Property Rights and Resource Utilization in the Context of the Outer Space Treaty" is indeed complex and multifaceted, encompassing legal, ethical, and practical considerations. The challenges stem from the Outer Space Treaty's (OST) ambiguity regarding private and commercial activities in space, particularly resource extraction and utilization. However, there are some proposed solutions and recommendations to address these issues.

2.1 Solutions and Recommendations

Given the OST's lack of clarity on private and commercial activities in space, one solution could be the development of supplementary international agreements or treaties. These should specifically address issues of property rights, resource utilization, and commercial activities in space, providing clear guidelines that are in harmony with the principles laid out in the OST. Moreover, the international community should work towards establishing clear legal definitions and standards for activities related to space resources. This includes defining what constitutes "the benefit and in the interests of all countries" and setting parameters for the equitable and sustainable use of space resources.

Creating an international regulatory body dedicated to space activities could also be a significant step forward. This body would oversee and regulate the exploitation of space resources, ensuring that activities are conducted in a manner that is equitable, sustainable, and by the OST's principle of space being the "province of all mankind". Furthermore, strengthening conflict resolution mechanisms in line with Article III of the OST is crucial. This could involve the creation of a space-specific arbitration or judicial body under the auspices of the United Nations or another international entity.

Moreover, there should be mechanisms in place for sharing the benefits derived from space resources. This could involve financial contributions, technology sharing, or capacity-building initiatives for countries that do not have the means to explore or utilize space resources independently. Policies and regulations should also be designed to prevent the monopolization of space resources by a few technologically advanced entities. This could include setting limits on the amount or type of resources that can be utilized by a single entity. In addition, encouraging public-private partnerships under clear guidelines can help balance private interests with the communal nature of space resources. Governments could play a pivotal role in overseeing and regulating these partnerships, ensuring adherence to international law and ethical standards.

As well, any framework or regulation should include strong environmental considerations to prevent the depletion or unsustainable use of space resources. This is crucial for ensuring that space activities do not harm the celestial environment or compromise future utilization opportunities. The process of developing these frameworks and agreements must be inclusive, involving a wide range of stakeholders, including countries with emerging space capabilities, private companies, and civil society groups. As space technology and our understanding of space resources evolve, the legal and regulatory frameworks should also be subject to continuous review and adaptation to ensure they remain relevant and effective.

These recommendations aim to balance the need for exploration and utilization of space resources with the principles of international cooperation, sustainability, and equitable benefit-sharing, as envisioned in the Outer Space Treaty.

3   Jurisdiction and Governance in Space Settlements

As humanity progresses towards establishing human habitats in space, a significant challenge arises in the form of jurisdiction and governance. The foundational Outer Space Treaty (OST) of 1967, while crucial in setting general principles for space activities, does not adequately provide for governance structures or legal jurisdictions in extraterrestrial habitats. This lack of clarity becomes a critical issue as multiple nations and private entities, each with their unique legal systems and interests, engage in space exploration and settlement.

The OST offers some foundational guidance on jurisdictional matters in space. Article VIII, for instance, provides that ownership of space objects, such as satellites and space stations, remains with the launching state regardless of their location in space or on celestial bodies. This article lays the groundwork for a form of extraterritorial jurisdiction by the launching state over its space objects. However, the application of this principle to space settlements, especially those involving a diverse array of public and private stakeholders, is not straightforward [9].

3.1   Limitations of Article VIII in Addressing Space Settlements

Article VIII's framework, while clear in terms of ownership and control, does not encompass the broader aspects of jurisdictional governance required for permanent human habitats in space. It does not address how laws would be enforced within these habitats or how legal disputes between inhabitants, who may be subject to different national jurisdictions, would be resolved. This gap in the OST becomes more evident as we consider the complex scenarios of space settlements, which may involve multinational crews, private companies, and potentially even tourists [10].

3.2   Risk of Jurisdictional Conflicts and Legal Vacuum

The transition from temporary space missions to permanent settlements necessitates a reevaluation of current governance structures. The OST, primarily drafted during an era when human spaceflight was in its infancy, does not provide explicit guidelines for the governance of long-term human presence in space. This omission raises the potential for jurisdictional conflicts and a legal vacuum, where the legal rights and responsibilities of individuals and entities in space are undefined [11].

Such a scenario could lead to disputes over resource utilization, criminal jurisdiction, and civil matters, potentially hindering the development and sustainability of space settlements. Moreover, the emphasis of the OST on international cooperation and the peaceful use of space, as highlighted in Articles I and III, underscores the need for collective management approaches and effective conflict resolution mechanisms. The treaty encourages states to work together in exploring and utilizing outer space, but it stops short of providing a mechanism for managing the day-to-day governance of space habitats. This lack of specificity in the OST points to the necessity for new international agreements or protocols that specifically address the governance of space settlements.

Another critical issue is the enforcement of laws and dispute resolution in space settlements. The unique conditions of space, detached from any Earth-bound nation, pose challenges in law enforcement and justice administration. Determining which legal system or combination of systems applies in space settlements, especially those involving international collaboration or private entities, remains an unresolved and complex question [12].

Given the OST's limitations in addressing the needs of permanent space settlements, there is a growing consensus among legal scholars and space policy experts that a new legal framework, or a significant expansion of the existing one, is required [13]. This framework should be designed to handle the unique challenges of space habitation, including the establishment of judicial systems, law enforcement mechanisms, and administrative structures suitable for the space environment. It should also consider the diverse legal backgrounds and cultural contexts of space inhabitants, ensuring that governance structures are inclusive, equitable, and representative of the international community.

An additional challenge in developing jurisdictional frameworks for space settlements is the integration of private sector interests and activities. Private companies are increasingly playing a significant role in space exploration and development. Their activities, ranging from satellite launches to proposed space tourism and resource extraction missions, add another layer of complexity to the jurisdictional landscape. The legal framework must delineate the boundaries of private sector activities, define their legal responsibilities, and establish mechanisms for oversight and accountability in line with international space law. While the OST lays an essential foundation for space law, its provisions fall short in addressing the complex jurisdictional challenges of space settlements. The transition to permanent human habitation in space necessitates the development of a more comprehensive legal framework. This framework should not only expand on the OST's principles but also provide practical solutions for governance, conflict resolution, and the integration of private sector activities in space. As humanity's presence in space evolves, so too must the legal structures that support and govern it, ensuring that space remains a domain for peaceful exploration.

3.3 Proposed Models for Governance

The proposed models for governance in the context of space settlements or similar scenarios address the challenges of jurisdiction, representation, and legal integration. Here's an expansion of the three models:

Extension of National Jurisdictions provides a clear legal framework based on existing national laws, ensuring familiarity and stability. It also simplifies legal processes for the citizens of the nation responsible for the settlement. However, this model can lead to jurisdictional conflicts in cases where multiple nations are involved in a settlement, leading to legal complexities. Also, it can raise significant concerns regarding the representation and rights of non-citizen residents. This can lead to a lack of inclusivity and potential human rights issues [14]. In addition, it may not adequately address the unique challenges and needs of living in space, as Earth-based laws might not be fully applicable or sufficient.

The Multinational Governance Model encourages cooperation and shared responsibility among nations, fostering international collaboration. It can also lead to the development of comprehensive laws that consider the perspectives and legal principles of multiple nations. Nevertheless, the Multinational Governance Model requires the harmonization of different national laws and interests, which can be a complex and time-consuming process. The need for a comprehensive international treaty means extensive negotiations and potential compromises, which can be difficult to achieve. Might struggle with enforcement issues if the participating countries have conflicting interests or priorities.

Independent Governance Structures allow for the creation of laws and governance structures specifically designed for space living, which can be more effective and relevant than Earth-based laws. It can also offer a high degree of autonomy and flexibility, enabling innovative and tailored approaches to governance. However, It Faces difficulties in gaining recognition and integration with Earth-based legal systems, which can impact international relations and cooperation. In addition, it might struggle with issues of legitimacy and enforcement, particularly in the early stages of establishment. There is, also, a risk of isolationism or the development of governance systems that diverge significantly from Earth's ethical and legal norms, potentially leading to conflicts.

3.4 Developing a Comprehensive Legal Framework

Applicability of earth-based laws includes analyzing current Earth-based laws to identify which can be directly applied, adapted, or are irrelevant in space, and developing new legal provisions specifically for space to address unique challenges such as microgravity effects, resource sharing, and space debris management. Challenges, however, include balancing the application of Earth laws with the need for new laws that cater specifically to the unique environment of space, and ensuring that space laws are in harmony with international laws and treaties on Earth.

Moreover, conflict resolution includes establishing judicial bodies within space settlements for immediate dispute resolution, and creating international arbitration panels, possibly under the aegis of an international space organization, to handle conflicts between entities from different nations. However, developing fair and equitable processes that are acceptable to all parties involved remains a challenge, considering the diverse legal systems of Earth, and ensuring the enforcement of decisions made by these bodies.

Furthermore, the protection of individual rights includes ensuring that human rights standards are maintained in space settlements, including rights to life, liberty, and security [15], and providing clear and accessible avenues for individuals to address legal grievances, possibly through ombudsman services or legal aid systems. Challenges involve adapting human rights norms to the space environment, where conditions are vastly different from Earth, and protecting rights in a high-tech environment where privacy and data security could be major concerns.

The establishment of Legal Institutions, also, encompasses setting up courts or tribunals within space settlements for law enforcement and adjudication of disputes, and integrating these institutions with Earth-based legal systems to ensure coherence and mutual recognition. Challenges involve establishing the authority and legitimacy of these institutions, especially in newly formed settlements, and training and appointing qualified legal personnel who understand both space-specific and Earth-based legal issues.

Additionally, international collaboration involves both space-faring and non-space-faring nations in the development of the legal framework to ensure a broad, inclusive perspective. Moreover, regularly revising and updating the framework to keep pace with advancements in space technology and exploration. It includes achieving consensus among a diverse group of stakeholders with varying interests and priorities, and ensuring that the legal framework remains adaptable and flexible to accommodate future developments in space exploration.

The development of such a comprehensive legal framework is crucial for the sustainability and success of space settlements. It involves a delicate balance between adapting Earth-based laws and creating new norms suited to the unique challenges of space. Moreover, the involvement of a wide range of international stakeholders is essential to ensure the framework's fairness, effectiveness, and wide acceptance.

To summarize, addressing jurisdiction and governance in space settlements is a complex but crucial task. It involves choosing an appropriate governance model and underpinning it with a comprehensive legal framework developed through international cooperation. As humanity ventures further into space, the establishment of fair, practical, and adaptable governance structures becomes increasingly important for the success and sustainability of space settlements.

4 Safety and Liability in the Context of Human Settlements in Space

The Liability Convention of 1972, a pivotal document in the fabric of space law, establishes that the launching state bears absolute liability for damage caused by its space objects on the surface of the Earth or to aircraft in flight (Article II). However, its scope is less clear when it comes to damage that might occur in the context of human settlements on celestial bodies or within the structures that support life in space. The provisions within Articles II and III address the nuances of fault and the processes for claims between states, but they stop short of detailing the complex scenarios that could arise from sustained human presence on another planet or moon [16].

Given the growth in space activities, the lack of specific guidelines for safety and liability in human settlements is a pressing concern. The Liability Convention articulates a framework for determining the launching state's liability, but it does not contemplate the intricate living conditions or the types of activities that would be commonplace in extraterrestrial settlements. For example, Article IV suggests that the convention applies to damage caused elsewhere than on the surface of the Earth, yet it does not elaborate on the practicalities of applying this to settlements, where the environment is unlike anything contemplated by the original drafters [16].

Safety standards for habitats in space are a critical consideration, yet current international space law provides no specific framework for the development and enforcement of these standards. The Rescue Agreement of 1968 highlights the responsibility of states to assist astronauts in distress, which implies a safety concern. However, this does not translate into a comprehensive set of safety protocols for long-term settlement efforts.

As we address the legal implications of safety and liability in space settlements, a multi-faceted approach is required. This approach must take into account not only the direct damage envisaged by Article II of the Liability Convention but also the broader implications of Article VII of the Outer Space Treaty, which holds states internationally responsible for their national activities in space, including the activities of non-governmental entities.

The complications of liability in space are accentuated by the multitude of actors involved in potential space settlements. Disputes arising from damage or accidents in space habitats could involve private individuals, corporations, and governments, each potentially from different countries. The Liability Convention's Article VIII provides for a claims process through diplomatic channels, but how this would function in a densely populated extraterrestrial environment is unclear [17].

Further complicating the matter is the enforcement of any liability determinations. On Earth, judicial decisions can be enforced by the coercive power of states, but no such mechanism currently exists in space. While Article IX of the Liability Convention discusses the settlement of disputes, the practical enforcement of such settlements is an open question that becomes even more complex in the context of human settlements in space.

Addressing these gaps requires an international consensus on expanded liability provisions, the development of safety protocols, and perhaps most critically, the establishment of a mechanism for enforcing legal norms in the space environment. Such efforts would build on the foundation provided by the Liability Convention but would extend and refine these principles to meet the unique challenges of life in space.

In conclusion, the Liability Convention of 1972 provides a starting point for addressing liability issues in space, but it does not offer a comprehensive solution for the challenges posed by human settlements on celestial bodies. As humanity's presence in space becomes more permanent, the development of an expanded legal framework becomes imperative. This framework must address not only the issues of liability and safety but also the practical enforcement of these norms, ensuring that space remains a domain characterized by order, safety, and justice.

To address the challenges of safety and liability in the context of human settlements in space, as highlighted above, several solutions and recommendations can be proposed. These proposals aim to extend and refine the principles of the Liability Convention of 1972 and other relevant space laws to meet the unique challenges of extraterrestrial life.

4.1 Solutions and Recommendations

The Development of specific safety protocols for space settlements establishes comprehensive safety standards and protocols tailored for habitats and human settlements in space. This would include guidelines for construction, life support systems, emergency procedures, and environmental controls. These standards should be developed through international collaboration and draw from existing knowledge in related fields such as aeronautics, engineering, and environmental science. Hence, promoting international cooperation in research and development could enhance safety technologies for space habitats and this could lead to advancements in life support systems, structural integrity, radiation shielding, and emergency response capabilities. This also requires creating a specialized international body to oversee and regulate human settlements in space. This authority would be responsible for enforcing safety standards, managing liability issues, and facilitating dispute resolution among parties in space settlements by implementing a clear and efficient process for resolving disputes that arise. This framework should also include provisions for mediation, arbitration, and, if necessary, adjudication by an impartial tribunal. Moreover, defining the legal jurisdiction and applicable laws, for human settlements on celestial bodies, would clarify which nation's laws apply in specific scenarios and how international law interacts with these national laws.

Expansion of the liability convention's scope is another solution, which would include scenarios pertinent to human settlements in space. This would involve detailing the liability issues that could arise in extraterrestrial habitats, such as accidents, environmental hazards, and other incidents that could cause harm to individuals or property. Besides, implementing a system for regular auditing of space settlements would ensure compliance with established safety standards, which would involve periodic inspections, reviews of procedures, and assessments of equipment and technologies used in space habitats. This also requires developing an enforcement mechanism for liability rulings related to space settlements. This could involve agreements that ensure compliance with decisions made by the proposed international space authority or an appointed judicial body. Nevertheless, space technology and human activities in space will continue to evolve, necessitating an adaptive legal framework. Regular reviews and updates to the laws and regulations governing space settlements should be a continuous process, accommodating new challenges and advancements in space exploration, and engaging the public and stakeholders, including space agencies, private companies, and scientific communities, in discussions about space settlement safety and liability [18]. Public awareness initiatives and stakeholder consultations can also help shape effective policies and ensure broad support for the regulatory framework, confirming that ethical and environmental considerations are integral to the governance of space settlements, which include addressing issues of sustainability, the impact of human activities on celestial bodies, and the moral implications of long-term human presence in space.

Since many space activities are likely to involve private corporations, it's essential also to incorporate these entities into the liability framework. This would ensure that private companies are also held accountable for their actions and bear responsibility for any damages they cause in space settlements. This also requires space settlement operators to have comprehensive insurance policies or financial guarantees, which would ensure that funds are available to cover potential damages or liabilities arising from accidents or other incidents in space.

Solutions also encompass developing training programs and certification requirements for individuals working in space settlements, which would ensure that all personnel are adequately trained in emergency procedures, safety protocols, and the operation of life-supporting systems in space environments.

By implementing these solutions and recommendations, the international community can create a robust and comprehensive framework for ensuring safety and liability in human settlements in space. This framework would not only address current gaps in space law but also pave the way for sustainable and responsible expansion of human activities beyond Earth.

5   Environmental Protection in Outer Space in the Context of Human Settlements in Space

The burgeoning interest in extraterrestrial settlements brings environmental protection in outer space to the forefront of space law. The Outer Space Treaty (OST) of 1967, along with subsequent treaties, provides a framework for space activities but does not offer exhaustive provisions specifically dedicated to the environmental protection of celestial bodies. Article IX of the OST stipulates that states must avoid harmful contamination of space and celestial bodies, signaling a foundational concern for extraterrestrial environmental impacts. However, the treaty falls short of detailing the measures required to prevent contamination and does not define the scope of "harmful contamination" [19].

As the potential for human settlements becomes more concrete, the lack of specific guidelines on environmental stewardship in space treaties is increasingly problematic. The ambiguity within the OST and related treaties leaves much room for interpretation, posing a significant challenge as human activities in space escalate. Activities such as planetary mining, construction of habitats, and even the introduction of Earth-based biological matter have the potential to disrupt extraterrestrial ecosystems or geological processes, should they exist.

Moreover, the absence of clear, actionable standards in the OST creates a vacuum where each spacefaring nation or entity might follow its unilateral measures, leading to inconsistent and possibly damaging practices. The need for an international consensus on environmental protocols is becoming urgent to ensure that the celestial bodies are explored and utilized sustainably.

The Moon Agreement of 1979 attempted to address this issue by elaborating on the principles in the OST and calling for the prevention of the disruption of the existing balance of the environment on celestial bodies. However, with limited ratification, the Moon Agreement's influence on space activities remains minimal.

Given the pace of technological advancement and the interest in resource utilization in space, it is imperative to develop a set of internationally agreed-upon environmental principles. These principles should aim to prevent biological contamination, preserve the scientific value of celestial bodies, and establish guidelines for the mitigation of debris and other forms of pollution that could result from human or robotic presence and activity.

The planetary protection policies of space agencies like NASA and ESA are a step in the right direction, offering frameworks for avoiding biological contamination during robotic and crewed missions. Nevertheless, these policies are not legally binding at the international level and may not be sufficient to address the scale of impact that could result from full-scale settlement efforts [20].

Legal scholars and policymakers are advocating for a new, comprehensive treaty or a set of amendments to existing ones to incorporate detailed environmental provisions.

This treaty would need to define terms like "harmful contamination," set out clear protocols for environmental impact assessments, and establish mechanisms for monitoring and enforcing compliance [21].

The creation of a dedicated international body to oversee and manage the environmental aspects of space activities could be instrumental. Such an entity could facilitate cooperation between nations, ensure transparency in the execution of space projects, and serve as a forum for addressing environmental disputes.

Understanding the existing challenges in environmental protection in space, particularly in the context of human settlements, it's clear that while there are treaties and protocols, they may not fully address the unique and emerging issues associated with long-term human presence in extraterrestrial environments. To bridge the gap between scientific innovation and legal frameworks in the context of environmental protection in space, especially for human settlements, we need to propose legal mechanisms that are adaptive to scientific advancements and challenges. Here are several integrated scientific and legal approaches to create a comprehensive and effective solution.

5.1 Scientific and Legal Approaches for an Effective Solution

The legal mandate for closed-loop life support systems legally requires the implementation of advanced systems in space habitats. This mandate would ensure that all space settlements utilize the latest technology in resource recycling and waste management, as guided by scientific advancements. This also suggests developing a legal framework that governs the creation and maintenance of artificial ecosystems in space habitats. This framework would be based on scientific principles of ecological balance and would set standards for the use of these ecosystems in supporting human life and maintaining environmental stability. Correspondingly, an international treaty that mandates the use of nano-technology for environmental monitoring in space can be proposed. This treaty would outline the responsibilities of spacefaring entities in continuously monitoring and mitigating environmental impacts, using the latest scientific tools and methodologies.

Legal guidelines for in-situ resource utilization (ISRU) is another approach, which establishes legal guidelines for the use of in-situ resources, such as lunar or Martian regolith. These guidelines would encourage scientific research and development in ISRU technologies, ensuring that space exploration and settlement are sustainable and reduce dependence on Earth-based resources. Moreover, regulations that govern the use of quantum computing in managing space habitats' environmental systems could be developed. These regulations would ensure that the deployment of this advanced technology aligns with legal standards for sustainability and environmental protection.

An international charter that promotes the research and implementation of regenerative life support systems should be also created. The charter would legally bind parties to integrate biological processes into their life support strategies, based on scientific evidence of their effectiveness and sustainability. Likewise, legal standing to an Environmental Ethics Charter specifically designed for space activities should be given. This charter would not only outline ethical principles but also have enforceable legal consequences for non-compliance, ensuring that space activities adhere to high environmental standards.

Approaches can also include implementing a certification process for space habitat designs that assesses their environmental efficiency. This process would be based on scientific criteria for material usage, energy efficiency, and sustainability, ensuring legal compliance with environmental protection standards. Moreover, integrating automated compliance systems into the legal frameworks governing space habitats would ensure that all activities and resource usage comply with environmental laws and regulations. Similarly, incorporating the space environment impact rating (SEIR) system into legal requirements for space activities would legally require space missions to assess and minimize their environmental impact, promoting sustainable practices based on scientific evaluation.

By integrating these scientific advancements with legal frameworks, we can ensure that space exploration and settlement are both legally sound and scientifically advanced, leading to sustainable and responsible activities in space environments.

6 Ethical and Social Considerations in Space Settlement

As we contemplate the settlements of celestial bodies, it becomes increasingly clear that the current framework of space law is not equipped to address the myriad of ethical and social issues that such a monumental step for humanity entails. Existing treaties, including the Outer Space Treaty (OST) of 1967, provide the legal foundation for space activities but remain silent on the profound ethical and social questions that arise when we consider the potential of living beyond Earth [22].

The rights of individuals in space, for instance, pose a significant ethical concern. The OST and related documents primarily address the activities of states and do not delve into the rights of private persons. As we move towards the establishment of space communities, it becomes critical to consider how individual rights will be protected. How will human rights be upheld in the vacuum of space? What legal status will be afforded to individuals who may be born on a celestial body? These questions demand careful consideration to ensure that space settlements are grounded in respect for human dignity and freedom [23].

The social structure of space communities is another area that is uncharted by current space law. The development of space settlements will require careful planning not only in terms of physical infrastructure but also social organization [24]. What kind of social contracts will bind these communities? How will decisions be made, and leadership be determined? The construction of equitable and functional social structures in space requires foresight and deliberate ethical reflection to avoid the replication of Earth's social and economic inequities in space.

Moreover, the ethical considerations in colonizing celestial bodies encompass a range of concerns, from the preservation of their pristine environments to the respect for their potential intrinsic value. The concept of 'astroethics' has been proposed to tackle such considerations, advocating for a moral framework that guides human conduct in space and the treatment of celestial bodies. The principles of astroethics call for responsible stewardship of space environments, ensuring that exploration and settlement efforts do not lead to irreversible harm.

Furthermore, as we extend our reach into the cosmos, the potential encounter with extraterrestrial life forms brings about profound ethical implications. How will we ensure that our actions do not disrupt potential ecosystems or harm unknown life forms? The principle of non-maleficence, a core tenet in ethics, must be at the forefront of our exploration to minimize harm to any form of life or environment we may encounter.

The equitable distribution of the benefits derived from space activities is also a pressing ethical issue. The OST articulates that space shall be the province of all humankind, implying that the benefits of space exploration should be shared broadly. However, the implementation of this principle is vague and does not account for the disparities in technological and financial capabilities between nations. The risk of space becoming a domain dominated by a few at the expense of the many is real and must be addressed through inclusive policies and international cooperation [25].

To summarize, the unique challenges of applying human rights and social protocols in the context of space settlements, a new and unique model is needed. This model should integrate existing Earth-based human rights principles while adapting them to the specific conditions and challenges of living in space. Here's a proposed model, titled the "Extraterrestrial Human Rights and Social Framework (EHRF)".

6.1 Extraterrestrial Human Rights and Social Framework (EHRF) Model

Extraterrestrial Human Rights Declaration (EHRD) is a comprehensive declaration that extends Earth-based human rights to space settlements. This declaration would adapt existing rights to the context of space, covering aspects unique to space life, such as rights related to the prolonged absence of a terrestrial environment, reproduction in space, and the potential statelessness of individuals born on celestial bodies.

A governing body established to oversee the application and adaptation of the EHRD in space settlements is The Space Settlement Governance Council (SSGC). The SSGC would consist of representatives from spacefaring nations, legal experts, ethicists, and representatives from space settlements. It would be responsible for interpreting the EHRD, resolving disputes, and updating policies as space settlement evolves.

Also, to reflect the unique needs and conditions of space living, The Extraterrestrial Social Contract (ESC) would be developed in consultation with future space settlers. The ESC is a social contract specifically designed for space communities, outlining the responsibilities and rights of settlers, governance structures, and decision-making processes.

Moreover, to integrate and harmonize space laws with Earth-based international law and to ensure consistency and address jurisdictional complexities, The Interplanetary Legal Integration Mechanism (ILIM) system would facilitate the legal recognition and enforcement of rights and obligations across Earth and space settlements.

Extraterrestrial Environmental and Cultural Preservation Protocol (EECPP) would also ensure that space settlements respect the intrinsic value of celestial bodies and their potential ecosystems. In addition, it would establish guidelines for the preservation of extraterrestrial environments and cultural heritage, including any interactions with potential extraterrestrial life.

A final entity that would offer guidance on complex ethical dilemmas and ensure that settlements operate within the agreed ethical boundaries would be The Space Settlement Ethical Oversight Panel (SSEOP) which is dedicated to monitoring and addressing ethical issues arising from space settlement.

The EHRF model would require ratification through a series of international treaties and agreements, ensuring global cooperation and adherence. The framework would also be subject to regular reviews, allowing for adaptation as technological, social, and ethical understanding evolves. The EHRF model would also require the implementation of education and training programs for space settlers, focusing on the principles of the EHRF, to foster a shared understanding and commitment to these principles.

To sum up, the Extraterrestrial Human Rights and Social Framework (EHRF) offers a comprehensive and adaptable approach, specifically tailored to the unique environment and challenges of space settlements. Integrating and expanding upon Earth-based human rights and social protocols provides a robust foundation for the governance and ethical management of human activities beyond Earth, ensuring the protection and well-being of individuals in these new frontiers.

7   Conclusion

In conclusion, as we enter a new era of space exploration and consider human settlements beyond Earth, addressing the legal challenges is imperative. The current space law framework, particularly the Outer Space Treaty (OST) of 1967, is inadequate for the complexities of extraterrestrial settlements. This necessitates evolving and expanding our legal frameworks to ensure peaceful and equitable utilization of space. The challenges include property rights, resource utilization, governance, liability, safety, environmental protection, and ethical and social considerations. Supplementary international agreements and a regulatory framework are essential to establish clear guidelines for the exploitation of space resources, ensuring equitable access and sustainability.

Governance and jurisdiction require new legal structures adapted to space environments. The Liability Convention of 1972 needs expansion to address safety and liability in space settlements. Environmental protection demands a comprehensive treaty and an international body to oversee space activities sustainably. Ethical and social considerations necessitate the development of an Extraterrestrial Human Rights and Social Framework (EHRF) to adapt human rights principles to space conditions.

The future of space law must be adaptable, inclusive, and sustainable. As humanity ventures into space, our legal frameworks must evolve to ensure responsible, ethical, and beneficial exploration and settlement for all. Establishing human settlements in space requires careful consideration and international cooperation to ensure success and sustainability for future generations.

References

1. Mosher, D.: 2 sentences in Starlink's terms of service show that SpaceX is serious about creating its 'own legal regime' on Mars. Business Insider (2020)
2. Gupta, B.: Liability for Commercial Outer Space Activities: Need for a Legal Framework in India. Nova Science, New York (2020)
3. SpaceX. Mars & Beyond (2021). https://www.spacex.com/mars
4. Miller, Z.: Space settlement and the celestial subjectivity model: shifting our legal perspective of the universe. In: Froehlich, A. (ed.), A Fresh View on the Outer Space Treaty, vol. 13, pp. 59–66. Springer International Publishing (2018)

5. Benkö M., Schrogl K. (eds.): Outer space - future for humankind: Issues of law and policy. Eleven International, The Netherlands (2021)
6. Kerkonian, A. D.: The legal aspects of permanent human settlement on celestial bodies. Master Thesis, McGill University (2017)
7. Li, X.: An analysis of the channels for accessing economic benefits in the commons governance regime of space resources. Int. J. Commons **17**(1), 787–186 (2023)
8. Miller, Z.: Space Settlement and the Celestial Subjectivity Model: Shifting Our Legal Perspective of the Universe. In: A. Froehlich (Ed.) A Fresh View on the Outer Space Treaty, vol. 13, pp. 59–66. Springer International Publishing (2018)
9. Hofmann, M., Bergamasco, F.: Space resources activities from the perspective of sustainability: legal aspects. Global Sustainability **3**, 1–7 (2020)
10. Bhat, S.: Legal issues surrounding human settlements on the moon and other celestial bodies. Inter. Instit. Space Law **63**(2), 91–106 (2020)
11. Svitlychnyy, O.: Ownership of space objects. Adv. Space Law **2**, 76–95 (2018)
12. Buchan, R., Franchini, D., Tsagourias, N. (eds.): The Changing Character of International Dispute Settlement: Challenges and Prospects. Cambridge University Press, New York (2023)
13. Miller, Z.: Space settlement and the celestial subjectivity model: shifting our legal perspective of the universe. In: Froehlich, A. (ed.) A Fresh View on the Outer Space Treaty, vol. 13, pp. 59–66. Springer International Publishing (2018). https://doi.org/10.1007/978-3-319-704 34-0_6
14. Tronchetti, F.: The PCA rules for dispute settlement in outer space: a significant step forward. Space Policy **29**(3), 181–189 (2013)
15. Ishola, F., Fadipe, O., Taiwo, O.: Legal enforceability of international space laws: an appraisal of 1967 outer space treaty. New Space **9**(1), 33–37 (2021)
16. Tăiatu, C.: legal implications for gender mixed human settlements on mars—preliminary thoughts on human reproduction and childbirth in space. In: Froehlich, A. (ed.) Assessing a Mars Agreement Including Human Settlements, Vol. 30, pp. 99–112. Springer International Publishing (2021). https://doi.org/10.1007/978-3-030-65013-1_9
17. von der Dunk, F.: 'For All moonkind'- legal issues of human settlements on the moon: jurisdiction, freedom and inclusiveness. Inter. Instit. Space Law **63**(2), 77–80 (2020)
18. Grepperud, S.: Individual or enterprise liability? the roles of sanctions and liability under contractible and non-contractible safety efforts. Rev. Law Econ. **16**(3), 1–28 (2020)
19. Kovudhikulrungsri, L., Nakseeharach, D.: Liability regime of international space law: some lessons from international nuclear law. J. East Asia Inter. Law **4**(2), 291–318 (2011)
20. Alshdaifat, S.: Who owns what in outer space? Dilemmas regarding the common heritage of mankind. Pécs J. Inter. European Law II, 21–43 (2018)
21. Chung, G.: Emergence of environmental protection clauses in outer space treaty: a lesson from the rio principles. In: Froehlich, A. (ed.) A Fresh View on the Outer Space Treaty, vol. 13, pp. 1–13. Springer International Publishing (2018). https://doi.org/10.1007/978-3-319-70434-0_1
22. Seasly, E.: NASA's planetary protection policies revised in response to changes in space exploration. In: Soares, C., Wooldridge, E., Matheson, B. (eds.) Space Systems Contamination: prediction, Control, and Performance 2022, vol. 12224, pp. 11489–11499. SPIE, California (2020)
23. Breccia, P.: The evolving role of the environmental impact assessment as a valuable legal tool for the protection of outer space and the sustainability of space activities. In: Proceedings of The International Institute of Space Law, pp. 41–52. Eleven, The Netherlands (2019)
24. Marboe, I.: Living in the moon village - ethical and legal questions. Acta Astronaut. **154**, 177–180 (2019)

25. Freeland, S.: The intersection between space law and international human rights law. In: Jakhu, R., Dempsey, P. (eds.) Routledge Handbook of Space Law, Routledge, pp. 245–258. Routledge, London (2016)
26. Lim, J.: Charting a human rights framework for outer space settlements. Inter. Institute Space Law **63**(2), 155–168 (2020)
27. Ferreira, G., Ferreira-Snyman, A.: The application of international human rights instruments in outer space settlements: today's science fiction, tomorrow's reality. Potchefstroom Electr. Law J. **22**(1), 1–43 (2019)
28. Avveduto, R.: Past, present, and future of intellectual property in space: old answers to new questions. Pac. Rim Law Policy J. **29**(1), 203–246 (2019)

A City-Wide Planetarium

Marwan A. Shwaiki[✉]

Sharjah Academy for Astronomy, Space Sciences and Technology, University of Sharjah Arab Planetarium Society (APS), Sharjah, UAE
mshwaiki@sharjah.ac.ae, m.shwaiki@gmail.com

Abstract. In this project, hundreds "could be thousands" of powerful drones to be used. The drones will fly high for about 0.25–1.0 km up and need to be stable for 30 min duration show called City Wide Planetarium Show. The drones will carry powerful LED light sources. LED light's intensity, color and luminosity can be controlled. All the drones will work together as one single structure. Drones will resemble the stars of the sky, spread out in the positions of the natural stars as seen from the center of the city where the public stargazing party is taking place.

Keywords: planetarium · drone · stargazing show · "large planetarium" · "external astronomy show"

1 Why Large and External Planetarium?

1.1 Dark Skies Are no Longer Available

Light pollution has turned the nights in our large cities into star-free places, which is the opposite case of what humans used to experience thousands of years ago. This is likely to have a negative impact on urban residents in the future. It may look like a good idea for the future to allow people to go out in the middle of their cities and be able to see star-like sources that beautify the sky [1].

1.2 Planetariums in Normal Sizes

Planetariums have been invented by Carl Zeiss in 1923. This year is the 100 years anniversary of this event. A planetarium is a large cylindrical room with seats where the audience can watch the sky and the stars on a dome screen above them. The stars are usually projected from a central StarBall projector to directly on the dome. Planetariums come in different sizes from 4 m to 30 m the maximum. Some proposals suggest a 50-m planetarium.

1.3 External and Huge

This proposal focusses on establishing a 500 m to 2000 m diameter planetarium in the middle of a city like Sharjah where thousands of people can watch the artificial sky and the constellations at once. This will give the city a distinctive position among the cities of the world and will enhance tourism in it.

H. M. K. Al Naimiy et al. (Eds.): AUASS-CONF 2023, SPPHY 420, pp. 248–254, 2025.
https://doi.org/10.1007/978-981-96-3276-3_18

1.4 Viewer Experience

Viewers within a certain circle close to the center of the "sky show" will be able to see the "stars" above them, but, in reality, those stars are nothing, but light emitted by drone's light sources.

2 The Idea

2.1 Flying Drones

In this proposal, hundreds of powerful drones that can fly as high as about one kilometer and stay stable for 30 m duration show are needed. These drones carry powerful LED light sources. LED light's intensity color and luminosity can be controlled. All the drones will work together as one single structure [2]. Drones will resemble the stars of the sky, spread out in the positions of the natural stars as seen from the center of the city where the public stargazing party is taking place (Figs. 1 and 2).

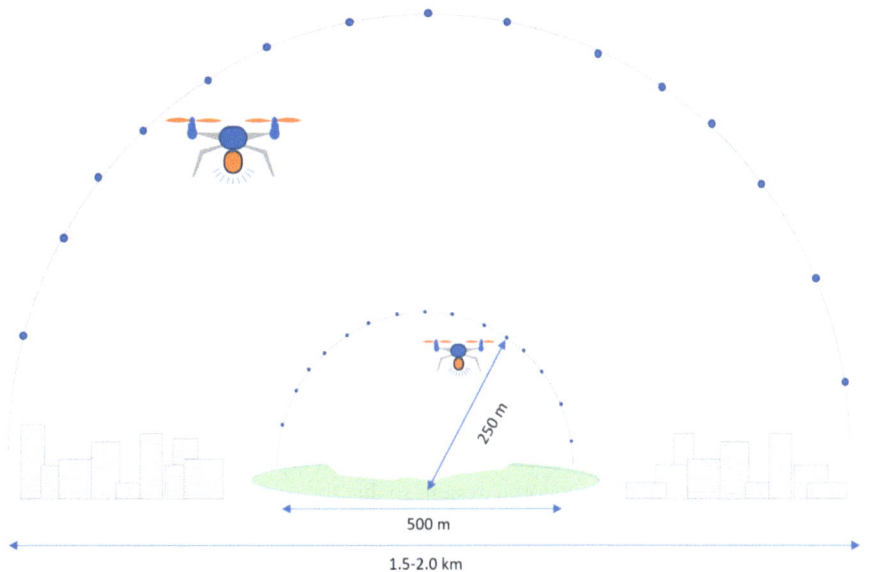

Fig. 1. An imaginary view that shows how the drones would be distributed on a 500 m dome above al Majaz Park in the middle of Sharjah City. A wider drone planetarium can be implemented in a place where buildings are shorter than the high ones there. (image out to scale).

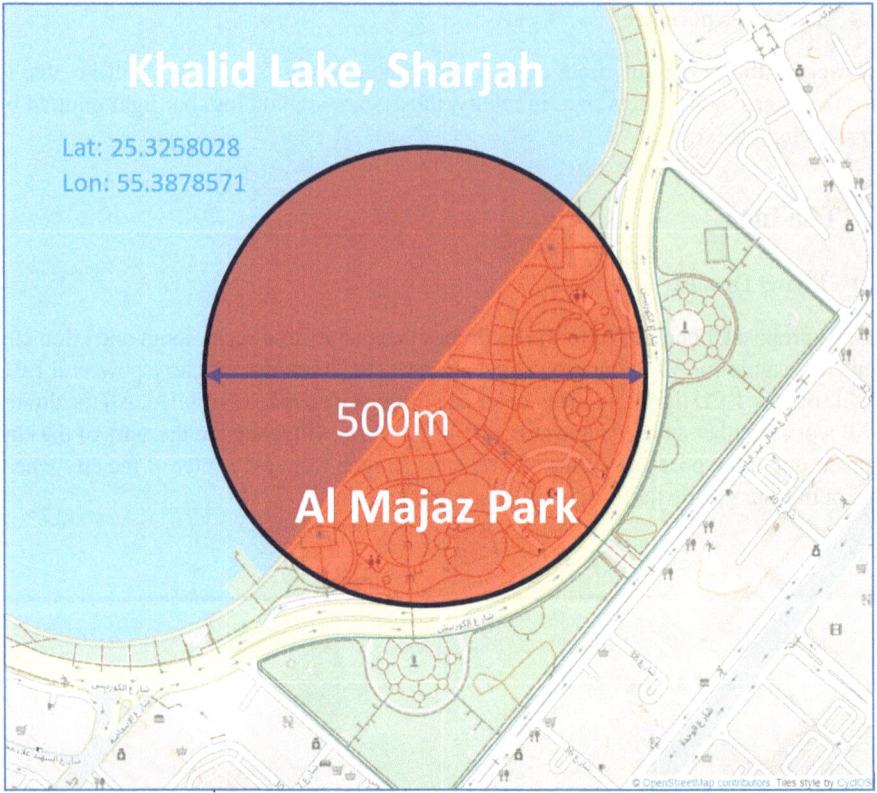

Fig. 2. An OpenStreetMapContributers map image with a top view of an imaginary 500-meter diameter dome of the drone planetarium in al Majaz Park in the middle of Sharjah City. https:// www.openstreetmap.org/copyright.

2.2 The Code and the Controlling Algorithm

The program code should be built with a 3D structural algorithm that can simulate the positions of the stars in the equatorial system so that the drones appear to move in their natural paths of the stars relative to the center of the celestial sphere in the center of the city where the audiences are gathered [2].

The algorithm will be designed to do the following tasks:

a- To simulate the star's diurnal motion from east to west.

b- To simulate the sky different city latitude in the world, from high North to high South.

c- A few hundred more drones can display the constellations stick lines to educate the audiences with the mythology of different cultures and civilizations.

d- The constellations can be dramatically animated to perform the historical drama of the sky mythology to become a giant screen that tells the stories of the ancients.

e- Can display some of the celestial sphere elements such as the ecliptic, equator and meridian circles in addition to the.

f- Another category of drones can be assigned to display sky objects labels such as stars names, constellations.

g- In the case of limited number of drones, the operators can focus on specific constellations every night.

h- Broadcasting a recorded scientific material or live show broadcast through channels that can be received through smartphone applications. People will be able to use their smartphone with headphones and follow the audio, and they can also run celestial apps that can support the show.

i- To simulate the magnitude of the stars after adding the city atmospheric brightness factor.

j- To simulate the star's brightness differences according to the star's height above the horizon.

k- Planets can be displayed by drones and real bright planets can be applied.

l- To simulate the retrograde motion of some planets like Mars.

m- Some drones could be developed to jet off gases around themselves to represent the Milky Way only on windless nights.

n- To simulate the transit of some artificial satellites like the ISS.

2.3 How Many Drones (Stars)?

Most of amateur astronomers who don't want to travel very far from their cities to see the fain stars, a place that can provide stars that are brighter than magnitude 3.5 is considered to be good enough to enjoy the stars and recognize the shapes of the constellations with the unaided eye observation. At this level of darkness, the number of stars can form about 300 stars. Therefore 300 drones that resemble the stars are enough for having an excellent star show for a general audience in this project.

Star Magnitude	Range	Number of Stars	Cumulative Stars
−1	−1.50 to -0.51	2	2
0	−0.50 to + 0.49	6	8
1	+0.50 to + 1.49	14	22
2	+1.50 to + 2.49	71	93
3	+2.50 to + 3.49	190	**283**
4	+3.50 to + 4.49	610	893
5	+4.50 to + 5.49	1,929	2,822
6	+5.50 to + 6.49	5,946	8,768

The number of stars that can be seen with the naked eye at each magnitude. [3].

Two facts will help in reducing the number of drones that are required for operating this city- wide artificial planetarium:

- Some stars are always hidden by the horizon depending on the city latitude. This can be estimated to; from 0% of the star at the latitude zero (at the equator) to about 50% at latitude + 90 or −90; (the poles).

- At any moment of time, about 50% of the stars are hidden below the horizon due to the diurnal motion, many of these drones will not be functional in the shows when diurnal motion is not simulated. Those drones can be used then to project fainter stars if required. More hundreds of drones can be used for the constellations stick figures and even for the mythological pictures (Fig. 3).

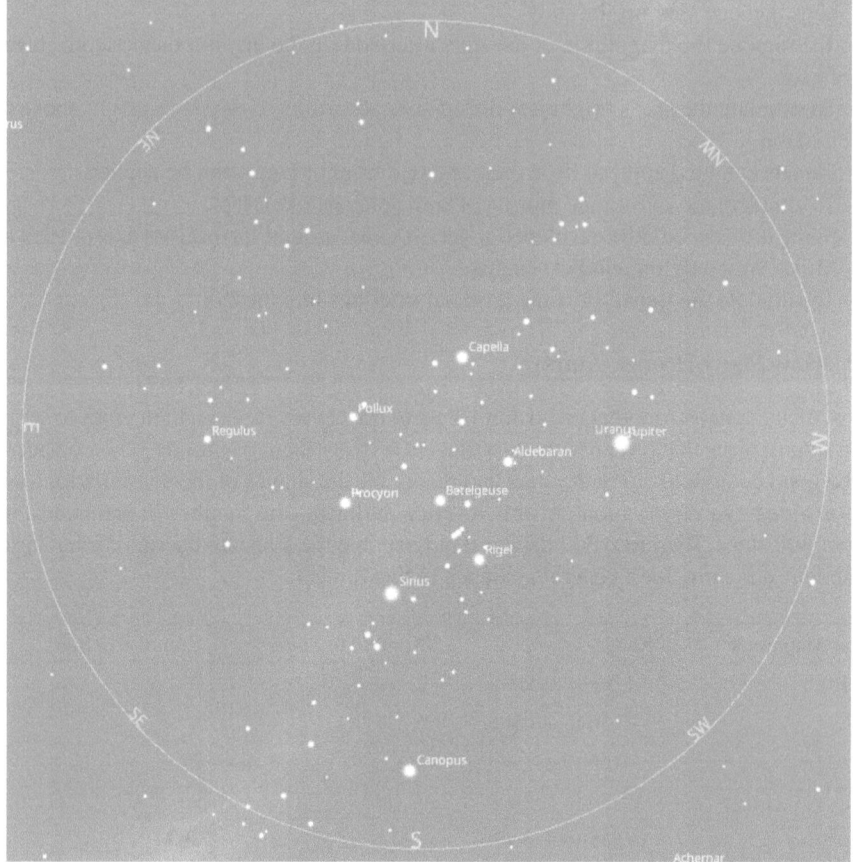

Fig. 3. The star at magnitude 3.5, (edited image, Stellarium software) [4]

2.4 City Light Pollution and Sky Limiting Magnitude

A person in al Majaz Park near Khalid Lake in Sharjah City, would see a very little number of stars with unaided eyes. Astronomically, this can be described the faintest stars magnitude that can be seen there which varies depending on many factors to about magnitude 1.5–2.0 [5]. This means most of the stars are almost invisible due to the bright sky which resulted by the huge light pollution there. To show stars up to magnitude 3.5

we need to calculate how bright the LED source of the lights that are carried by the drones should be. Obviously, this depends on market availability.

On the other hand, we need to consider the city light pollution limit on the drone's bulbs. To put it simply, a normal dinner candle at a distance of 1km can shine light as much as a star of magnitude 2 in an ideal dark sky. However, as a stimulation, the brightness of the artificial stars of the drones will be as bright as double or triple of the natural stars in order to be seen comfortably. This can be controlled by a factor to be implemented for all bulbs of the drone (Fig. 4).

Fig. 4. A side view photo of al Majaz Park and the imaginary 500-meters dome above it. Photo is taken by the author.

3   Challenges Need to Be Considered

In general, the climate in Sharjah and UAE as whole, is sunny with low-speed winds for most of the year [6]. This provides a suitable place for external sky shows at night for the public. Nevertheless, here are some challenges that need to be put into consideration during the design of this project:

3.1   Wind Speed

The algorithm will empower the drones to counteract low speed winds to retain their positions in relation to the virtual digital planetarium dome above the city.

3.2 Rain and Bad Weather

Stars shows will be terminated in rainy sky, low clouds, or nights with strong winds.

3.3 Airplane Path

Needless to mention that such drones will be a real hazard for the airplanes that are flying across the city. At this moment cities where airplanes are flying across them below 5km, this project is not applicable.

3.4 Light Source Power

For a drone-planetarium dome of diameter estimated to 0.5–2.0 km, the light source that a drone needs to carry must emit a powerful light that can resemble the bright stars like Sirius (mag $= -1.5$) in a light polluted city. The brighter the light required, the stronger and drone is needed due to the more batteries that the drone should carry.

Nevertheless, such a project will encounter a few difficulties and challenges but with the rapidly improving technology and some facilities that can be provided by the hosting city, I believe most of these challenges will be resolved.

References

1. Sky & Telescope Magazine, What's My Naked-eye Magnitude Limit? By: Roger W. Sinnott July 19 (2006)
2. Huang, J., Tian, G., Zhang, J., Chen, Y.: On unmanned aerial vehicles light show systems: algorithms, software and hardware. Appl. Sci. **11**, 7687 (2021). https://doi.org/10.3390/app 11167687
3. Article, How Many Stars You Can Observe. www.stargazing.net
4. Stellarium astronomy Software. www.stellarium.org
5. By direct estimations of experts "Al-Wardat, M. A.", "Talafha, M. F." and Mr. Marwan Shwaiki, director of Sharjah Planetarium and astronomy educator
6. Website "weatherspark.com", link:https://weatherspark.com/y/105471/Average-Weather-in-Sharjah-United-Arab-Emirates-Year-Round

Controlling Robot Through Satellite: Telesurgery or Remote Surgery

Muna Alqam$^{(\boxtimes)}$ and Samir Al-Busaidi

Department of Electrical and Computer Engineering, Sultan Qaboos University, Muscat, Oman
w01434@iu.edu.jo, albusaid@squ.edu.om

Abstract. Remote surgery is a promising medical technology that has the potential to revolutionize healthcare with the potential to provide life-saving care to patients in remote areas. However, one of the challenges facing this technology is the time delay associated with data transmission over conventional network infrastructures which makes it difficult to perform delicate procedures. CubeSats are a type of small satellite that can be used to overcome this challenge. CubeSats can be used to reduce the time delay by providing a low-cost, high-bandwidth communication link between the surgeon and the patient. In this report, we present the design and simulation using General Mission Analysis Tool (GMAT) software of a CubeSat constellation for 7 h remote surgery connecting two hospitals, Sultan Qaboos University Hospital (SQUH) in Sultanate of Oman and OSF children hospital in Illinois USA. We discuss the orbital design of the constellation to ensure that the satellites are always in view of both hospitals. The constellation must be designed to provide continuous coverage of the two hospitals, even as Earth rotates, as well as the challenges and solutions associated with this application.

Keywords: CubeSat · Remote surgery · Formation · Constellation · GMAT

1 Introduction

The exploration and manipulation of physical objects from a distance represent a promising dimension in networked services within the interconnected global landscape [1]. The IEEE P1918.1 Working Group (WG) has established a definition for this operational approach, terming it the Tactile Internet: "A network (or network of networks) for remotely accessing, perceiving, manipulating, or controlling real or virtual objects or processes in perceived real time by humans or machines" [2]. This definition underscores the potential applications, with a notable emphasis on facilitating remote surgeries across geographically linked locations, overcoming considerable distances. Minimally Invasive Surgery (MIS) technology emerges as a highly sought-after advancement in terms of Quality of Life (QoL). MIS is renowned for alleviating patients' physical strain, reducing hospital stays, and lowering medical costs. The contemporary landscape witnesses numerous studies on robotic systems designed for MIS applications, with several products already commercialized. Moreover, with the capability of remote surgery, it is possible to realize the real time collaboration of highly specialized doctors to an ongoing surgery, where all parties concerned are geographically dispersed. Such possible interaction will facilitate teamwork between doctors from around the world for all kinds of medical treatments.

© The Author(s) 2025
H. M. K. Al Naimiy et al. (Eds.): AUASS-CONF 2023, SPPHY 420, pp. 255–265, 2025.
https://doi.org/10.1007/978-981-96-3276-3_19

To support the advances of QoL, a strong communications link must be established between involved medical institutions. One greatly promising communications candidate is the use of nano-satellites, CubeSats, a field of communications that has been evolving from its inception in great leaps. The first group of CubeSats was launched in June of 2003 from Plesetsk, Russia, on a Eurokot, utilizing the Russian Multiple Mission Orbit Service. The involved cost of launching the 1U (1 unit) CubeSat was $30,000 [3]. Despite the advantages of the CubeSat, which can be summarized as: low cost, small size [4], modular design [5] short developing time [6] and educational value [7], there are some disadvantages that must be taken into account. The relatively high price for highly capable CubeSats, which stands at an average price of $200,000 [8], and its short average lifespan of 2.2 years are two main drawbacks. To date, the oldest operational CubeSat is 6.5 years old, but most CubeSats last much shorter [8].

2   Telesurgery (Remote Surgery)

The first remote surgery was performed in 2001 [9], when Professor Jacques Marescaux and his team removed a patient's gallbladder using a robotic system called the DaVinci in France. The surgery was controlled from New York, and the only feedback that the surgeon had from the operative field was verbal and visual. The surgeon saw the operative field through a high-definition camera and heard the comments of the assistant surgeon who was in direct contact with the patient. The surgeon also received information from the robotic arms, which provided feedback on the forces that were being applied to the tissue. The time delay between the surgeon's actions and the patient's response, which is a major challenge for remote surgery, can be as long as 200 ms, a time that is enough for the patient to feel pain or discomfort. However, the use of high-speed data connectivity supporting two-way transmission can help reduce the time delay [10]. Fundamentally, a remote surgery setup involves patients and medical professionals who are not physically present in the same operating room. Through the utilization of display screens and tactile sensors, surgeons receive live updates regarding the patient's status. Haptic technology facilitates the tracking of the doctor's arm posture, position, and movements. This information is then sent to the patient's location. Subsequently, the mechanical arm of the surgical console mimics the actions of the surgeon to perform the necessary procedures on the patient. Simultaneously, audiovisual data and tactile sensations detected by the surgical consoles are relayed back to the operating doctor [11].

3   Cubesat

The smallest size CubeSat standard specification is defined as a single U (1U) with dimensions of 10cm x 10cm x 10cm. This represents the unit of physical measurement of a CubeSat limited to an associated maximum mass of one kilogram. Given the limitations imposed by the standard, satellite developers need to streamline both the structural mass and volume to accommodate the payload effectively. Aluminum 7075 or 6061 are frequently preferred materials for constructing the primary structure of satellites. The primary structure usually accounts for 15% to 20% of the total mass of a CubeSat [12].

Most CubeSats are deployed into a Low Earth Orbit (LEO) with an expected operational life span of two years, before being placed in a post-mission disposal orbit. During its operational lifetime, each CubeSat will use its own collision avoidance maneuvers to avoid collisions with other objects in space. These maneuvers are based on CubeSat's own orbit and the predicted orbits of other objects in its vicinity. The CubeSat will also be able to receive updates from ground stations to provide more accurate information about the predicted orbits of other objects. After the CubeSat's operational lifetime, it will be placed in a post-mission disposal orbit where it will decay within 25 years. This ensures that the CubeSat will eventually deorbit and burn up in the atmosphere, minimizing the risk of space debris [13]. The Japanese CubeSat, CUTE-1, lasted for over five years [14], becoming the longest-lived CubeSat [3]. CubeSats have proven reliability in space exploration due to its real low-cost missions. A basic 1U CubeSat can cost as little as $50,000 to develop and launch. A more complex 6U CubeSat can cost up to $250,000 [15]. For smaller nations, educational institutions, and businesses worldwide, cost-effective space missions play a crucial role, allowing them to develop and launch their spacecraft [16]. These endeavors offer various benefits, including rapid development cycles and the facilitation of innovative scientific missions through distributed space system architectures like fractionated spacecraft, satellite swarms, and constellations, leading to unprecedented temporal and spatial coverage. When compared to conventional large satellites, CubeSats exhibit reduced complexity in both development and launch processes. This reduction in complexity has enabled the deployment of large constellations comprising small satellites, potentially achieving missions comparable to/or surpassing those of conventional spacecraft in certain circumstances [16]. One significant advantage of CubeSats lies in their lower susceptibility to individual failures due to their relatively low cost and short development times, in contrast to larger, more sophisticated satellites. This characteristic enhances their value in scientific exploration, particularly for targeted missions, multipoint observations, and high-risk, high-value endeavors. Additionally, CubeSats can complement larger missions effectively, filling specific niches or bridging gaps between traditional large-scale missions. The reconsideration of space exploration within the CubeSat framework opens up possibilities for novel and revolutionary mission concepts in the future. Despite these considerable advantages, it is improbable that CubeSats will entirely replace their larger spacecraft counterparts. Challenges inherent in deep space missions, for example, strongly favor the use of large satellites over CubeSats [17].

3.1 Cubesat Communication Systems

Satellites offer three main categories of communication services: broadcasting, telecommunications, and data communications. Broadcasting encompasses direct delivery of radio and television signals to consumers, as well as mobile broadcasting services like satellite television (DTH). Telecommunications services encompass a wide range, including telephone calls and services for telephone companies, as well as support for wireless, mobile, and cellular network providers. Data communications involve the exchange of data between different points. Businesses and organizations seeking to exchange financial and other information across various locations often utilize satellite technology, leveraging very small-aperture terminal (VSAT) networks to facilitate data transfer [18]. For telesurgery, an uninterrupted high data rate link is essential to achieve success in time critical and life-threatening operations which require highly skilled operation techniques. Using higher frequency bands offers greater bandwidth, which can be used to transmit more data. However, higher frequency bands also suffer from more attenuation, so they are not always the best option for CubeSats [19]. CubeSats can alternatively transmit data over a wide geographical area, which could be translated to an improved QoS data link. With multiple ground stations, the ground infrastructure and support will be relatively more complex when compared to a single ground station scenario [20].

3.2 Cubesat Constellation

The potential for adaptability and flexibility within a CubeSat constellation is vast. When considering safety and security, CubeSat technology offers inherent advantages. By employing CubeSats, the risks associated with intermediaries in data transmission can be minimized, ensuring the confidentiality and integrity of sensitive information exchanged during surgical procedures. CubeSats also mitigate potential security vulnerabilities associated with transmitting data through satellites in countries with inadequate data protection laws. This strategic advantage enhances the security of patient data and reduces the risk of geostrategic and commercial tensions [21]. The closed-circuit system approach facilitated by CubeSat technology ensures a direct and secure exchange of information between operating surgeon sensors and remote operating theater sensors, minimizing the risk of information leaks. Additionally, CubeSat technology provides extended coverage time, enhancing the reliability and continuity of data transmission during surgical procedures.

3.3 Cubesat Antenna

The most popular antenna types for CubeSat applications were found to be planar (patch), monopole/dipoles, reflectors, reflect-arrays, helical and horn antennas [22]. Each antenna type is suitable for different missions. For remote surgery application, it is necessary to have high speed link where the most suitable antenna type is reflectors. Figure 1 shows the commonly used antenna types for different CubeSat applications. Using large parabolic antennas can aid to receive weak signals, which reciprocates to transmit signal with a relatively high gain. However, large antennas are also more expensive and could become difficult to deploy on the relatively small sized CubeSats [23]. As an alternative to parabolic reflector antennas, phased array antennas can be used to focus the signal at specific locations. However, the drawbacks of phased array antennas can be summed as complex, expensive and will require power for array signal processing [24], all of which negates the attractiveness of CubeSats for such missions. For an uninterrupted data link, it could be possible to employ multiple ground stations.

Fig. 1. Suitability of each antenna type to the CubeSat applications

4   Design Assumptions

The adopted remote surgery design assumptions based on research of applications which require similar satellite designs are given as follows:

Satellite Specifications

- The satellite is a 3U CubeSat (100 x 100 x 340.5 mm) operating in the X-band frequency range. While most CubeSat-class radios function in UHF or S-band frequencies, a few operate in the X-band (8–12 GHz), offering higher speeds of approximately 50–100 Mbp [25]. Although Ka-band provides significantly more bandwidth than X-band, NASA is shifting to Ka-band to enhance radio communication speeds. However, Ka-band is susceptible to higher attenuation due to atmospheric conditions and rain fade, making it undesirable for this application.
- The satellite features two physical channels: one for receiving video and audio from the patient's side and another for positioning and controlling robotic arms from the doctor's side. These channels correspond to right-hand circular polarization (RHCP) and left-hand circular polarization (LHCP), with each utilizing 100 MHz of total bandwidth. The satellite employs a dual polarization antenna to support these channels.
- An adaptive coding and modulation (ACM) scheme dynamically adjusts the modulation and coding for each channel based on the available link margin.
- Channel specifications include a symbol rate of 76.8 Msps, channel spacing of 100 MHz, and a channel bandwidth of 96 MHz. Channel 0 is RHCP, while Channel 1 is LHCP.
- The onboard antenna provides a gain of 15 dBi per polarization. Modulation options include QPSK, 8PSK, 16APSK, and 32APSK, with variable data rates ranging from 37 Mbps to 336 Mbps.

Ground Station Details

- The system comprises two ground stations: SQUH and OSF Children Hospital Illinois stations.
- Each station is equipped with two S/X-band antennas, facilitating telemetry, tracking, and command (TTC) operations. Additionally, UHF band support is provided. The receiver chain for all S/X-band antennas includes a left-hand polarized feed and a low noise block down converter (LNB) to support the second polarization.
- Two demodulators interface with the two-channel transmitter.

- Ground dish sizes range from 4.5 m to 5 m in diameter, with X-band gains reaching 49 dBi. The beamwidth (half power) is 0.55 degrees, with a G/T (gain over noise temperature) of 29 dB/K and cross-polarization of 25 dB.

5 GMAT Simulation

GMAT stands as the sole enterprise-level, open-source software system globally for designing, optimizing, and navigating space missions. It caters to missions across various flight regimes, spanning from low Earth orbit (LEO) to lunar, liberation point, and deep space missions. Developed collaboratively by a team comprising members from NASA, private industry, and both public and private contributors, GMAT serves practical mission support purposes, aids engineering studies, acts as an educational tool, and fosters public engagement [26]. GMAT can also give an estimation for the onboard satellite power consumption and eclipse events providing an overall illustrative image of CubeSats behavior in orbit. For remote surgery application using CubeSat, we decided to go with a constellation of 30 CubeSats in 2 formations orbiting at 500km altitude at LEO elliptical orbit with eccentricity of 0.024. The simulation is run on GMAT platform using locator conductor to estimate the coverage time for both ground stations. We started the simulation by entering one CubeSat orbital parameters and cloned the CubeSat to obtain 15 CubeSat to make the first formation. Figure 2 shows the first three CubeSats parameters.

The first formation started from true anomaly (TA) $= 0°$ and applied phase difference of $24°$ between the 15 CubeSats to obtain complete orbital coverage. This formation gave approximately 3 h and 30 min coverage time for the first station located at SQUH. The coverage time had been calculated using:

$$Coveragetime = \frac{(2 * Cube\ Satorbital\ period)}{(360\ degrees)} * \left[1 - \left(\frac{500\ km}{6371\ km}\right)\right]^2 \qquad (1)$$

where 500km is the altitude of the CubeSat and 6371 km is the Earth radius.

The next step was to form another formation of 15 CubeSats with the same parameters and phase shifts but with different right ascension of the ascending node (RAAN), specifically it is set to $70°$ instead of $20°$ for the first formation. This step increases the coverage time. We managed to obtain a continuous 6 h and 55 min, which is a sufficient duration for moderately complex surgery. For more complicated surgeries, however, require longer continuous duration links, that can be done by adding a third formation with different RAAN. The same step needs to be repeated to ensure between the two ground stations for the duration of the surgery a continuous uninterrupted connection. To verify the satellite constellation parameters, the default propagate parameters were set in GMAT, and the propagator was set over a one-day period of propagation. This allows for the estimated coverage time for one day.

5.1 Ground Station and Coverage Time Simulations

The locations for both ground stations were chosen based on the two hospitals locations. Ground station 1: SQUH, location latitude 23.5877° N and longitude 58.1707° E. Ground

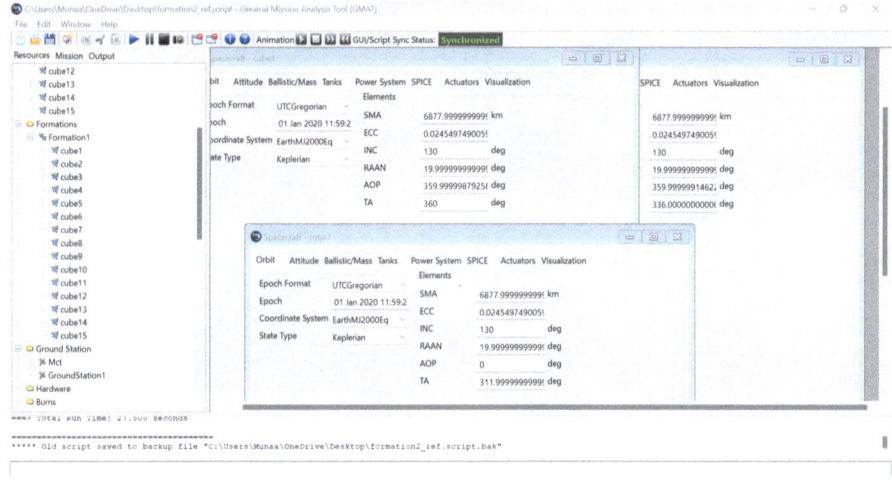

Fig. 2. First three CubeSats in the first formation orbital parameters

station2: OSF children Hospital location latitude 37.77° N and longitude 89.32° W. Figure 3 shows the GMAT parameters for both ground stations, and Fig. 4 shows the locations of both ground stations and the ground track of one CubeSat.

From the simulation, using a single constellation, the continuous coverage time achieved using a constellation of 15 CubeSats was 3 h and 35 min (sufficient only for short operations).

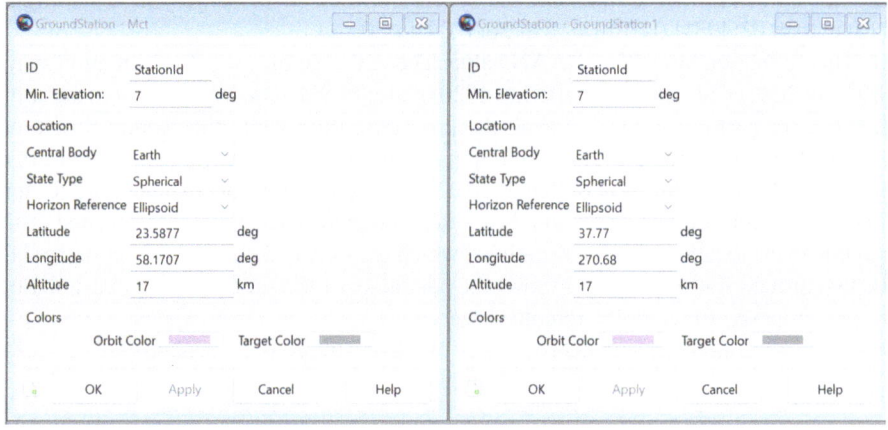

Fig. 3. Ground stations parameters

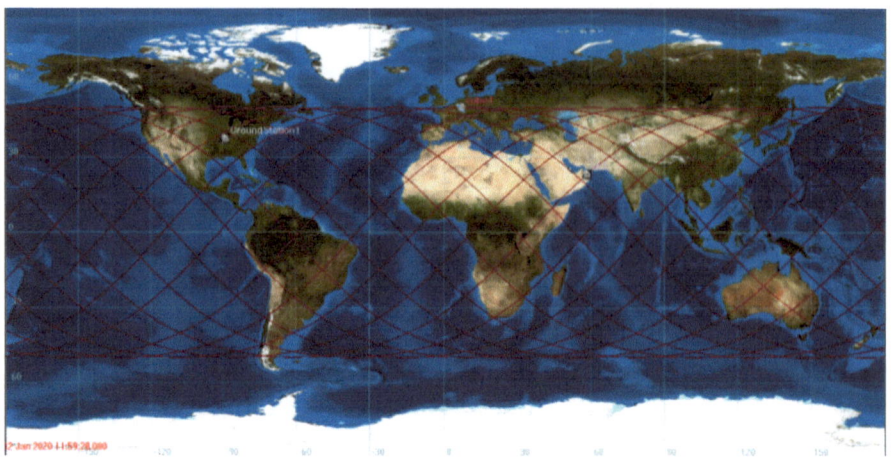

Fig. 4. Ground stations1 and 2

Applying another 15 CubeSat formation to the constellation into the simulator, it was possible to achieve 6 h and 55 min of full and continuous coverage for the first ground station. For longer surgeries duration other formations are required. The coverage time based on the chosen Epoch (01 Jan 2020 11:59:28.000) started for the first ground station of the first formation at 02 Jan 2020 02:49:04.047 and ended at 02 Jan 2020 06:24:37.950 (approximately 3 h and 35 min of coverage) the second formation continued the coverage starting at 02 Jan 2020 06:23:36.687 and finishing at 02 Jan 2020 09:44:50.124 (approximately 3 h and 20 min of coverage). For ground station 2, a third formation was introduced to ensure the synchronization of the surgery time starting at 02 Jan 2020 02:49:04.047 till 02 Jan 2020 09:44:50.124.

This formation has a 220° RAAN value. The coverage time for this formation starts at 02 Jan 2020 02:44:56.560 till 02 Jan 2020 09:46:05.551. The first formation duration didn't cover ground station 2 during the procedure time while the second formation of RAAN = 70° gave the first continuous duration needed starting at 02 Jan 2020 02:44:10.236 and finishing at 02 Jan 2020 02:47:55. It is noticeable that the third formation was enough for ground station 2 coverage, no extra formations were needed. This was due to the location of ground station 2 which is closer to the North pole compared to the first ground station. Figure 5 shows the complete 3-formation CubeSat constellation system proposed for complex remote surgeries.

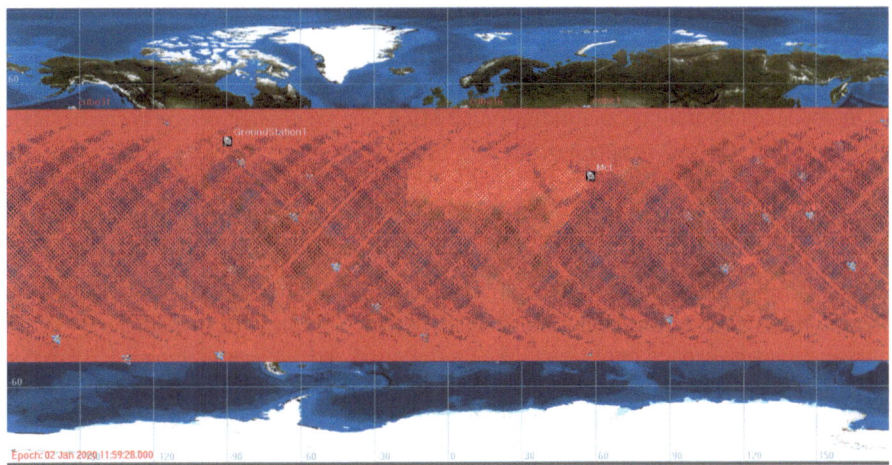

Fig. 5. Coverage of Proposed CubeSat Constellation using three Orbital Formations

6 Comparative Analysis with Related Work

In comparing our work with a notable study on remote surgery between Japan and Thailand [27], we highlight key differences. While the Japan-Thailand study relied on conventional internet infrastructure, our approach utilizes a CubeSat constellation for dedicated, high-bandwidth communication, minimizing latency issues. This CubeSat system offers advantages in signal transmission delays inherent to LEO satellites, ensuring consistent low latency data transfer between the hospitals regardless of terrestrial internet limitations. Additionally, our study proposes a three-formation CubeSat constellation for redundancy, enhancing communication reliability. Unlike the Japan-Thailand study, which faced internet stability issues, our approach provides continuous coverage. Overall, our innovative use of CubeSat technology addresses challenges in remote surgery, whilst outlining the potential advantages in latency, coverage, and reliability. While another research effort [1], explored the broad challenges of network infrastructure for remote surgery using Tactile Internet, focusing on latency and reliability issues, our work delves deeper into the practicalities. We investigate the technical feasibility of utilizing CubeSat formations for remote surgery applications. Our research details specific aspects like satellite design while leveraging upon GMAT simulations to analyze coverage time. In essence, our focus complements the existing research by providing a technical solution (CubeSat constellations) that tackles the network infrastructure challenges highlighted in [1], ultimately, bridging the gap between surgical implementation expertise and robust network infrastructure design in realizing remote surgery via a space-based network. A separate study explored the concept of a highly articulated MicroSat for on-orbit construction and repair [1], focusing on the robotic arms and its ability to manipulate items. Our study takes a complementary approach, our research takes a complementary approach. Their work delves into the intricate design of the MicroSat itself, including its robotic arms, tools, and human-machine interface for remote control. In contrast, our research focuses on the communication infrastructure needed to support such a MicroSat. We investigate the feasibility of utilizing a network of CubeSats to provide crucial real-time, high-bandwidth connection for remote surgery applications. By combining expertise in network infrastructure (CubeSats) with the in-space manipulation capabilities of the MicroSat, significant progress can be made towards achieving remote space surgery.

7 Conclusion

The idea of remote surgery through satellite communication has been investigated considering the coverage time requirement for relatively complex life critical surgery. The investigation was carried out using GMAT simulation and was based upon a 15 CubeSat formation. From the results, it was observed that for complex surgeries, a 3-formation CubeSat constellation system is required for a link between 2 locations, one in Oman and the other in USA. It was further observed that such a CubeSat constellation can be successfully employed for this relatively novel application under the coverage time for complicated procedures. The proposed application using 3-formations of 15 CubeSats could be relatively costly, but the benefits of having this application will overcome the cost in a short period of time, especially when counting the number of souls that can be saved. Further research involving the collaboration of surgeons to assess the system implementation possibilities and address practical concerns would be valuable for future work.

References

1. Kolovou, G., Oteafy, S., Chatzimisios, P.: A remote surgery use case for the IEEE P1918.1 tactile internet standard. In: ICC 2021 - IEEE International Conference on Communications, Montreal (2021)
2. Holland, O., et al.: The IEEE 1918.1 "tactile internet" standards working group and its standards. Proc. IEEE **107**(2), 256–279 (2019)
3. Cappelletti, C., Battistini, S., Malphrus, B.: CubeSat Handbook: From Mission Design to Operations. Elsevier, London (2020)
4. Tumlinson, J., Postman, M.: The CubeSat Revolution: A New Platform for Space Science and Technology Development. STScI Newsletter, Baltimore (2019)
5. Johnson, C., Spencer, D.: CubeSats: a tool for education and research. Acta Astronaut. **108**, 26–37 (2015)
6. Weeden, B., Spencer, D.: CubeSats: a review of their past, present, and future. Acta Astronaut. **133**, 247–266 (2017)
7. Spencer, D., Weeden, B.: The future of cubesats: a review of current trends and future prospects. Space Policy **33**(1), 1–12 (2017)
8. Weeden, B., Spencer, D.: The cost of cubesats: a preliminary analysis of the 2020s. Acta Astronaut. **177**, 135–146 (2022)
9. Brower, V.: The cutting edge in surgery. Telesurgery has been shown to be feasible–now it has to be made economically viable. EMBO Rep. **3**(4), 300–301 (2002)
10. Marescaux, J., et al.: Transcontinental robot-assisted remote telesurgery: feasibility and potential applications. Ann. Surg. **235**(4), 487–492 (2002)
11. Chen, M., et al.: A 5G cognitive system for healthcare. Big Data Cognitive Comput. **1**(2), 1–15 (2017)

12. Toorian, A., Diaz, K., Lee, S.: The cubesat approach to space access. In: 2008 IEEE Aerospace Conference, Big Sky, MT, USA (2008)
13. Matney, M., Vavrin, A., Manis, A.: Effects of cubesat deployments in low-earth orbit. In: 7th European Conference on Space Debris, Darmstadt, Germany (2017)
14. Yamakawa, H., Imai, M., Suzuki, O.: The CUTE-1 cubesat: a technology demonstrator for future small spacecraft. Acta Astronaut. **51**(11), 1055–1062 (2002)
15. Spencer, D., Weeden, B.: The cost of cubesats: a preliminary analysis of the 2020s. Acta Astronaut. **177**, 135–146 (2022)
16. Poghosyan, A., Golkar, A.: CubeSat evolution: analyzing CubeSat capabilities for conducting science missions. Prog. Aerosp. Sci. **88**, 59–83 (2017)
17. Spencer, D., Weeden, B.: The future of deep space exploration. Acta Astronaut. **177**, 147–158 (2022)
18. Labrador, V.: satellite communication, Britannica. https://www.britannica.com/technology/wireless-communications. Accessed 6 Oct 2022
19. Spencer, D., Weeden, B.: CubeSat communication systems. Acta Astronaut. **177**, 150–170 (2022)
20. Spencer, D., Weeden, B.: CubeSat communication systems for high data rate applications. Space Policy **34**(2), 113–124 (2018)
21. Soencer, D., Weeden, B.: CubeSat constellations: a review. Acta Astronaut. **177**, 171–182 (2022)
22. Abulgasem, S., et al.: Antenna designs for cubesats: a review. IEEE Access **9**, 45289–45324 (2021)
23. Brown, C., Weeden, B.: A review of cubesat communication systems. J. Spacecr. Rocket. **55**(3), 741–752 (2018)
24. Spencer, D., Weeden, B.: Phased array antennas for high data rate cubesat communication systems. Space Policy **35**(1), 49–58 (2019)
25. Devaraj, K., et al.: Planet High Speed Radio: Crossing Gbps from a 3U CubeSat. In: 33rd Annual AIAA/USU Conference on Small Satellites, Utah (2019)
26. GMAT user guide R2020a, SourceForge, 29 September (2023). https://sourceforge.net/projects/gmat/
27. Weeden, B., Spencer, D.: The challenges of cubesat launches: a review of current trends and future prospects. Acta Astronaut. **140**, 225–246 (2018)

Investigating the Effects of Ionizing Radiation on C-V Characteristics of SOI-MOS Capacitors in Harsh Environmental Conditions

Fawzi A. Ikraiam[✉]

Physics Department, Faculty of Science, University of Omar Al-Mukhtar, El-Beida, Libya
fawzi.ikraiam@omu.edu.ly

Abstract. The performance and reliability of electronic devices in harsh environments depend on the response of these devices when they are inevitably subjected to different radiation effects in applications as complex as space-borne systems. Electronics devices, thus, may experience degradation which may damage and/or result in device failure. This paper presents the effects of ionizing radiation on the C-V characteristic, as a diagnostic tool, of the SOI-MOS capacitor (SOI-MOS-C) at room temperature. These effects are examined before and after exposing the capacitor to radiation at high frequency (HF). The radiation effects influence on the C-V curves is reviewed under different radiational dose conditions. The irradiated capacitors exhibit changes in the capacitance. Increasing irradiation dose results in increased capacitance values and a shift in voltage values. The change in the C-V curves is ascribed to variations of charge accumulation and the generated electron-hole pairs as well as the increase of interfacial states at the buried oxide (BOX)-Si interface. Appropriate protection of these devices in these harsh environments is also essential to ensure proper operation of the devices and to exclude undesirable effects. It is concluded that C-V characteristic of SOI-MOS-C is a suitable tool for characterizing electronic devices in harsh environments, such as space applications, and SOI technology offers an alternative choice for radiation-hardened devices. The usefulness of HF SOI-MOS-C C-V characteristics is also discussed for device modeling in such cases. The effect of voltage sweep direction during C-V measurement is also discussed.

Keywords: SOI Technology · Device Reliability · SOI-MOS Capacitor · C-V Characteristic · Radiation Effects · Space applications

1 Introduction

Radiation spectrum changes considerably in various environments. These environments can be either space oriented such as space applications and aviation or artificial ones such as nuclear applications, high energy physics experiments as well as device processing. Space is a harsh environment for humans as well as electronic devices where both are exposed to conditions harsher than the usual circumstances. Astronauts and spacecraft involved in space missions are subjected to different harmful radiation consisting of

H. M. K. Al Naimiy et al. (Eds.): AUASS-CONF 2023, SPPHY 420, pp. 266–276, 2025.
https://doi.org/10.1007/978-981-96-3276-3_20

galactic cosmic radiation (GCR) and solar particle events (SPEs). GCRs are energetic particles which come from outside our solar system. These are the most hazardous type of space radiation because they are most energetic. Solar Energetic Particles (SEPs) are energetic particles that come from the sun. The overwhelming energetic ions generated in an SPE are protons. Protons are, therefore, the main concern when studying possible SPE radiation effects. Designing and performing these missions necessitate precise and appropriate knowledge about this environment. This knowledge requires evaluation of occurrence time, spatial distribution of the radiation and the type and energy spectrum of the particles involved. All space applications inevitably involve exposing electronic-systems to radiation where its effects cannot be ignored. Conventional advanced silicon ICs are vulnerable to single-event upsets (SEUs) either near earth or in low altitude missions because of cosmic rays as well as being sensitive to soft errors [1, 2]. These radiation-induced effects were investigated and studied extensively throughout decades of device development [e.g., 3–5]. Silicon-on-insulator (SOI) emerged as one of the most promising present-day silicon technologies. SOI based devices offer considerable advantages when compared conventional bulk silicon technology in such environments as well as allowing the miniaturization of these devices into the nanometer range [6–8]. It is, thus, anticipated that with the reduction of device dimensions and the increase of ICs complexities, radiation-hardened devices will be needed for reliability require-ments for future technologies. To guarantee the compliance of electronic devices with the functions expected of them and to increase their immunity to radiation influences, electronic devices are placed in ceramic hardened cells to bar the absorption of light at the metallurgical junctions, which may cause the biasing properties of the device to be altered. Conventionally, insulating coatings like ceramic packages of aluminum and bond wires of gold are customary substances for IC packaging. These may include trace quantities of radioactive elements which may emit alpha particles [9]. SOI technology inherently reduces or eliminates these effects [10]. The term ionizing radiation belongs to the radiation which is able to remove electrons from atoms lattice and break chemical bonds between atoms. These ionizing radiations exist in the cosmic background radi-ation. Examples of radiation effects that may occur are neutron effects, Single Event Effects (SEEs), transient dose effects and Total Ionizing Dose (TID) effects. It is, thus, essential to assess device operation at different frequency ranges, particularly low and high frequencies conditions, and to estimate the influence of radiation-induced interfa-cial states that may be present on the different interfaces of the electronic device using the SOI-MOS capacitor (SOI-MOS-C) as a characterizing tool. SOI devices offer bet-ter circuit isolation, reduced leakage currents, eliminated parasitic capacitance, reduced power consumption, improved radiation protection, faster switching speed and others when compared to conventional bulk MOS devices. SOI based devices have also better SEU protection. The main idea behind SOI technology is the ability of manufacturing devices using a thin Si film electrically isolated from the bulk substrate using a buried SiO_2 oxide (BOX). To understand the physics of these devices and analyze their func-tions, modeling device electrophysical parameters is necessary [11, 12]. For example, threshold voltage (V_{Th}) shift in C-V characteristics is a principal element to characterize SOI based devices. These characteristics depend on a number of parameters such as: Si film thickness, BOX thickness, impurity concentrations, temperature effects, interface states and so on. Scaling down device dimensions also provoked quantum mechanical effects, particularly for SOI with ultrathin channel [13]. Using C-V measurements for characterizing different biasing conditions of the SOI based devices is a well- founded mean [11, 14]. SOI technology is, hence, an alternative option for space applications as opposed to conventional bulk MOS technology, mainly as a result of its superior SEE tolerance. The BOX, which is inherent to the SOI-MOS based devices, is the principal merit in the radiation tolerance for SOI technology when compared to conventional Si technology. This BOX has direct effects on reducing leakage currents that may be caused by ionizing radiation where charge centers accumulate in the SiO_2 region and interfacial states are generated at the Si-SiO_2 interface. In particular, TID irradiation produces charge traps on defect sites in the oxide and generates new interfacial states at the interface resulting in a shift in the capacitance value proportional to the accumulated dose [2]. TID effect is, accordingly, an essential concern for circuits' operation used in space applications. However, TID effects in SOI devices, and due to the existence of the BOX, are minimized. Interfacial properties characterization in SOI devices still continues to be a challenge and an on-going task. The C-V characteristic for SOI devices is a reliable method to examine and investigate the behavior of the electrophysical parameters when devices are subjected to radiational effects [11, 15].

In this paper, radiation effects on electronic devices are examined and the use of SOI technology to mitigate their effects is reviewed. Effects of irradiating SOI-MOS capacitors and the consequent effects on the C-V characteristics are illustrated and explained. The usefulness of SOI-MOS based devices for high-frequency and high-radiation environment applications is validated. This paper focuses on how the SOI-MOS-C, from technological point view rather than the mathematical foundations behind it, is used as a tool for studying radiation effects in harsh environment applications and its use in radiation-tolerant circuit applications.

2   Device Description and Properties

Figure 1 shows a schematic diagram for the SOI-MOS capacitor with two terminals, while Fig. 2 shows the equivalent capacitor model for the structure neglecting the interface-traps capacitance. The total capacitance per unit area, C_T, is obtained by the series combination of these capacitances as [11, 16]:

$$\frac{1}{C_T} = \frac{1}{C_{GOX}} + \frac{1}{C_{BOX}} + \frac{1}{C_{SCR1}} + \frac{1}{C_{SCR2}} + \frac{1}{C_{Sub}} \tag{1}$$

where:

C_{GOX} and C_{BOX} are gate and BOX capacitances, respectively. C_{SCR1} and C_{SCR2} are space charge region capacitances in SCR_1 and SCR_2 in the body, respectively. C_{Sub} is the substrate capacitance.

Calculating C_{SCR1} and C_{SCR2} is the most important in finding C_T. All five capacitances are given as follows:

$$C_{GOX} = \frac{\varepsilon_{GOX}}{t_{GOX}}, \; C_{BOX} = \frac{\varepsilon_{BOX}}{t_{BOX}}, \; C_{SCR1} = \frac{\partial Q_{SCR1}}{V_{SCR1}}, \; C_{SCR2} = \frac{\partial Q_{SCR2}}{V_{SCR2}} C_{Sub} = \frac{\partial Q_{Sub}}{V_{Sub}} \tag{2}$$

where:

Q_{SCR1} and Q_{SCR2} are the space charge densities for the front and back regions in the Si body, respectively. V_{SCR1} and V_{SCR2} are the voltage drops in these two space charge regions, respectively. V_{Sub} is the voltage drop in the space charge in the bulk Si substrate. C_{GOX} and C_{BOX} are the gate oxide and the BOX capacitances, respectively. t_{GOX} and t_{BOX} are the gate and the buried oxides thicknesses, respectively. ε_{GOX} and ε_{BOX} are the permittivities for the gate oxide and the BOX, respectively.

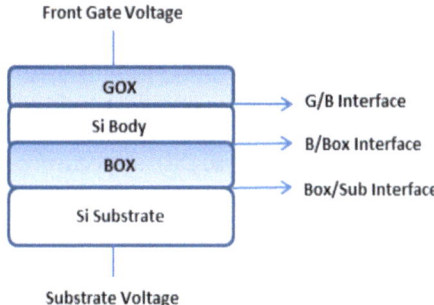

Fig. 1. A schematic diagram of the two-terminal SOI-MOS Capacitor.

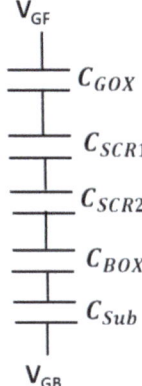

Fig. 2. The equivalent capacitor model. V_{GF} and V_{GB} are the voltages for the front and the back gates, respectively. C_{GOX}, C_{SCR1}, C_{SCR2}, C_{BOX} and C_{Sub} are defined in the text above.

3   Experimental Description

Zhongshan *et al.* [16] carried out C-V measurements for three p-type SIMOX SOI capacitors, fabricated on three SOI wafers, at 1 MHz high frequency (HF) before and after irradiation with different doses with a ^{60}Co source. The SIMOX SOI devices were fabricated using three p-type (100) silicon wafers with a nominal resistivity 20 Ωcm with a dose of 2×10^{18} cm^{-2}. Oxide layers were deposited by magnetron sputtering followed by a post-implantation 1300 °C annealing. A quasi-reactive gas mixture of Ar_2 and O_2 was employed to achieve the anticipated stoichiometry. The top silicon region was calculated to be 200 nm and the BOX thickness was 375 nm. In order to reduce defect concentrations of SIMOX SOI wafers and to improve the radiation hardness of the BOX region in two of the SOI wafers (2 and 3), nitrogen ions, with 160 keV, were then implanted into the BOX layer with doses of 2×10^{15} and 3×10^{15} cm^{-2}, respectively, followed by an annealing step in a nitrogen ambient at 1200 °C. One of the three wafers (wafer 1) was used as a control wafer and, thus, was not given any nitrogen implantation. The SOI-MOS capacitors were fabricated on the three SOI wafers, with a 1.96×10^{-3} cm^{-2} Al-gate by electron beam evaporation, after removing the top Si layers using ion reactive etching. In order to investigate the influence of gamma ray irradiation on the SOI-MOS capacitors, the fabricated capacitors were irradiated using the ^{60}Co source at a dose rate of 1.38×10^4 rad (Si)/min. The three SOI-MOS capacitors were irradiated with doses of 5×10^4 and 1×10^5 rad (Si), respectively, with a gate voltage bias of 6.5 V during irradiation. To make the comparison with a zero-gate voltage bias, the capacitors were irradiated with a dose of 5×10^5 rad (Si).

4   Analysis and Discussions

Limited studies have been published on the effect of ionizing radiation using SOI-MOS-C C-V characteristics, which is an established tool for process characterization and extracting device parameters in bulk MOS technology [17]. This is possibly because of the fact that using C-V measurement is not straightforward in SOI technology due to the multi-interface presence in SOI devices leading to competing biasing at these three interfaces. Zhongshan *et al.* [16] performed C-V measurements for the three p-type SIMOX SOI capacitors at 1 MHz high frequency (HF) before and after irradiation with different doses with the ^{60}Co source. These C-V characteristics were analyzed before and after exposure to gamma irradiation at HF and room temperature. These characteristics of the irradiated capacitors exhibited changes in the capacitance values as gamma irradiation dose is varied. The capacitance values increased as well as a shift in voltage values was detected when the irradiation dose increased. Our previous model for the partially and fully depleted SOI-MOS-C demonstrated the effects of changing different electrophysical parameters, such as doping concentration, interfacial traps, on the C-V curves of unirradiated capacitors [11]. From the experimental data on the effect of radiation doses on the C-V characteristics by [16], our model [11] is used to shed new light on these curves and to make the correspondence between these parameters and the effect of the radiational doses. From these data, it is seen that the annealing of the samples in nitrogen results in C-V curves where the shift in flatband voltages (V_{FB}) is less negative. This is probably because the radiation produced a poor Si-buried oxide interface which leads to reducing the positive charge at the interface. It is also noted that, after first irradiation, the value of this shift does not vary when radiation dose is increased due to the radiation hardening in examined capacitors. Figure 3 presents the HF *C-V* curves of SOI-MOS capacitors (1, 2 and 3) measured at a frequency of 1M Hz before and after 1×10^5 rad (Si) irradiation. A gate bias of 6.5 V was applied during irradiation. Figure 4 presents HF C-V characteristics of these capacitors at a frequency of 1M Hz before and after 5×10^5 rad (Si) irradiation, where a gate bias of 0 V was employed during irradiation. While, Fig. 4 presents the HF C-V characteristics for the SOI-MOS-C (1) measured at 1MHz before and after 5×10^4 and 1×10^5 rad (Si) irradiation during a 6.5 V gate bias [16].

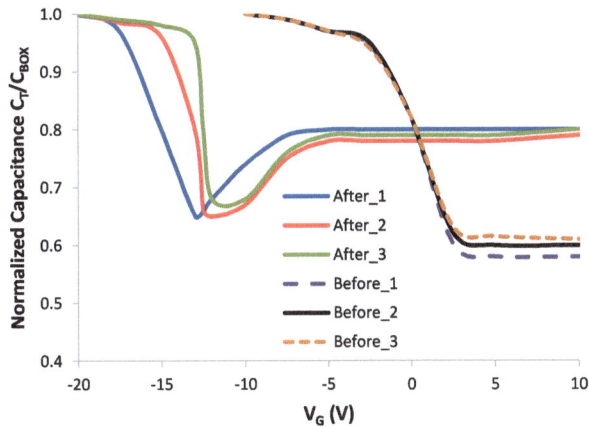

Fig. 3. HF C-V characteristics of the three SOI-MOS capacitors measured at a frequency of 1MHz before and after 1×10^5 rad (Si) irradiation doses with a gate bias of 6.5V during irradiation [16].

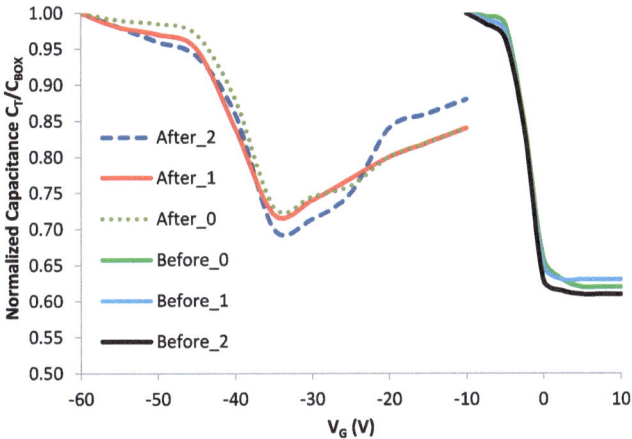

Fig. 4. HF C-V characteristics of the three SOI-MOS capacitors measured at a frequency of 1MHz before and after 5×10^5 rad (Si) irradiation doses with a gate bias of 0 V during irradiation [16].

The ionizing radiation gives rise to depositing energy at $Si-SiO_2$ interface resulting in oxide-trapped charge accumulation producing interfacial traps at this interface degrading the expected performance of the devices. Interface-traps build-up, as a process, is more complicated than oxide charge trapping. Since there is lattice parameters incompatibility between the Si and oxide which makes interface traps presence intrinsic in the fabrication process of Si technology [18]. These interfacial traps can emit and/or capture either

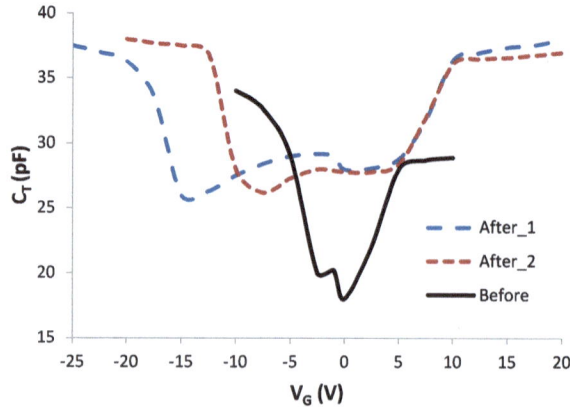

Fig. 5. HF C-V characteristics of the SOI-MSOS capacitor (no. 1) measured at 1MHz before and after 5×10^4 and 1×10^5 rad (Si) irradiation doses with a gate bias of 6.5V during irradiation [16].

electrons or holes depending on the voltage biasing. These traps are then attributed to the increase in dopants, mismatch in dielectric-silicon lattice, break in chemical bond or to radiation effects, etc. Thus, as expected, the defects induced due to the radiation affect and shift the *C-V* curves of SOI-MOS-C. Hence, to accurately investigate these parametric shifts, the increase in interfacial traps must be precisely understood as in MOS devices [5].

From Fig. 5, the HF C-V characteristics of the SOI-MOS-C (no. 1) exhibit atypical behavior as expected of high frequency curves. The behavior exhibited is similar to that of low frequency (LF) C-V characteristics. These resemble to a large degree the LF curves presented in [11]. To understand this difference, the behavior of the device in inversion under HF and LF conditions should be emphasized. Since the SOI-MOS capacitor is p-type then electrons in the inversion layer must be generated somehow since there are not adequate free electrons in the substrate to form the inversion layer without some excitation. For the SOI-MOS capacitor, the electrons are generated slowly by thermal excitation and/or applying external voltage. In the case of HF, the time is not enough for this generation to take place and the depletion capacitance becomes the leading capacitance. However, in Fig. 5, thus, instead of the capacitance remaining at its minimum depletion value, the capacitance attains a value similar to the LF case when the device goes into inversion [11]. Because of the radiation dose, the trapped holes will be generated in the BOX region attracting more electrons into the inversion layer leading to LF resembling curves. These trapped carriers follow the applied ac signal and may not be in equilibrium state during the whole time of C–V measurements. Another important point to mention is the voltage sweep direction. Since the SOI-MOS capacitor has three Si-SiO$_2$ competing interfaces, then each of these interfaces will be in any of the different cases, accumulation, inversion, depletion or even deep depletion subject to the value of the applied voltage as well as the sweep direction of the voltage. Due to this increased number of interfaces, when compared to the conventional MOS capacitors, the precautions usually taken when measuring conventional C-V curves to avoid driving test capacitors into deep depletion state, instead of strong inversion, are no more feasible. For conventional MOS capacitors and to avoid deep depletion (non-equilibrium) state, the HF C-V measurements are usually measured from strong inversion to accumulation states. In this case and when applying a potential on conventional MOS capacitors, the relatively slow recombination-generation process is not able to provide or eliminate minority carriers as a result of this applied potential. On the other hand, in the LF case minority carriers are generated or annihilated in response to the applied voltage. This is, however, experimentally impossible to achieve in SOI-MOS capacitor with its three Si-SiO$_2$ interfaces. Because of this, any direction voltage sweep direction must inevitably lead to driving at least one of the three interfaces from accumulation towards the inversion condition which may result in a non-equilibrium state unless this interface is of super high quality. Figure 6 illustrates the experimental data of HF C-V characteristics dependency on the voltage sweep direction for the SOI-MOS capacitor. This is even more difficult for thinner body regions where longer relaxation times are required to reach equilibrium where majority carrier charges react to voltage changes at approximately the dielectric relaxation time. This means that that the measurements are not only voltage dependent but also time dependent where not only recombination-generation processes are to be considered but time delay between the voltage step and the measurement time has to be taken into account. Therefore, non-equilibrium states, at least on one of the three interfaces, are expected which affects the measured values depending on the voltage sweep direction. The main problem in the non-equilibrium state is the Fermi-level splitting into two quasi-Fermi levels for the majority and minority carriers, respectively. To add to this difficulty, the value of the total capacitance (C_T) is a result of a number of individual capacitances making C_T value relatively small. Therefore, to make accurate and reliable measurements, very small displacement currents must be measured where noise, parasitic capacitances and leakage currents become a major challenge. With thinner and thinner gate areas measuring both LF and HF C-V characteristics for SOI-MOS transistors is experimentally difficult because of these small gate areas. Measuring these devices, with their specific carrier lifetimes and low carrier generation rates, at low frequencies may produce HF characteristics. This may necessitate specially fabricated SOI-MOS capacitors with large gate areas to allow reliable C-V measurements.

Interface states are able to trap charges in the channel region leading to a shift in the threshold voltage. This shift influences carriers' mobility in the channel. Generally, there are two types of consequences as a result of this; holes trapping and interfacial- state defects production. These two effects follow different mechanisms. These two effects share their dynamics in two ways: formation of Si inversion layer under the oxide developing a conductive channel and reducing device gain. Trapping of holes occurs rather fast. Detrapping these holes can be achieved by annealing. Formation of interface states, on the other hand, arises relatively slowly. Annealing of these states takes place at temperature above 400 °C. Since these effects are time-dependent, methodologies are required that permit reliable assessments of the device operation in real radiation environments. In conventional MOS technology, this radiation-induced charge may change polarity type, particularly at the lower surface of the channel, giving rise to a conductive back channel connecting both source and drain terminals in undesirable manner. Like the I-V characteristics and in line with the methodologies employed in conventional MOS

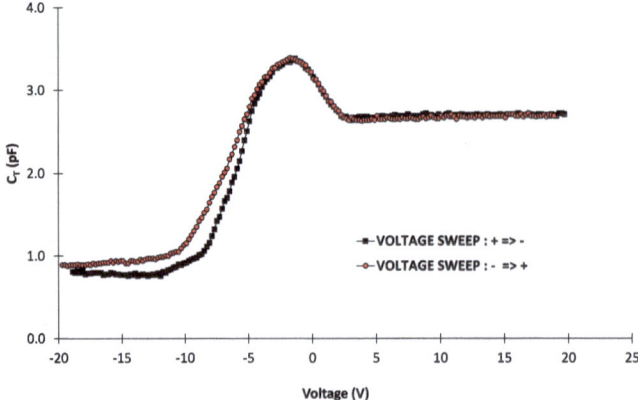

Fig. 6. HF C-V characteristics dependency on the voltage sweep direction for the SOI-MOS capacitor.

capacitors, extracting electrical and physical parameters related to the buildup of charges in the oxides may be deduced using C-V curves [8].

The body film of the capacitor can also be partially or fully depleted. If this Si film is not fully-depleted, the gate-potential bias influence on the leakage current is limited. In this case, the BOX triggers, relatively independent of gate bias, a constant increase in leakage current. However, if the body film is fully-depleted, the leakage current is significantly affected by gate voltage. The device is also temperature dependent which must be taken into account in future modeling for these devices [19].

5 Conclusion

Ionizing radiation influences on C-V characteristic of the SOI-MOS capacitor were investigated. The effects were examined before and after exposure to gamma irradiation at high frequency and room temperature under different radiational dose conditions. These C-V curves of the irradiated capacitors exhibited changes in the capacitance amount with gamma irradiation dose. A raise in the capacitance and a shift in voltage values were detected when the irradiation dose increased. The change of the C-V characteristics was credited to changes in the charge accumulation at the BOX-Si interface and electron-hole pair formation as well as the increase of interface states in the BOX during irradiation process. This was due to charge transport and trapping in the near interfacial states in the BOX. The impacts of this radiation on C-V characteristics were a negative V_{Th} shift and a rise total capacitance, mainly due to the inversion region. Experimental dependency of HF C-V characteristics on the voltage sweep direction for the SOI-MOS capacitor was also illustrated. Including effective carrier mobility that depends on charge position and/or energy state may also be necessary. As this understanding evolves, modeling the charge exchange between the Si layer and BOX interfacial states requires deeper knowledge of the interfacial defects involved. This understanding of the physical phenomena, caused by ionizing radiation affecting the physical and electrical parameters of SOI-based devices, is essential to predict potential malfunction and/or failure in the device operation in the harsh environments. It was concluded that using C-V curves of SOI-MOS capacitor is a useful option for characterizing electronic devices for harsh environments such as space applications with appropriate modeling and experimental techniques. However, appropriate protection of these devices in these environments is essential to ensure proper operation of the devices and to exclude undesirable effects such as SEEs and total dose effects.

References

1. National Research Council (NRC): Managing Space Radiation Risk in the New Era of Space Exploration. The National Academies Press, Washington, DC (2008). https://doi.org/10.17226/12045
2. Karmakar, A., Wang, J., Prinzie, J., De Smedt, V., Leroux, P.: A review of semiconductor based Ionising radiation sensors used in harsh radiation environments and their applications. Radiation 1, 194–217 (2021). https://doi.org/10.3390/radiation1030018
3. Ibarra, M.L., Barrera, M., Alurralde, M.: Preparation and characterization of MOS capacitors for in situ measurement during radiation damage studies. Procedia Mater. Sci. 9, 319–325 (2015). https://doi.org/10.1016/j.mspro.2015.04.041
4. Esqueda, I., Barnaby, H., Holbert, K., El-Mamouni, F., Schrimpf, R.: Modeling of ionizing radiation-induced degradation in multiple gate field effect transistors. IEEE Trans. Nucl. Sci. 58(2), 499–505 (2011)https://doi.org/10.1109/TNS.2010.2101615
5. Esqueda, I., Barnaby, H., King, M.: Compact modeling of total ionizing dose and aging effects in MOS technologies. IEEE Trans. Nucl. Sci. 62(4), 1501–1515 (2015). https://doi.org/10.1109/TNS.2015.2414426
6. Rudenko, T.E., Nazarov, A.N., Lysenko, V.S.: The advancement of silicon-on-insulator (SOI) devices and their basic properties. Semicond. Phys., Quantum Electron. Optoelectron. 23(3), 227–252 (2020). https://doi.org/10.15407/spqeo23.03.227
7. Ikraiam, F.A.: An analysis of radiation effects on electronics and SOI-MOS devices as an alternative. In: 6th Annual International Conference on Sustainable Development through Nuclear Research and Education (Nuclear 2013), Piteşti, Romania, May 22–24 (2013)
8. Cristoloveanu, S., Bawedin, M., Ionica, I.: A review of electrical characterization techniques for ultrathin FDSOI materials and devices. Solid-State Electron. 117, 10–36 (2016)https://doi.org/10.1016/j.sse.2015.11.007
9. May, T.C., Woods, M.H.: Alpha-particle induced soft errors in dynamic memories. IEEE Trans. Electron Dev. ED-26. 2 (1979)https://doi.org/10.1109/T-ED.1979.19370
10. Lagaev, D.A., Klyuchnikov, A.S., Shelepin, N.A.: Prospects for applying FD-SOI technology to space applications. J. Phys.: Conf. Ser. 2388, 012135 (2022). https://doi.org/10.1088/1742-6596/2388/1/012135
11. Ikraiam, F., Beck, R., Jakubowski, A.: Modeling of SOI-MOS capacitors C-V behavior: partially- and fully-depleted cases. IEEE Trans. Electron Devices 45(5), 1026–1032 (1998)https://doi.org/10.1109/16.669517
12. Li, Y., et al.: Applications of direct-current current-voltage method to total ionizing dose radiation characterization in SOI NMOSFETs with different process conditions. Electronics 10, 858 (2021). https://doi.org/10.3390/electronics10070858
13. Omura, Y., Horiguchi, S., Tabe, M., Kishi, K.: Quantum-mechanical effects on the threshold voltage of ultrathin-SOI nMOSFETs. IEEE Electron Device Lett. 14(12), 569 –571 (1993)https://doi.org/10.1109/55.260792
14. Afzal, B., Zahabi, A., Amirabadi, A., Koolivand, Y., Afzali-Kusha, A., Nokali, M.E.: Analytical model for C-V characteristic of fully depleted SOI-MOS capacitors. Solid-State Electron. 49(8), 1262–1273 (2005)https://doi.org/10.1016/j.sse.2005.06.006

15. Rustagi, S.C., Mohsen, Z.O., Chandra, S., Chand, A.: C-V characterization of MOS capacitors in SOI structures. Solid-State Electron. **39**(6), 841–849 (1996)https://doi.org/10.1016/0038-1101(95)00395-9

16. Zhongshan, Z., et al.: Sensitivity of total-dose radiation hardness of SIMOX buried oxides to doses of nitrogen implantation into buried oxides. Chin. J. Semiconductors **26**(5), 862–866 (2005)

17. Sze, S., Ng, K.: Physics of Semiconductor Devices, 3rd edn. John Wiley & Sons Inc, Hoboken, NJ, USA, April (2006)

18. Fleetwood, D.M., et al.: Effects of oxide traps, interface traps, and "border traps" on metal-oxide-semiconductor devices. J. Appl. Phys. **73**(10), 5058–5074 (1993)https://doi.org/10.1063/1.353777

19. Kim, S., Shim, T., Park, J.: Electrical behavior of ultra-thin body silicon-on-insulator n-MOSFETs at a high operating temperature. J. Ceram. Process. Res. **10**(4), 507–511 (2009)https://doi.org/10.36410/jcpr.2009.10.4.507

A Complete Analysis of Subgiant Stellar Systems III: Hip70868

Hoda E. Elgendy[1](✉), Mashhoor A. Al-Wardat[1,2,3], Hassan B. Haboubi[1],
Lin R. Benchi[1], Abdallah M. Hussein[3], and Hussein M. Elmehdi[1]

[1] Department of Applied Physics and Astronomy, University of Sharjah, Sharjah 27272, UAE
U20104142@sharjah.ac.ae, Hoda.elgendy@ese.gov.ae
[2] Sharjah Academy for Astronomy, Space Sciences and Technology, Sharjah 27272, UAE
[3] Department of Physics, Faculty of Sciences, Al Al-Bayt University, PO Box: 130040,
Mafraq, Jordan

Abstract. This study utilizes "Al-Wardat's method for analysing binary and multiple stellar systems" to estimate a set of parameters for the binary system Hip70868. The method compares the system's observational magnitudes, color indices, and spectral energy distribution (SED) and synthetic SEDs generated through atmospheric modeling of each component. Feedback-adjusted parameters and an iterative approach were employed to achieve the best fit between observational (including the latest measurements of Gaia, release DR3) and synthetic spectral energy distributions. The findings were completed using the distance of 71.89 pc given by Hipparcos 2007's new reduction. The individual components' parameters for the system were derived afterward. The parameters obtained for the individual components are as follows: Component A has an effective temperature (T_{eff}) of 6072 ± 50 K, a surface gravity (log g_a) of 4.50 ± 0.05, and a radius (R_a) of 1.40 ± 0.07 $R_\odot$. Component B has a (T_{eff}) of 5887 ± 50 K, a (log g_b) of 4.40 ± 0.05, and a radius (R_b) of 1.326 ± 0.07 $R_\odot$. The spectral types of the components were found to be F9 IV and G1 IV, respectively. The findings from this analysis were utilized to accurately position the two system components on the Hertzsprung-Russell (H-R) diagram and the evolutionary tracks. The analysis shows that the components have transitioned from the main sequence to the sub-giant stage of their evolution.

Keywords: binary · stars · visual · fundamental parameters · individual component · individual: Hip70868

1 Introduction

Numerous studies indicate that over half the stars in our solar region and galaxy belong to the binary or multiple-star group systems [1]. This fascinating discovery highlights the structures and levels of organization that can be observed within the cosmos. The way these stellar systems form and evolve their architecture is closely linked to factors that include their period ratios, orbit orientation, and mass distribution [2]. This information provides insights into the process of star formation in general. Understanding

H. M. K. Al Naimiy et al. (Eds.): AUASS-CONF 2023, SPPHY 420, pp. 277–286, 2025.
https://doi.org/10.1007/978-981-96-3276-3_21

the characteristics of higher-order systems is significant as most stars are born within groups sharing similar properties [3]. The presence of companions has impacts on the life and eventual death of stars [4]. Thus, understanding the formation, evolution, and behavior of stellar groups requires a thorough analysis of star systems and their binary features. This is widely backed by scientific research papers [5].

The physical and geometrical parameters of binary and multiple star systems are analyzed using "Al-Wardat's method for analyzing binary and multiple stellar systems." This method blends atmospheric modeling with spectrophotometry. The geometrical and physical properties of the components of stars are predicted by this method for several stars (for example see [6–12]). It uses the estimated metallicities and ages of binary and multiple stellar systems (BMSSs) to help explain the formation process [4]. The technique applies to all kinds of binaries, including eclipsing, and visual ones that have not yet reached the giant stage [8, 13].

To apply this method, one must determine the color indices and visual magnitude in addition to the magnitude differences (obtained via measurements made using speckle interferometry). Numerous binary systems have used this method since 2002 [10].

The two components that make up the Hip 70868 system, A and B, are presented in this document alongside their geometrical and physical specifications. The selection of this system was based on its representation of a sub-giant star system, about which there appears to be little information because it is in between the main system and the red giant stages of growth [14]. Studying sub-giant systems is crucial to understanding the transition from main sequence to red giant stages. Further, it helps clarify the unique characteristics of stars in this "intermediate" level [12].

Table 1. Basic data of the system Hip 70868.

Properties	Parameters	Value	Ref
Position	α_{2000}	$14^h\ 29^m\ 32^s\ .61$	(*SIMBAD*)
	δ_{2000}	$-37°\ 02'\ 22''\ .33$	
Magnitude [mag]	m_v	7.84	[15]
	$(B-V)_J$	0.626 ± 0.015	[15]
	AV	0.0248 ± 0.002	[16]
Parallax [mas]	π_{Hip1}	14.79 ± 1.10	[15]
	π_{Hip2}	13.91 ± 0.91	[17]
	π_{GDR2}	12.5757 ± 0.5952	[18]

1.1 Hip 70868

Hip 70868, or HD 126935, belongs to the G3V spectral classification type in the Henry Draper Catalog [19]. The components of binary stars are distinguished by the suffixes "A" (for the brighter or primary component) and "B" (for the faint secondary component in the system). The fundamental parameters of hip 70868 are presented in Table 1.

The system's parallax comes in three different values. The first parallax, with a value of 14.79 mas and a distance of 67.61 pc, is derived from literature catalogs [15]. The second value, 13.91 mas, equivalent to a distance of 71.89 pc, is taken from the Hipparcos reduced catalog [17]. The third value, 12.5757 mas, corresponds to a distance of 78.38 pc and was taken from the Gaia Release 2 [18]. In this paper, the parallax of the reduced Hipparcos catalog is used for the analysis to calculate the distance of the system.

2 Atmospheric Modeling

The components of hip 70868's atmospheric properties were identified by applying "Al-Wardat's method for analyzing binary and multiple stellar systems". Using grids of line-blanketed model atmospheres ATLAS 9, this method focuses on modeling entire synthetic spectral energy distributions (SEDs) using preliminary parameters. The "Al-Wardat" atmospheric modeling approach requires additional data to analyze the components of the hip 70868 system.

The magnitude difference between the components $\Delta m = 0.323$ (from *Fourth Catalog*) is used as the average of all measurements done with visual filters 550 ± 90 nm in Eq. 1, as well as the visual magnitude $m_v = 7.84$ from Table 1. We employed Eq. 2 to get the preliminary magnitudes for each component. The magnitudes of the primary and secondary components of the system were found to be $m_v^A = 8.46$ mag and $m_v^B = 8.74$ mag respectively.

$$\frac{f1}{f2} = 2.512^{-\Delta m} \tag{1}$$

$$m_v = -2.5 \log(f1 + f2) \tag{2}$$

Additionally, we can compute the initial input parameters for each component, including bolometric magnitudes, luminosities, and effective temperatures using Eq. 3, Eq. 4, and Eq. 5.

$$M_v = m_v + 5 - 5\log(d) - A_v \tag{3}$$

$$\log\left(\frac{R}{R_\odot}\right) = 0.5\log\left(\frac{L}{L_\odot}\right) - 2\log\left(\frac{T}{T_\odot}\right) \tag{4}$$

$$\log(g) = \log\left(\frac{M}{M_\odot}\right) - 2\log\left(\frac{R}{R_\odot}\right) + 4.43 \tag{5}$$

The preliminary input parameters were calculated with $T_\odot = 5777$ K, of $R_\odot = 6.69$ x 10^8 m, $A_v = 0.2268$, the dynamic parallax 13.91 ± 0.91 (d = 71.89 pc) from Table 1 (to calculate the magnitude) and finally, the bolometric correction from [20].

Employing solar metallicity model atmospheres as the ATLAS 9 initial parameter output, the total luminosities of components A and B, which are located at a distance d (pc) from Earth, combine to form the total energy flux from the system as follows [6, 21]:

$$F_\lambda.d^2 = H_\lambda^A R_A^2 + H_\lambda^B R_B^2 \tag{6}$$

Rearranging

$$F_\lambda = \left(\frac{R_A^2}{d^2}\right)^2 \left[H_\lambda^A + H_\lambda^B \cdot \left(\frac{R_B}{R_A}\right)^2\right] \qquad (7)$$

where H_λ^A and H_λ^B are fluxes from the unit surface of each component. F_λ is the total spectral energy density of the system. R_A and R_b are the radii of primary and secondary components in solar units. "Al-Wardat's method" was used to create their synthesized Spectral Energy Distributions to calculate the binary system's magnitudes and color indices.

This technique was also applied as a calibration technique, whereby the synthetic magnitudes of a whole stellar system were compared with available observational data. The precision of the observations determines how accurate the results are.

3 Results and Conclusion

The resulting whole synthetic SED is not going to fit the first trial's observational spectral energy density. To find the best fit between the synthetic and observational total absolute fluxes, numerous attempts were conducted. This resulted in the construction of hundreds of synthetic SEDs using various parameter settings and comparisons with the observational SED. Efforts were made to alter the radii, parallax, effective temperatures, and surface gravity acceleration. As our objective preliminary parameters, we consider the values of the total observational V_J, B_T, V_T, and Δm. To sum up, the parameters were selected for the best fit (Table 2).

Table 2. Parameters of the system Hip70868 components as estimated using Al-Wardat's Method.

Physical Parameter	Value
T_A	6072 ± 50 K
T_B	5887 ± 50 K
$\log g_A$	4.50 ± 0.05
$\log gB$	4.40 ± 0.05
R_A	1.40 ± 0.07 $R_\odot$
R_B	1.326 ± 0.07 $R_\odot$
d	71.89 pc

The synthetic data for each component and the entire data for the system are shown in Table 3. The spectral energy distributions obtained by the synthetic analysis are shown in Fig. 1. According to the tables in [20], the spectral types of the components are F9-IV and G1-IV, respectively.

The luminosity of each component is calculated using Eq. 4 and the error of luminosity values is calculated through the following method:

$$\sigma_L = \pm 2L \sqrt{\left(\frac{\sigma_R}{R}\right)^2 + 4\left(\frac{\sigma_{Teff}}{Teff}\right)^2} \qquad (8)$$

Table 3. Magnitudes and color indices of the composed synthetic spectrum and individual components of Hip70868 as a result of applying the method. The entire values were used for the best fit.

Sys	Filter	Entire Obs	Entire Synt	HIP70868	
			$\sigma = \pm 0.03$	A	B
Joh- Cou	U	8.48	8.60	9.17	9.57
	B		8.47	9.06	9.34
	V	7.84	7.84	8.46	8.74
	R	0.626 ± 0.015	7.50	8.13	8.39
	U – B		0.13	0.10	0.17
	B – V		0.63	0.60	0.66
	V – R		0.34	0.33	0.35
Stromgren	u		9.74	10.32	10.72
	v		8.80	9.39	9.75
	b		8.19	8.80	9.11
	y		7.81	8.43	8.71
	u – v		0.94	0.92	0.97
	v – b		0.61	0.59	0.64
	b – y		0.38	0.37	0.397
Tycho	B_T	8.631 ± 0.010	8.62	9.20	9.56
	VT	7.912 ± 0.008	7.91	8.53	8.82
	B_T - V_T	0.719	0.71	0.68	0.75
Gaia	G	7.722	7.72	8.35	8.61
	Bp	8.001	8.07	8.68	8.98
	Rp	7.211	7.20	7.83	8.07
Infrared	J	6.709	6.71	7.37	7.56
	H	6.479	6.42	7.09	7.25
	K	6.361	6.38	7.06	7.22

Yielding $L_A = (2.39 \pm 0.25)$ $L_\odot$ and $L_B = (1.896 \pm 0.210)$ $L_\odot$ luminosities for each component. This shows that our system has begun evolving into the sub-giant stage according to the evolutionary tracks shown in Fig. 2 [22].

Overall, we estimated the atmospheric parameters of each component of HIP70868 using "Al-Wardat's method for analyzing BAMSs". Using the estimated parameters, "Al-Wardat's method" builds a synthetic spectral energy distribution for each component (SED) and provides the position of the components' system on the H-R diagram with the evolutionary tracks. The parameters of the components are shown in Table 4.

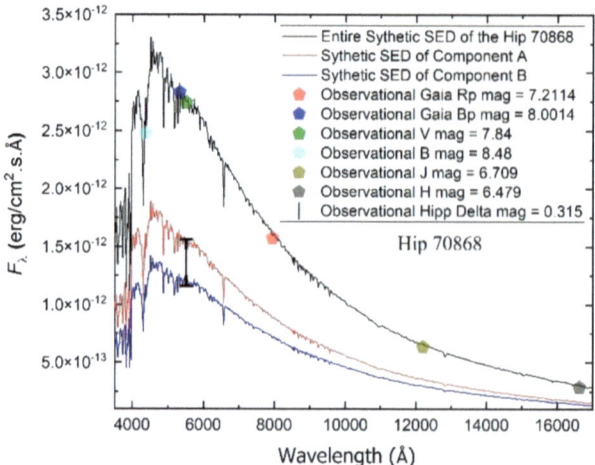

Fig. 1. Synthetic Spectral Energy Distribution of the entire Hip 70868 system and its individual components. The complete synthetic SED aligns well with the observed magnitudes from various sources, including the latest Gaia DR3 data.

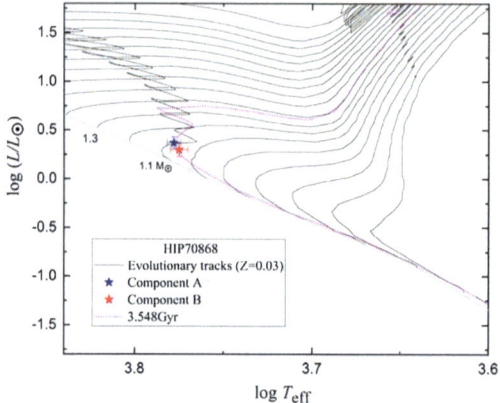

Fig. 2. The components of Hip 70868 on the evolutionary tracks and isochrons of (Girardi et al., 2000).

Table 4. Parameters of the individual components of Hip70868 (This work).

Parameter	Unit	Component A	Component B
T_{eff}	K	6072 ± 50	5887 ± 50
logg	cm/s^2	4.50 ± 0.05	4.40 ± 0.05

(continued)

Table 4. (*continued*)

Parameter	Unit	Component A	Component B
R	$R_\odot$	1.40 ± 0.07	1.326 ± 0.07
L	$L_\odot$	2.39 ± 0.25	1.89 ± 0.21
M_{bol}	Mag	3.80	4.05
M	$M_\odot$	1.22	1.16
d	pc	71.89	
Age	Gyr	3.548	

4 Orbital Analysis

Using the IDL code ORBITX developed by Tokovinin [23, 24], and inputting the orbital data shown in Tables 5 and 6, we get an improved orbit of our system shown in Fig. 3.

Table 5. New orbital elements of Hip70868.

Parameters	Units	Values
P [Period]	[Yr]	14.18830 ± 0.01593
T [Epoch of passage]		2008.6061 ± 0.0024
e [Eccentricity]	-	0.4398 ± 0.0012
a [Semi-major]	[Arcsec]	0.1173 ± 0.0002
ω [Longitude]	[deg]	167.10 ± 0.16
Ω [Position angle]		125.17 ± 0.27
i [Inclination]		128.15 ± 0.18

According to Fig. 3 we notice that all 12 points are aligned in the orbit starting from 1991 to the 2021 observations. Using this data, we can use Eq. 9 and Eq. 10 to find the mass and error of the system.

$$\mathrm{Mdyn} = \mathrm{MA} + \mathrm{MB} = \left(\frac{a^2}{\pi^3 P^2}\right) M_\odot \tag{9}$$

$$\frac{\sigma_M}{M} = \sqrt{9\left(\frac{\sigma_\pi}{\pi}\right)^2 + 9\left(\frac{\sigma_a}{a}\right)^2 + 4\left(\frac{\sigma_p}{p}\right)^2} \tag{10}$$

Yeilding a total dynamical mass of (2.48 ± 0.55) $M_\odot$ using parallax $\pi_{Hip1} = 14.79$ mas and a total dynamical mass of (2.98 ± 0.58) $M_\odot$ using parallax $\pi_{Hip2} = 13.91$ mas from Table 1.

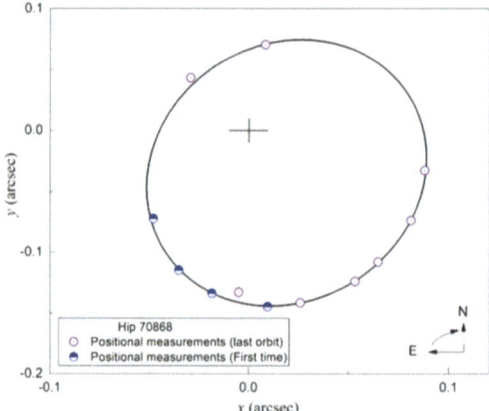

Fig. 3. New orbit of Hip 70868.

Table 6. The relative positional measurements of the system as of the Fourth Catalog of Interferometric Measurements of Binary Stars.

Date	θ (deg)	ρ (deg)
1991.250	178.0 ± 0.133	0.133
2008.542	34.1 ± 0.052	0.052
2009.260	353.4 ± 0.071	0.071
2013.132	249.8 ± 0.094	0.094
2014.303	227.8 ± 0.110	0.11
2015.497	211.1 ± 0.126	0.126
2016.134	203.4 ± 0.135	0.135
2017.430	190.5 ± 0.144	0.144
2018.162	183.9 ± 0.145	0.145
2019.372	172.3 ± 0.135	0.135
2020.117	163.1 ± 0.120	0.12
2021.078	146.6 ± 0.087	0.087

5 Conclusion

The method of analyzing binary or multiple stellar systems, which was developed by Mashhoor Al- Wardat, and the orbital analysis method, which was developed by Adrei A. Tokovinin, were applied to the visually close binary star HIP 70868. Based on the best fit between the system's synthetic SED and observational photometry, the atmospheric, fundamental, and orbital parameters have been computed. The results indicated a 3.548 Gyr old subgiant star, with F7 IV component A and a G1 IV component B. Furthermore, the system's total and partial synthetic magnitudes and color indices were calculated using the Hipparcos new reduction parallax of 13.91 mas. Finally, using orbital data, the dynamical mass of the system was calculated as well.

Acknowledgement. This work has made use of data from the European Space Agency (ESA) mission Gaia (https://www.cosmos.esa.int/gaia), processed by the Gaia Data Processing and Analysis Consortium (DPAC; https://www.cosmos.esa.int/web/gaia/dpac/consortium). This work has made use of the Multiple Star Catalog, SIMBAD database, and Al-Wardat's method for analyzing binary and multiple stellar systems with their codes.

References

1. Tanineah, D.M., Hussein, A.M., Widyan, H., Al-Wardat, M.A.: Trigonometric parallax discrepancies in space telescopes measurements I: the case of the stellar binary system Hip 84976. Adv. Space Res. **71**(1), 1080–1088 (2023). https://doi.org/10.1016/j.asr.2022.09.025

2. Abushattal, A.A., et al.: The 24 AQR triple system: a closer look at its unique high-eccentricity hierarchical architecture. Adv. Space Res. **73**(1), 1170–1184 (2024). https://doi.org/10.1016/j.asr.2023.10.044

3. Al-Wardat, M.A., et al.: Physical and geometrical parameters of CVBS XIV: the two nearby systems HIP 19206 and HIP 84425. Res. Astron. Astrophys. **21**, 161 (2021)

4. Hussein, A.M., Abu-Alrob, E.M., Alkhateri, F.M., Al-Wardat, M.A.: Atmospheric parameters of individual components of the visual triple stellar system HIP 32475. (2023). arXiv:2304.03604. Accessed 13 Oct 2023. http://arxiv.org/abs/2304.03604

5. Docobo, J.: Precise orbital elements, masses and parallax of the spectroscopic–interferometric binary HD 26441. **469**, 1096–1100 (2017). https://doi.org/10.1093/mnras/stx906

6. Al-Wardat, M.A.: Spectral energy distributions and model atmosphere parameters of the quadruple system ADS11061. Bull Spec. Astrophys. Obs. **53**, 51–57 (2002)

7. Al-Wardat, M.A.: Physical parameters of the visually close binary systems hip70973 and hip72479. PASA **29**, 523–528 (2012)

8. Abu-Alrob, E.M., Hussein, A.M., Al-Wardat, M.A.: Atmospheric and fundamental parameters of the individual components of multiple stellar systems. Astron. J. **165**(6), 221 (2023)

9. Hussein, A.M., Abu-Alrob, E.M., Alkhateri, F.M., Al-Wardat, M.A.: Atmospheric parameters of individual components of the visual triple stellar system HIP 32475. ArXiv Prepr. ArXiv230403604 (2023)

10. Hussein, A.M., Abu-Alrob, E.M., Mardini, M.K., Alslaihat, M.J., Al-Wardat, M.A.: Complete analysis of the subgiant stellar system: HIP 102029. Adv. Space Res. (2023). https://doi.org/10.1016/j.asr.2023.07.045

11. Al-Wardat, M., et al.: Physical and geometrical parameters of CVBS XIV: the two nearby systems HIP 19206 and HIP 84425. Res. Astron. Astrophys. (2021). http://www.raa-journal.org/docs/papers_accepted/2020-0472.pdf

12. Al-Wardat, M.A., Hussein, A.M., Al-Naimiy, H.M., Barstow, M.A.: Comparison of gaia and hipparcos parallaxes of close visual binary stars and the impact on determinations of their masses. Publ. Astron. Soc. Aust. **38** (2021). https://doi.org/10.1017/pasa.2020.50

13. Hussein, A.M., et al.: Atmospheric and fundamental parameters of eight nearby multiple stars. Astron. J. **163**, 182 (2022)

14. Masda, S., et al.: Physical and dynamical parameters of the triple stellar system: HIP 109951. Astrophys. Bull. **74**(4), 464–474 (2019). https://doi.org/10.1134/S1990341319040126

15. ESA: The hipparcos and tycho catalogues (ESA), vol. 1239 (1997)

16. Lallement, R., et al.: Three-dimensional maps of interstellar dust in the local arm: using Gaia, 2MASS, and APOGEE-DR14. Astron. Astrophys. **616**, A132 (2018)
17. van Leeuwen, F.: Validation of the new Hipparcos reduction. Astron. Astrophys. **474**, 653–664 (2007). https://doi.org/10.1051/0004-6361:20078357
18. Gaia, C., et al.: Gaia data release 2 summary of the contents and survey properties. Astron. Astrophys. **616**(1), A1 (2018)
19. Fourth catalog of interferometric measurements of binary. http://www.astro.gsu.edu/wds/int4.html
20. Mamajek, E.: A modern mean dwarf stellar color and effective temperature sequence (2013). https://www.pas.rochester.edu/~emamajek/EEM_dwarf_UBVIJHK_colors_Teff.txt
21. Al-Wardat, M.A.: Model atmosphere parameters of the binary systems COU1289 and COU1291. Astron. Nachrichten **328**, 63–67 (2007). https://doi.org/10.1002/asna.200610676
22. Girardi, L., Bressan, A., Bertelli, G., Chiosi, C.: Evolutionary tracks and isochrones for low- and intermediate-mass stars: From 0.15 to 7 Msun, and from Z = 0.0004 to 0.03. Astron. Astrophys. Suppl. Ser. **141**, 371–383 (2000). https://doi.org/10.1051/aas:2000126
23. Tokovinin, A.: Low-mass visual companions to nearby G-dwarfs. Astron. J. **141**(2), 52 (2011). https://doi.org/10.1088/0004-6256/141/2/52
24. Tokovinin, A.: Speckle spectroscopic studies of late-type stars, vol. 32, p. 573 (1992)

Author Index